INSTRUMENTATION IN INDUSTRY

HAROLD E. SOISSON

Electrical System Design & Development;
General Electric Company, RESD.

Instrumentation
in Industry

A WILEY-INTERSCIENCE PUBLICATION

John Wiley & Sons
New York · London · Sydney · Toronto

Library of Congress Cataloging in Publication Data:

Soisson, Harold E.
   Instrumentation in industry.

   "A Wiley-Interscience publication."
   Includes bibliographies and index.
   1. Engineering instruments. I. Title.

TA165.S725   681   74-23222
ISBN 0-471-81049-5

Printed in the United States of America

10 9 8 7 6 5 4 3 2 1

Preface

Automation and continuous process operations are expanding both the scope and the use of individual instruments and instrument systems for automatic control and measurement of the variables encountered in our manufacturing and process facilities. The increased scope and expanded use of instruments has made the business of building, operating, maintaining, and calibrating these instruments a large and vital part of our national economy. This development has underlined the need for personnel at a level below that of the professional engineer trained in the application, operation, maintenance, and calibration of instruments and instrument systems. Such personnel would provide sorely needed assistance for manufacturing and process engineers.

The growth of instrumentation has also resulted in a need for a specific study of measurement and control with the associated instruments and instrument systems. *Instrumentation in Industry* represents an attempt to meet this need. It is intended for use by engineers and engineering support personnel as both a reference handbook and an introduction to automatic control instrumentation. It is a broad survey of industrial instrumentation, including a discussion of general principles of operation and a brief coverage of operational characteristics.

The material is developed by showing the broad industrial requirements and the problems encountered in selecting instruments to meet specific requirements, and by describing the types of instruments used to meet specific requirements. Primary emphasis is placed on the instrument and its essential accessories and their relationship to the system that requires measurement and/or control.

Instruments and instrument systems for static and/or dynamic measurements are examined. Mechanical, hydraulic, electrical, electronic, electromechanical, electroelectronic and other combinations essential to accomplishing the measurement and/or control function are also considered and discussed.

Mass production techniques require the standardization of components to meet specific tolerances, and instruments have been developed that meet these tolerances. This book provides a basic description of the theory and instrument function, a discussion of measurement and response limitations

v

and calibration of the instruments, and an analysis of the success of the industrial process or operation in maintaining a quality product.

The material presented requires a basic understanding of mathematics through trigonometry, electricity, and physics. In most cases, readers with a high school background in these areas will have no difficulty understanding the material presented. In some instances, a knowledge of differential and integral calculus would be helpful, but not absolutely essential. The nonrigorous approach has been adopted so that the book is accessible to a broad range of readers.

Extensive use has been made of sketches and diagrams to illustrate principles and systems. Photographs have been used to give the reader a better understanding and recognition of particular instruments as they existed at the time the book was written. Review questions and problems have been included in most of the chapter to emphasize the theory or function and to challenge the reader's imagination in devising an approach to a problem.

To produce a book that covers such a broad scope of information requires the use of a large number of sources of information. The author is indebted to many individuals and instrument manufacturers for their assistance in making the material available. The author is also grateful for the encouragement of the management of the General Electric Company.

Contributing companies include, but are not limited to: Helicoid Gage Division of American Chain and Cable, Baird Atomic, Inc., Barnes Engineering Company, Beckman Instruments, F. W. Bell, Inc., CGS/Thermodynamics Division of CGS Scientific Corporation, Chemquip Company, Cohu Electronics, Inc., Columbia Research Corporation, Centraves Viscometer Division of Consolidated Vacuum Corporation, Olkon Corporation, Corning Glass Company, Dresser Industrial Valve and Instrument Division of Dresser Industries, Inc., Ellison Instrument Division of Dietrich Standard Corporation, Fischer and Porter Company, Ford Motor Company, General Electric Company, General Radio Company, Gulf Oil Corporation, Gulton Industries, Inc., Hays Corporation, Hewlett Packard Company, Honeywell, Inc., ITT Barton, King Engineering Corporation, Kistler Instrument Corporation, Kulite Semiconductor Products, Inc., Leeds & Northrup Company, Marshalltown Manufacturing Company, Micro-Measurements, Microdot, Inc., Monitor Manufacturing Company, National Instrument Company, National Research Corporation, Nuclear Chicago Corporation, Nuclear Measurements Corporation, Pace Engineering Company, Penn Airborne Products Company, Penn Meter Company, Pyrometer Instrument Company, Inc., Jules Racine & Company, Inc., Randolph Company, RFL Laboratories, Inc., Ruska Instrument Corporation, Schaevitz Engineering Company, Herman H. Sticht Company, Inc., Tabor Instrument Corporation, Taylor Instrument Companies, Texas Nuclear Corporation, Trapelo Division of LFE Corporation, Henry Troemner, Inc., Victoreen Instrument Company, Charles F. Warrick Company, West Virginia Pulp and Paper Company, United States Steel Corporation, Yellow Springs Instrument Company, Inc., Dwyer Instruments, Inc., ITT Marlow, Hersey Products, Inc., Ametrol Division, and Neptune Meter Co.

HAROLD E. SOISSON

Strafford-Wayne, Pennsylvania
September 1974

Contents

CHAPTER 1   *HOW INSTRUMENTS ARE USED IN INDUSTRY*   1

1.1   Ceramics, 2
1.2   Iron and Steel, 6
1.3   Chemicals, 10
1.4   Petroleum Products, 10
1.5   Pulp and Paper, 11
1.6   Food, 11
1.7   Electrical, 12
1.8   Nuclear Reactors, 13
1.9   Automotive, 15
1.10 Appliances, 20
1.11 Instrument Selection for the Application, 23

CHAPTER 2   *INSTRUMENT STANDARDS AND CALIBRATION*   28

2.1   Types of Reference Standards for Calibration, 29
2.2   Calibration of Temperature Instruments, 30
2.3   Pressure Calibration Standards, 32
2.4   Flow Calibration Standards, 40
2.5   Weight Calibration Standards, 41
2.6   Time Calibration Standards, 42
2.7   Electrical and Electronic Calibration Standards, 43
2.8   Radioactive Sources, 49
2.9   Velocity or Speed Standards, 49
2.10 Frequency Standards, 51
2.11 Summary, 52

CHAPTER 3   *PRESSURE AND VACUUM*   59

3.1   Pressure and Force Balance Gages, 61
3.2   Piston Pressure Gages, 71
3.3   Diaphragm Pressure Gages, Nonmetallic, 72

3.4   Elastic Membrane Gages, 76
3.5   Electromechanical Pressure Gages, 85
3.6   Vacuum Gages—Mechanical, Electrical and Electronic, 92
3.7   Pressure Gage Calibration, 99
3.8   Summary, 102

CHAPTER 4   THERMOMETERS   109

4.1   Temperature Scales, 109
4.2   Liquid-in-Glass Thermometers, 111
4.3   Liquid-in-Metal Thermometers, 117
4.4   Vapor Actuated Thermometers, 120
4.5   Gas Actuated Thermometers, 122
4.6   Speed of Response in Bulb-Thermometer Systems, 125
4.7   Bimetallic Thermometers, 126
4.8   Resistance Thermometers, 128
4.9   Summary, 147

CHAPTER 5   PYROMETRY   151

5.1   Thermocouples, 151
5.2   Potentiometric Pyrometers, 165
5.3   Radiation Pyrometers, 171
5.4   Two-Color Pyrometry, 189
5.5   Optical Pyrometry, 192
5.6   Summary, 202

CHAPTER 6   LIQUID AND DRY LEVEL INSTRUMENTATION 207

6.1   Mechanical Level Measuring Instruments, 207
6.2   Pressure Drop Systems, 215
6.3   Electrical Level Measuring Instruments, 221
6.4   Electroelectronic Level Measuring Instruments 225
6.5   Nuclear Level Measuring Instruments, 231
6.6   Level Measurement by Weighing, 232
6.7   Summary, 233

*CHAPTER 7   FLOW INSTRUMENTATION AND
             MEASUREMENT*   240

7.1   Head Flowmeters, 240

7.2   Area Flowmeters, 260
7.3   Electromagnetic Flowmeters, 267
7.4   Mass Flowmeters, 269
7.5   Positive Displacement Flowmeters, 274
7.6   Open Channel Flowmeters, 281
7.7   Summary, 283

*CHAPTER 8   AUTOMATIC MEASUREMENT AND CONTROL
              CONCEPTS AND SYSTEMS*   290

8.1   Common Basic Characteristics of Automatic Measurement and
      Control Systems, 290
8.2   Basic Characteristics of Measuring Devices, 291
8.3   Basic Process Characteristics, 301
8.4   Basic Characteristics of Automatic Control, 307
8.5   Computer Control Systems, 316
8.6   Summary, 317

CHAPTER 9   ANALYTICAL INDUSTRIAL INSTRUMENTATION 323

9.1   Stress and Strain, 324
9.2   Transducers, 344
9.3   Piezoresistive Transducers, 345
9.4   Piezoelectric Accelerometers, 346
9.5   Linear Variable Differential Transformers (LVDT), 359
9.6   Optical Electronic Chemical Analyses, 370
9.7   pH Measurements, 377
9.8   Viscosity, 396
9.9   Moisture Measurements, 393
9.10 Chromatography, 395
9.11 Summary, 398

*CHAPTER 10   RADIATION MEASUREMENT AND
               INSTRUMENTATION*   404

10.1   Radioactivity Detectors, 405
10.2   Readout Instruments and Accessories, 424
10.3   Reactor Radiation Instrumentation, 440
10.4   Process Radiation Instrumentation, 444
10.5   Calibration of Radiation Instrumentation, 452
10.6   Summary, 453

CHAPTER 11 *NONDESTRUCTIVE TESTING EQUIPMENT*   457

11.1   Magnetic Particle, 457
11.2   Dye Penetrants, 459
11.3   X Rays, 461
11.4   Gamma Rays, 466
11.5   Neutron Radiography, 468
11.6   Ultrasonics, 468
11.7   Eddy Current Testing, 477
11.8   Infrared as an NDT Tool, 487
11.9   Microwaves as an NDT Tool, 489
11.10 Signature Analysis as an NDT Tool, 493
11.11 Ultrasonic Holography, 495
11.12 Ultrasonics in Bond Testing, 497
11.13 Summary, 499

CHAPTER 12 *ENVIRONMENTAL AND POLLUTION*
            MEASUREMENTS   502

12.1   Meteorological Instrumentation and Measurements, 503
12.2   Air Pollutant Measurements, 510
12.3   Environmental and Air Monitoring System Equipment for
       Remote and Central Station Locations, 525
12.4   Industrial Air Measurement and Control, 528
12.5   Noise Measurement, 536
12.6   Liquid Pollution Measurements and Control, 543
12.7   Solid Wastes and Pollutants, 548
12.8   Summary, 552

INDEX   555

CHAPTER 1

How Instruments Are Used in Industry

This chapter is designed to show the broad scope instrumentation plays in a key number of industries that manufacture products essential to human welfare, comfort, and safety.

Instruments are indispensable tools used to establish and maintain the quality values that identify a product being manufactured. They are used to control the variables in a process or system as accurately as necessary to meet product specifications for composition, form, color, or finish.

The instrument or instrument system may be mechanical, pneumatic, hydraulic, electrical, electronic, or a combination of any two or more of the basic forms, such as electromechanical. Each instrument or system contains three basic functions. They are:

1. Detector
2. Intermediate transfer device
3. End device

They are shown in the block diagram of Figure 1.1. The input device must detect the incoming signal and transfer it to some type of output. The type of instrument or system depends on the variables to be controlled or measured, and how rapid and accurate the measurement or control must be.

Automation with requirements for computer control and data logging has expanded the use of instruments for single-station or system measurements and control in all modern industries. These range from a simple manually controlled station to a complex computer actuation and control center. For every application there must be a clear and concise understanding of the function of each instrument and its limitations in the

1

Figure 1.1   *Three basic instrument or instrument system functions.*

measurement and control system. It is essential that adequate theory, functional operation, and interactions between the components in the process to be measured or controlled are known.

The usefulness of an instrument in any measuring and control system depends on how well it can initiate a control device and how reliably it can reproduce the initiation of the control. Both the accuracy and reliability of an instrument depend on its construction and how well it holds its calibration. An uncalibrated instrument is a measurement hazard and is not a true measuring device. To be a useful industrial tool for process use, an instrument must be calibrated to some acceptable standard. Standards and calibration are discussed in Chapter 2.

Instrumentation makes mass production possible and allows upper (plus) and lower (minus) limits to be set and maintained. The use of calibration standards establishes reliable measurements and controls at the point of manufacture, and permits one manufacturer to specialize in a product and act as a supplier to other manufacturers or assembly groups.

Industrial measurement and control, with its associated instrumentation, is a multimillion dollar operation and has shown good annual growth.

To illustrate how instruments are used in modern industry, we briefly look at some of the industrial applications in which instruments are essential to the success of the industry and its products.

1.1   CERAMICS

In the ceramics industry of manufacturing bricks and tiles, the pressure of the press or the extrusion pressure of the pug mill determines the density of brick, tile, or other clay products. The temperature of the drying oven establishes the drying cycle, and the temperature of the burning kiln determines the hardness and sometimes the color of the product. A block diagram of brick manufacturing is shown in Figure 1.2*a*, and a pictorial schematic in Figure 1.2*b*.

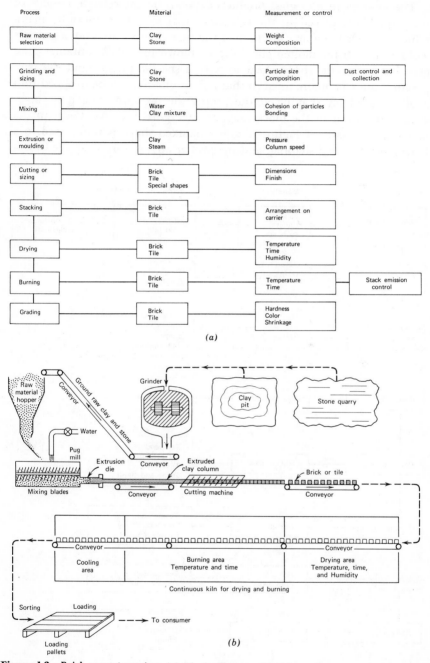

Figure 1.2 *Brick manufacturing. (a) Block diagram. (b) Schematic process flow.*

The manufacture of glass products depends on the relative weights of the ingredients placed in the refractory furnace which must be maintained at a fixed temperature to melt the ingredients into a homogeneous molten mass. If the molten glass is to be blown into glass containers, the glass has to be cut with a shear to have a specific weight, and the temperature must be maintained so that a glass globule can be dropped into a blank mold for shaping at the proper air pressure. The basic operations are shown in the block diagram of Figure 1.3*a*. As shown in the pictorial schematic of Figure 1.3*b*, the blanked shape is transferred to a blow mold for final shaping, and the proper air pressure is applied for a

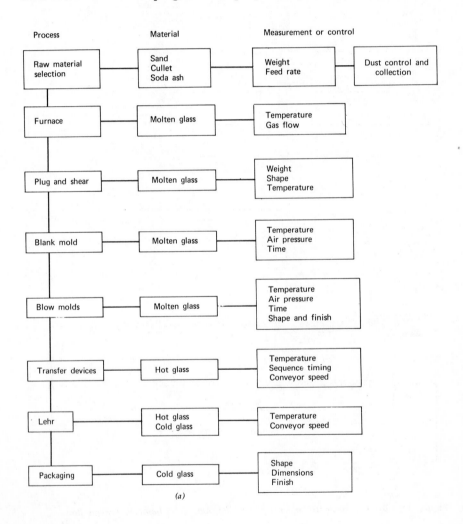

(a)

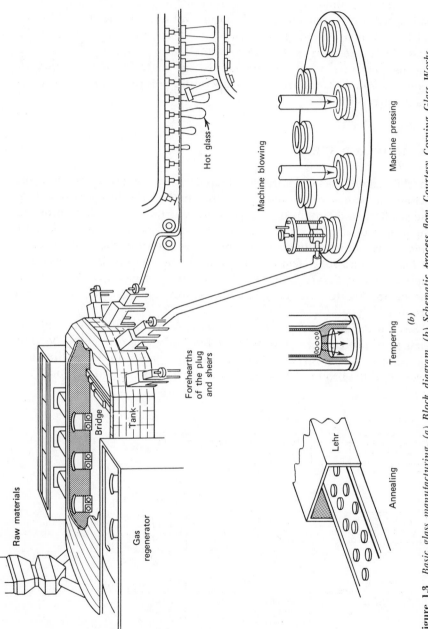

Raw materials

Gas regenerator

Bridge

Tank

Forehearths of the plug and shears

Hot glass →

Machine blowing

Machine pressing

Tempering

Annealing

Lehr

(b)

Figure 1.3 Basic glass manufacturing. (a) Block diagram. (b) Schematic process flow. Courtesy Corning Glass Works.

specific interval of time. The hot glass container is then transferred to a lehr for annealing. The lehr is maintained at the proper temperatures and speed to accommodate the output volume of the glass blowing machine. When molten glass is made into sheets, the temperature of the glass, the temperature and pressure of the forming rolls, and the annealing temperatures must be maintained to make a quality product. Sequence timing, pressure, and temperature controls make these operations successful.

1.2 IRON AND STEEL

In the ferrous metal industry, materials must be accurately weighed before they are smelted, and the temperature of the furnace, as well as the gas flow to maintain the temperature, has to be controlled to obtain the proper quality of iron from iron ore, limestone, and coke. The basic operations and measurements or controls are shown in the block diagram of Figure 1.4a. Proper temperatures must be maintained in the blast furnace shown in Figure 1.4b, and the molten iron has to be sampled as shown in Figure 1.4c to guarantee sound castings and other iron products.

Steel products depend on the materials used in the melt and the method of processing, as shown in the block diagram of Figure 1.5a. The temperature of the furnace is important in preparing for the rolling and shaping of the product. An open hearth furnace is shown in Figure 1.5b.

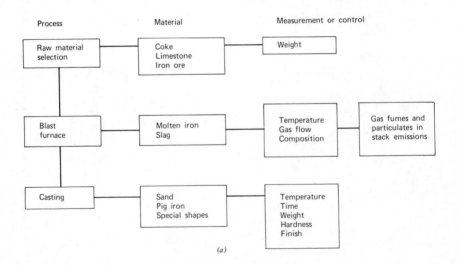

(a)

(b)

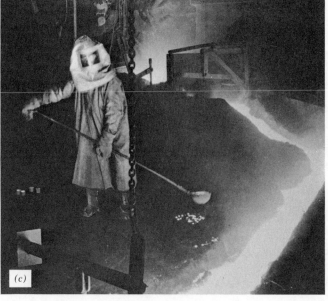

(c)

Figure 1.4 *Basic iron smelting and casting industry. (a) Block diagram. (b) "Dorothy" blast furnace at United States Steel Duquesne Works towers 284 ft above the west bank of the Monongahela river. (c) A sample of molten iron is taken from the "Dorothy," United States Steel's newest and Pittsburgh's largest blast furnace. Courtesy. U. S. Steel Corp.*

Open hearth ingots must be soaked in temperature pits to reach the proper rolling temperature, and the soaked ingots must be stripped from their casting molds before they can be rolled. Control of the pressure on the rolls determines the thickness and width of the sheet or bar of steel produced from a billet or slab, as shown in Figure 1.5c. The machine-ability, hardness, and weldability of steel are determined by the composition and the heat treatment given the product. Therefore temperature, weight, pressure, and time controls contribute to the success of the iron and steel industry.

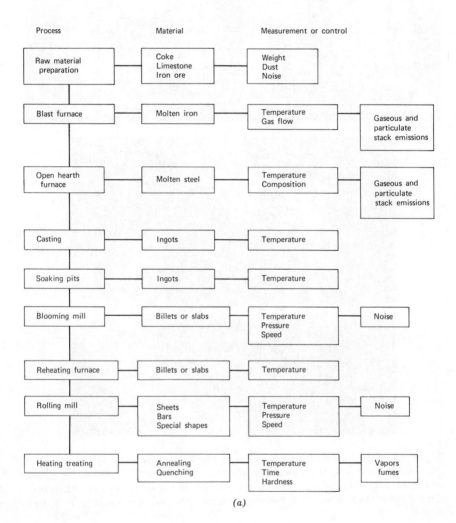

(a)

Figure 1.5 *Steel manufacturing. (a) Block diagram. (b) Open hearth furnace. The charging machine used for adding scrap and other materials is in the foreground, and a ladle making a hot molten addition is in the background. (c) This is a 36-in. bloom mill at United States Steel Duquesne Works rolling a steel bloom. Courtesy U.S. Steel Corp.*

1.3   CHEMICALS

The chemical industry has to control accurately the flow of liquids and the pressure and temperature at which chemical reactions take place, and to measure accurately the amount of each substance used. The basic requirements are shown in the block diagram of Figure 1.6. Inaccurate measurement and control of the variables can and have been hazardous. Chemical explosions can and have caused the loss of life, an entire plant, or a small community. Accurate controls are just as necessary to produce the right color and consistency of lipstick or nail polish as to produce a good weather-resistant paint or silicone plasticizer for use with epoxy resins.

1.4   PETROLEUM PRODUCTS

Petroleum products require close and accurate controls. Temperatures and pressures are critical during the refinement process of cracking and distillation. Problems involving the flow of volatile liquids are always

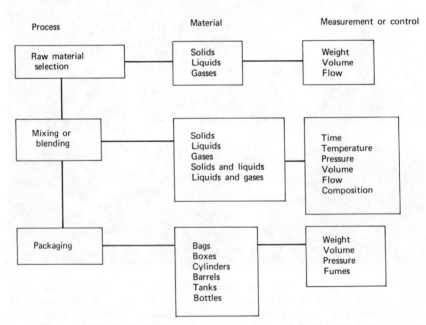

Figure 1.6 *Basic chemical process measurement and control requirements.*

present. The measurement and control instrumentation must not only be accurate, but it must also be reliable so that the process is maintained within the allowable safe limits to produce a specific grade of motor oil, diesel fuel, or gasoline. Accurate measurement of flow, pressure, and temperature make this industry relatively safe. Some basic requirements are shown in the block digram of Figure 1.7a. A typical modern fractionating tower is shown schematically in Figure 1.7b, in which crude oil is broken down into each of its useful components.

1.5 PULP AND PAPER

In the pulp and paper industry, logs have to be demarked and cut into small chips. These chips are then mixed with chemicals for a digestion process and maintained at the proper temperature until the fibers are separated from the other ingredients. The wood fibers are then sent into the drying and pressing cycles, and the chemicals are sent through a recovery process. After drying and pressing, the pulp is rolled into paper. Control of the rolling pressure and the speed of the roll determine the paper thicknesses. Speed regulation on the take-up roll is important or the paper will be torn, because as the roll increases in size the speed of the roll has to be reduced. Thus control of weight, flow, pressure, and speed of rotation are essential to the success of the paper industry, as shown in the block diagram of Figure 1.8.

1.6 FOOD

Selection of food for processing, such as sorting of beans, peas, and other regularly shaped or colored fruits or vegetables, can be done automatically by using an electric eye and a reject system. After foods are sorted and washed, they are cooked for canning or blanched for freezing. In either case the cooking or blanching vessel must be kept at a specified temperature and the food processed for a specified time. After cooking or blanching, the food is packaged in cans, jars, or waxed boxes, and labeled. In packaging, an exact weight or volume is placed in each container, and the container is sealed. In many cases the sealed and labeled food is inspected by an instrument system before it is sent to the shipping container or freezing unit. The basic requirements in food processing are shown in the block diagram of Figure 1.9.

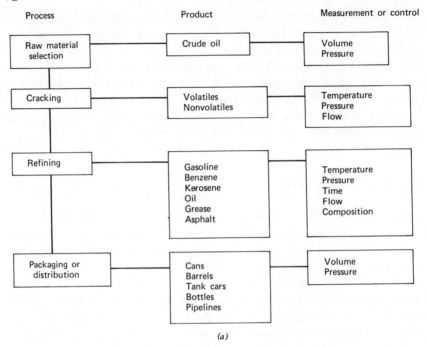

(a)

Preparation of cereals also makes use of instruments in sorting, rolling, exploding, baking, and packaging. Pressure, temperature, timing, weight, flow, and sequence operational control instrumentation are essential to supply us with prepared foods.

1.7 ELECTRICAL

The generation of electrical power and the manufacture of generating and propulsion equipment make up one of our largest economic areas; this industry could not exist as it is today without instrumentation. In the manufacturing of turbines and generators to produce both dc and ac electricity, the modern factory makes use of numerically controlled machines, balancing equipment, nondestructive testing equipment, and accurate dimensioning instruments. This phase of the industry is shown in the block diagram of Figure 1.10.

In the generation of dc electricity, the speed of the generator has to be measured and controlled. It is also necessary to measure the output current, voltage, and power generated. The block diagram of Figure 1.11 shows the basic requirements for both generation and consumption.

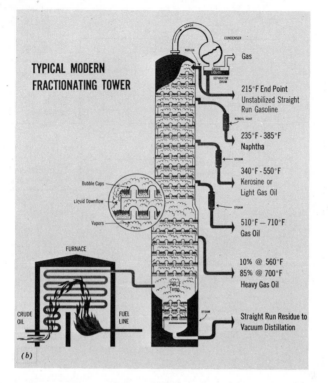

Figure 1.7 *Basic measurement and control in the petroleum industry. (a) Block diagram. (b) Typical modern fractionating tower. Courtesy Gulf Oil Corp.*

Generation of ac electricty requires measurement of the frequency (25, 50, or 60 Hz), and the phase relationships on three phase systems, as well as measurement of current, voltage, and power. Generation, distribution, and consumer requirements are shown in the block diagram of Figure 1.12.

The measurement of electrical power consumed by industrial propulsion and heating equipment depends on ammeters to measure current, voltmeters to measure voltage, and wattmeters or watt hour meters to measure power at the point of consumption.

1.8 NUCLEAR REACTORS

The building of nuclear reactors requires the best of manufacturing techniques and precision measurements. The components of a reactor

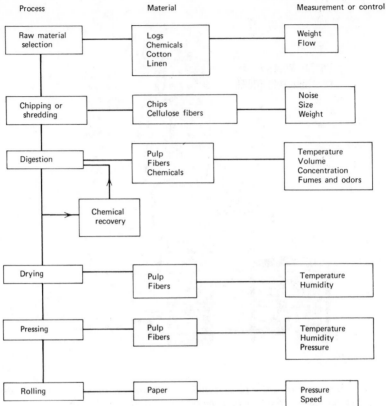

Figure 1.8 *Basic pulp and paper manufacturing. Block diagram.*

have to withstand the normal wear of equivalent conventional equipment, as well as the effects of radiation. Some of the components are exposed to high temperatures, and they have to be cooled so that they do not melt or deform while in service or during the period immediately after the reactor is shut down. Close tolerances are required to meet operational conditions safely. Also, accurate controls and measurements are necessary to ensure safe operation. Reactor components and assemblies probably receive the most thorough inspection of any manufactured product for tolerances, cleanliness, and soundness of specific materials. Materials must be clearly and positively identified, from the raw stock to the completed unit. Nondestructive, hardness, machineability, and weldability tests are commonly performed on each type of material used.

Nuclear reactors are presently used as power sources for the generation of electricity, as shown in Figures 1.11 and 1.12, and for the propulsion of naval vessels. As additional knowledge is gained of the metallurgical

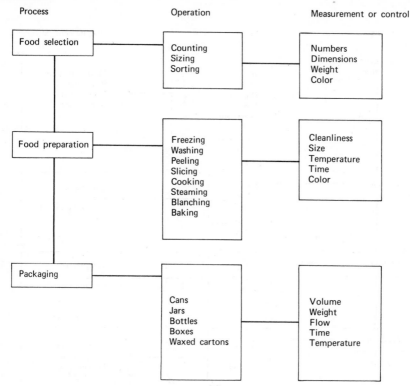

Figure 1.9 *Food industry basic instrument and control requirements.*

properties of materials for application in the nuclear reactor field, the industry for fabrication and reclamation of fuel will grow and develop standards that will require instruments for both measurement and control in addition to those now used in the fabrication and assembly operation. The basic requirements for this industry are shown in the block diagram of Figure 1.13.

1.9 AUTOMOTIVE

The manufacturing of automotive parts, and the assembly of manufactured parts depends on mass production techniques in which measurements and control are vital. Automotive companies depend on many suppliers and have set up many of their own parts and assembly operations in widely separated locations. In their designs they have established tolerances for parts, which must be maintained in order that the components

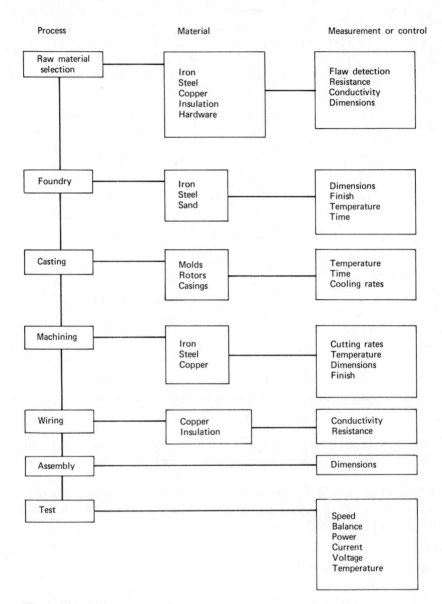

Figure 1.10 *Basic generator manufacturing.*

Process   Measurement or control

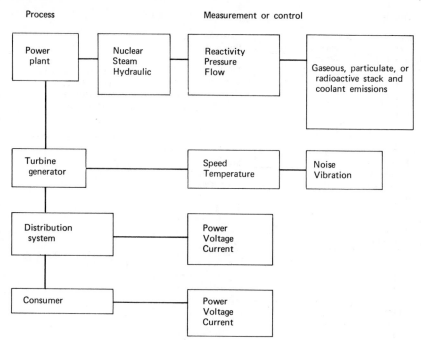

Figure 1.11 *Dc power generation and distribution.*

can be assembled into a unit that will function properly and have sales appeal. Automobile parts include castings of plastic, aluminum, iron, and steel, fabrics, glass, rubber, and other materials such as paint.

Machining of motor blocks, bearings, crankshafts, valves, valve springs, gears, and shafts requires use of the latest numerically controlled equipment to keep the manufacturer or his supplier economically competitive. Tolerances are usually automatically checked, and signals are activated when dimensions are either above or below the established tolerance, or if a tool breaks. Speeds and feeds are set by a taped control, and finishes are checked for defects and uniformity. All finished units receive a final inspection, usually on a statistical basis, before they are placed on an assembly line operation.

Assembly operations are timed and, if any operation fails, the faulty unit is removed or the entire assembly line is stopped until corrections are made. When faulty units are removed, they must be replaced from a standby supply and the faulty unit repaired or junked. Feed lines to each assembly station must be kept operative and timed for the final assembly

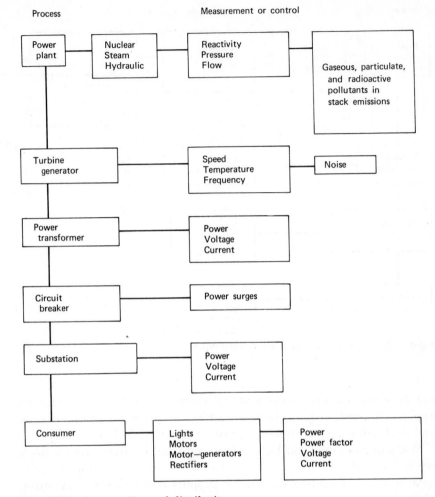

Figure 1.12 *Ac generation and distribution.*

operations. The final products of each assembly line are checked and tested for appearances and performance.

Automobile design features are tested for performance, vibration, shock, handling ability, roadability, and sales appeal. Intricate and complex measurements, using all types of instruments, are made for the production of the vehicle and its performance. These measurements include the pressure needed for hydraulic brakes or power steering units, power transfer efficiencies of automatic transmissions, brake horsepower

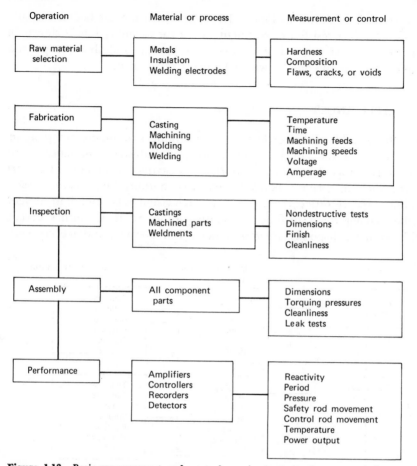

Figure 1.13 *Basic measurement and control requirements in the reactor industry.*

of the engine, revolutions per minute of the generator or the alternator for light and power, shock absorption for the weight of the vehicle, and wind resistance effects on body airflow design. A representative grouping for the automotive industry is shown in the block diagram of Figure 1.14a. The Ford Motor Company Rouge manufacturing complex layout is shown in Figure 1.14b. This complex starts with the raw materials and transforms them into completed automobiles. The Ford Motor Company operates its own blast furnaces and rolling mill operations. A computer control center for the electronic control of a Ford Motor Company plate glass facility is shown in Figure 1.14c. An electrocoating process tank is shown in Figure 1.14*d*, in which the car bodies are submersed for rust resistance treatment. A tractor assembly line is shown in Figure 1.14*e*. At the end of the assembly line an underbody inspection station such as the one shown in Figure 1.14*f* is used.

1.10   APPLIANCES

The household appliance field is high-volume and highly competitive, and is one of the larger economic areas in industry. Appliances must have buyer appeal based on comfort, labor savings, or necessity. Comfort items include electric blankets, electric toothbrushes, heating pads, radios, tape recorders, record players, television sets, and hair dryers. Labor saving items include automatic clothes washers, clothes dryers, dish washers, disposals, freezers, mixers, can openers, and ice crushers.

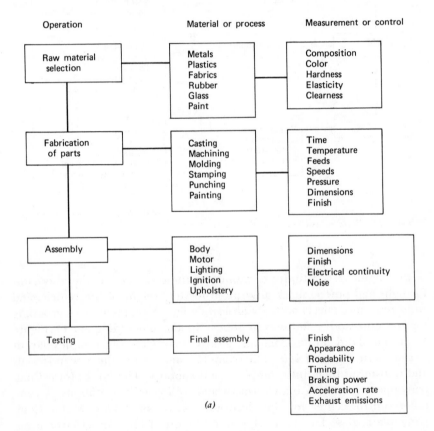

(a)

Figure 1.14 *Basic automotive industry measurement and control requirements.. (a) Block diagram. (b) Ford Motor Company's Rouge manufacturing complex. (c) Computer control center for electronic control of a plate glass faciltiy. (d) Electrocoating dip tank to apply rust resistant primer to car bodies. (e) Tractors on line in the assembly plant at Highland Park. (f) Underbody inspection station near the end of the assembly line. Courtesy Ford Motor Co.*

Necessity items include fans, motors, refrigerators, regular washing machcines, air conditioners, vacuum cleaners, stoves, and clocks.

Manufacturing controls in the appliance industry include pressure for forming metal parts; temperature for casting, curing of plastics, and drying of paints; positioning of parts for punching, stamping, and machining operations; speed of the assembly line operations; and quality measurements for finish, performance, and appearance of the final product. Instruments are needed for each of the measurement and control functions, as shown in the block diagram of Figure 1.15.

1.11 INSTRUMENT SELECTION FOR THE APPLICATION

To enhance the competitive position of an industry, instruments or instrument systems must be carefully chosen to meet a specific application. For an instrument manufacturing or process engineer and his trained assistants to make the best application choice, they have to have a sound knowledge of both the instrument and the operating system. Some very fundamental concepts are essential in making the best selection. They are types and sources of error, time lag, dead time, and frequency response of both the instrument and the system.

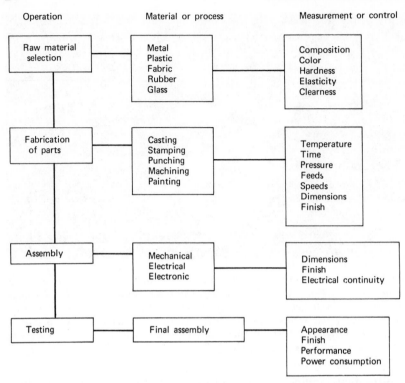

Operation	Material or process	Measurement or control
Raw material selection	Metal Plastic Fabric Rubber Glass	Composition Color Hardness Elasticity Clearness
Fabrication of parts	Casting Stamping Punching Machining Painting	Temperature Time Pressure Feeds Speeds Dimensions Finish
Assembly	Mechanical Electrical Electronic	Dimensions Finish Electrical continuity
Testing	Final assembly	Appearance Finish Performance Power consumption

Figure 1.15 *Basic measurement and control requirements in the appliance industry.*

Error.   In making a measurement the *true value* of an object of measurement can never be *exactly* established. This statement is now generally recognized by those in the business of making measurements, because if several independent measurements are made on one physical
quantity under presumably identical operating conditions, all with equal
care and making full use of the sensitivity of the measuring instruments,
the measurement values will not be exactly alike. There is no such thing
as two identical items. There are always some differences. When two
items appear to be identical, it is because the measuring device used does
not have adequate precision to reveal the actual differences that exist.
These differences exist because of limitations in the measuring device
itself, the accuracy with which the measuring scale can be read, the vision
angle of the individual making the observation, and indeterminate variations in the operating conditions.

These sources of variation, or so-called errors of measurement, either
inherent in the instrument or the system, do not represent a mistake in
making the measurement or in carrying out the control action, but are a

measure of the degree of uncertainty that exists in the method employed for the specific application. The greater the number of measurements taken on a quantity or process, the more will be known about the unknown quantity, and it will be possible to arrive at a more precise estimate of its true value than can be obtained through any single measurement.

There are two sources of variance of error, commonly called accidental or random error and systematic error.

Random errors are due to irregular causes, too many in number and too complex in nature for their origin to be traced. The chief characteristic of random errors is that they are likely to have a negligible effect on the value of the arithmetic mean of a set of measurements. This simply means that, when a set of measurements is recorded, random error makes equal contributions to the values above and below the average for the set of measurements and effectively cancels out on the average. This is not true, and should not be assumed to be true, for any single measurement.

Systematic errors enter into measurement records in a given pattern; they are generally in the same direction (having the same sign) and are of the same order of magnitude. These errors may be due to instrumental errors, such as a faulty scale gradation or an inaccurate standard, or they may be due to observer error, such as always responding too soon to a signal, reading values on the lower third of the scale on the low side, or reading values on the upper third of the scale on the high side. When the magnitude of a systematic error can be determined, it can be included as a correction factor in the measurement value and be added in the direction opposite the error contribution.

The ideal objective in good instrumentation is to reduce all errors to the random type, so that they can be treated by the probability theory to obtain the best estimate of the true value of the measured quantity. In practice, error is reduced to the minimum consistent with the economics of the industrial application to produce an object or device that has sales appeal.

Time Lag. Time lag can be defined as the time interval or length of time that elapses between the time a signal is generated and the time the measuring or control instrument or instrument system indicates, records, or actuates a control to correct an error or change the function.

The shorter the time lag, the better the dynamic measurement of the function. Time lag can become quite serious if rapid response is needed in the process or if the action is frequency dependent, because it represents a delay in action. If the time lag is of sufficient length, a change in the variable may not be made in time to prevent a process loss. When frequency is a part of the measurement or control cycle, a time lag may be responsible for a phase shift, so that the measurement or control action is out of phase with the process.

Time lag is caused by resistance to the flow of air in tubing connecting pneumatic instruments, by resistance in wires carrying current, by the moment of inertia in the mechanical movement of a meter or gage, and by any other factors that slow up the transporting time for a signal to accomplish the objective it was designed to perform.

Dead Time.   Dead time is one type of time lag or delay in a measuring instrument or system. Dead time in an instrument or process system is the time during which a new signal or variation in a signal cannot be detected. It is serious in all types of dynamic operations, because no action can be taken during dead time, and a variation occurring during dead time is not detected. In other words, one just does not know whether or not a change has taken place during the dead time of an instrument unless another instrument with a much shorter dead time verifies the condition. The choice of an instrument should be made so that the dead time does not constitute a measurement or process hazard. Dead time of the instrument should be chosen to be less than 10% of the time lag involved in the total measurement.

Frequency Response.   The manner in which an instrument or circuit handles the frequencies falling within its operating range is known as frequency response. One method of measuring this response is to measure the variation in gain or loss in output as a function of frequency. Frequency response is usually specified by the manufacturer as being flat over a range of frequencies, which means there is essentially no change in response characteristics for that range. The specifications may indicate an amplitude loss as being down so many decibels above the flat range and down an additional decibel amount beyond the second range, where a decibel is defined as 10 times the logarithm of the power rates.

Frequency response is important in dynamic measurements, both in phase relationships and in amplitude. The maximum phase shift is 180°, which represents a complete reversal in the direction of the signal, and phase shifts can range from 0 to 180°. When the dead time of an instrument is less than one-tenth of the total response time, the phase shift can be neglected in the analysis of the measurement in choosing the best instrument.

A frequency measurement that is now considered important, although it has been neglected in the past, is noise. While noise has more than a frequency content, it is normally broken down for analysis, as discussel in Chapter 12.

With this general and brief background showing how instruments are used in industry and the knowledge required for their selection for use in industrial processes, we now look more closely at specific instruments and their functions, applications, accuracy, and limitations.

Review Questions

1.1 How many basic functions does an instrument or instrument system have? Name them.

1.2 Upon what factors does the usefulness of an instrument depend?

1.3 Why are standards needed where instruments are used?

1.4 What are the main areas of measurement and control in the ceramics industry?

1.5 Why is temperature control important in the iron and steel industry?

1.6 Why is measurement and control needed in the chemical industry?

1.7 Do you consider precision measurements more essential in the petroleum industry than the chemical industry? Why?

1.8 What controls, if any, does the paper industry have that are not needed in the ceramics industry?

1.9 How does instrumentation help to feed you better?

1.10 What are the basic instrument functions used in power generation? Power consumption?

1.11 Why are measurement and control considered absolute essentials in nuclear reactor operations?

1.12 How has instrumentation and control contributed to the comfort and performance of automobiles?

1.13 What part has instrumentation played in mass production? In automation?

1.14 What does an engineer need to know in order to make the best application choice to perform a specific measurement or control function?

1.15 Name the types of errors always present in a measurement. Discuss each.

1.16 What effect does lag time have in a control operation?

1.17 What is dead time in an instrument or measuring system?

1.18 Where is frequency response important?

1.19 What is the maximum possible phase shift? How many degrees does this represent?

CHAPTER 2

Instrument Standards and Calibration

An uncalibrated instrument is a hazard and is not a measuring and control device that can be relied on.

Calibration is an essential part of industrial measurement and control. In fact it is an essential part of any measurement and control operation.

Calibration is not a glamorous operation. However, it must be performed properly if a product is to have controlled quality. Calibration is the only guarantee that industrial instruments have the accuracy and range required to maintain operating systems under economically controlled conditions. Calibrated instruments permit a manufacturer or processor to produce quality merchandise with desirable or demanded customer specifications.

Calibration must be performed periodically and requires the use of a standard for comparison values. Thus calibration can be simply defined as the comparison of specific value inputs and outputs of an instrument and a reference standard. These comparisons require operator skill and care, and good reference standards.

Calibration does not guarantee the performance of an instrument, but is usually a good indication whether or not its performance can meet the accuracy and range specifications for which the instrument is to be used. When the instrument has been designed to meet accuracy and range specifications but does not meet them during calibration, it must be repaired and/or adjusted so that it does meet them. Repairs and adjustments are usually made by skilled operators or mechanics. Recalibration is always performed after an instrument has been adjusted, repaired, modified, or abused. Calibrations are the manufacturer's verification that an instrument is capable of indicating, recording, or controlling system variables at values established for a specific industrial application. He maintains records and normally files a copy of each calibration for each instrument or control system. These records are often required as evidence of compliance by some customers.

2.1 TYPES OF REFERENCE STANDARDS FOR CALIBRATION

Industrial users normally maintain one or possibly two general types of instrument standards, *primary* and *secondary*.

A *primary* standard is an extremely accurate, absolute value unit certified by the National Bureau of Standards (NBS) to be within the allowable tolerances for the absolute units of measurement of that parameter maintained at the NBS in Washington, D.C. Normally, instrument manufacturers and a few of the larger users are the only groups who maintain primary standards. These standards are both expensive to own and maintain. Such standards are certified by the NBS and are used to calibrate instruments sold as secondary standards for the calibration of industrial instruments.

The calibration interval for *secondary* standards depends on accuracy and the type of standard being maintained. The calibration period or interval for industrial instruments varies from weeks to years, depending on the type of service in which they are used and the type of instrument construction. In some cases in which accuracy is very important, it may be more expensive to calibrate certain types of instruments at more frequent intervals than to purchase better instruments initially, which require less frequent calibration for good quality control.

A well-equipped industrial instrument calibration facility should maintain standards and parameter generating equipment for temperature, pressure, flow, weight, time, voltage, current, power, resistance, capacitance, velocity, frequency and radioactivity, as required for the industry being served. It is well to remember that a standard should be at least a factor of 10 more accurate than the instrument being calibrated, and that the reference standard need be only as reliable and accurate as required for the application. Economics is important in calibration, because it is a necessary overhead operating expense and should be held to a minimum.

In all calibration procedures it is advisable to make readings going both upscale and downscale. In mechanically operated meters this procedure normally discloses losses due to friction, hysteresis, spring set, and similar types of phenomena. During calibration never rub the glass or plastic cover over an electrical meter movement or pointer to clean it.

When cleaning a meter surface over the meter movement, use an anti-static technique and make certain that there is no charge left on the clean surface. Static charges affect both normal readings and calibration readings.

With digital, alphanumerical, and decimal readout instruments using cold cathode readout units or the latest light emitting diodes (LEDs), the readout should be plus or minus the least significant digit or letter in either the ascending or descending order.

2.2 CALIBRATION OF TEMPERATURE INSTRUMENTS

Temperature measuring instruments may range in span from nearly absolute zero in cryogenic work to thousands of degrees in plasma jet applications. The standards used are calibrated in the temperature reference scale most applicable, or in several reference scales such as Celsius and Fahrenheit.

Methods used to generate temperatures and the standards used for calibration purposes are shown in Table 2.1.

When a thermometer or thermocouple is being calibrated, there should be at least three calibration points; and preferably these values should be spaced such that one is below the values being measured, one is near the values being measured, and one is above the values being measured. Thermometers and some thermocouples are immersion units, and during the calibration procedure this immersion must be carefully observed. If immersion cannot be accomplished either in the application or in the calibration, a correction can be made if the accuracy is required.

For the glass used for practically all thermometers made for temperatures up to 450°C (842°F), the immersion stem correction can be computed closely by using the following equation:

$$\text{Stem correction} = 0.00016°C\ (0.000089°F) \times N(T - t) \qquad (2.1)$$

where N = number of degrees emergent above the level of the bath in degrees Celsius or degrees Fahrenheit

T = temperature of bath in degrees Celsius or degrees Fahrenheit

t = average temperature of emergent stem in degrees Celsius or degrees Fahrenheit

The correction value is added to the thermometer scale reading if the ambient temperature is lower than the bath temperature, and is subtracted from the thermometer reading if the ambient temperature is

Table 2.1   Temperature Generators and Standards

Generating Facility[a]	Standard	Temperature
Water bath	Thermometer	32 to 212°F
Salt bath	Thermometer	0 to 750°F
	Thermocouple	0 to 750°F
Sand bed	Thermometer	200 to 980°F
	Thermocouple	200 to 2000°F
Glass bead bed	Thermometer	200 to 980°F
	Thermocouple	200 to 1200°F
Liquid nitrogen	Thermometer	−40°F
	Thermocouple	−40°F
Liquid Nitrogen plus		
Dry Ice (CO_2)	Thermometer	−80°F
	Thermocouple	−80°F
Dry Ice (CO_2 sublimation)	Thermocouple	−109.3°F
Ice point	Thermometer	32°F, 0°C
	Thermocouple	
Liquid oxygen (bp)	Thermocouple	−297.364°F
Liquid helium	Thermocouple	
Sulfur (fp)	Thermometer	832.280°F
	Thermocouple	832.280°F
Gold (fp)	Thermocouple	1945.4°F
Palladium (fp)	Thermocouple	2826°F
Platinum (fp)	Thermocouple	3216°F
Blackbody	Infrared pyrometer	−40 to 9000°F
	Optical pyrometer	1000 to 20,000°F
Plasma jet	Optical pyrometer	3000 to 30,000°F

[a] fp, freezing point; bp, boiling point.

higher than the bath temperature. When accuracy is of the highest importance, the ambient temperatures are often specified for the precision calibration standards. Where temperature conditions vary widely from the average ambient temperatures used during calibration of the standard, corrections must be made for the highest order of accuracy.

Resistance thermometers or resistance temperature detectors (RTDs) change resistance with a temperature change, so that a good resistance bridge is an essential part of the calibration equipment. RTDs exhibit an increase in resistance with a temperature increase, and a calibration bridge may be calibrated in either resistance or temperature units. The calibration temperature may be produced by a suitable bath or an equilibrium mass that can be maintained at the freezing or boiling point of a suitable metal or alloy.

Thermistors are also resistance thermal detectors, and exhibit a decrease in resistance with a temperature increase. Bridges or other resistance type measuring equipment are an essential part of the calibration system just as in the case of RTDs.

Thermocouples are usually considered contact temperature detectors that exhibit an increase in emf as a result of increasing temperature. Therefore, in addition to temperature generating equipment and an appropriate standard, a precision millivolt indicator and a reference point are needed to carry out thermocouple calibration. While any ambient temperature can be chosen as a reference, in calibration work the ice point is a favorite choice. In fact, all standard reference tables base their conversion values for the various types of thermocouples on the ice point reference. To avoid matching the resistance of the thermocouple wires to the indicating meter, use is made of a self-balancing potentiometer or null meter where a local emf value is used to balance out or null the value generated by the thermocouple at any given temperature. Care must be exercised so that the thermocouple being calibrated has the same contact pressure and the same surface area contact as the calibration standard. For the most accurate measurements, the calibration setup should be as near the application setup as possible. For example, if the application requires a protective well or an insulation penetration with the thermocouple under spring pressure, or the thermocouple is welded to a tab fastened to the object whose temperature is to be indicated, measured, or controlled, both the unit under calibration and the standard should be placed in a simulated setup to obtain the most accurate calibration. When boiling or freezing points of metals are used in calibration practices, it is not necessary to use a measurement standard if pure metals or alloys are used. These points are recognized as standard values for calibration purposes.

Two precision test thermometers are shown in Figure 2.1, a resistance thermometer in Figure 2.2, and a noble metal thermocouple in Figure 2.3. Precision mercury-in-glass thermometers, resistance thermometers, and noble metal thermocouples are all good acceptable secondary standards for industrial calibration applications.

2.3 PRESSURE CALIBRATION STANDARDS

Standards for the calibration of pressure indicating and recording instruments cover the range from 10^{-11} mmHg, to tons per square inch. The range from atmospheric pressure to 10^{-11} mmHg is normally considered

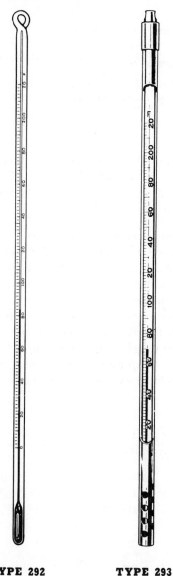

TYPE 292   **TYPE 293**

Figure 2.1 *Precision test thermometers. Type 292 glass thermometer for total immersion up to the reading point. Type 293 thermometer for highly precise temperature readings in a stainless steel sheath guard for maximum protection. Courtesy Dresser Industries.*

a vacuum, and all pressures above atmospheric pressure are considered pressure.

An industrial or laboratory instrument calibration facility should have adequate standards to cover the range applicable for the operation. The common standards employed are ionization gages for the lowest pressures (10^{-6} to 10^{-11}), McLeod or McCloud gages from 0 psi to 10^{-6} mmHg,

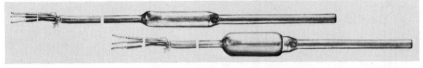

Figure 2.2 *Resistance thermometer. Platinum Thermohm detector using a platinum wire resistor adjusted precisely to 25 Ω at 32°F and hermetically sealed in a stainless steel case. Courtesy Leeds and Northrup Co.*

manometers from 0 lb/in.² to a maximum of 100 lb/in.² and deadweight testers, or precision pressure gages from 5 to 100,000 lb/in.² All these gages are covered in detail in Chapter 3, and the pressure standards most used in industrial applications are presented here.

Liquid Column Gages.   The liquid column gage shown in Figure 2.4 is used as a primary standard for all the secondary standards used for calibration of pressure measuring instruments in its range. The accuracy of this type of gage for measuring pressure, under laboratory conditions, is 1 part in 10,000. The gage consists primarily of a glass or metal tube filled with a liquid of known density that exerts a hydrostatic pressure which varies directly with the height of the liquid in the column. A mercury column 15 ft high has a pressure range up to 88.408 lb/in.² at 20°C (68°F) under standard pressure and temperature conditions. The pressure in the liquid column can be calculated by use of the equation

$$P_{psi} = Dh \qquad (2.2)$$

where P = pressure in pounds per square inch
       D = density of liquid in pounds per cubic inch
       h = height of column in inches

At 20°C the density of mercury is 0.491157 lb/in.³, and 15 ft equals 180 in. This gives a pressure of 0.491157 lb/in.³ × 180 = 88.40826 lb/in.²

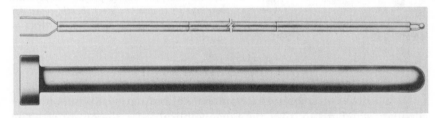

Figure 2.3 *Noble metal thermocouple. For high precision, using ice bath reference junction. Courtesy Leeds and Northrup Co.*

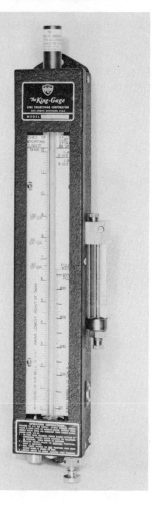

Figure 2.4 *Liquid column gage. An absolute pressure manometer available in ranges up to 100 in. of liquid acting on the same principle as a barometer. Courtesy King Engineering Corp.*

Manometers. Most industrial instrument shops employ a manometer, shown in Figure 2.5, instead of a straight liquid column, because manometers are less cumbersome to use and generally satisfy accuracy requirements for calibration purposes. The manometer pressure or vacuum standard normally requires only two corrections, except for very precise calibrations. These two corrections are for temperature and gravity errors. Other errors that may call for corrections are scale, capillary, compressibility, and the effects of absorbed gases.

Temperature error is introduced when a liquid column or manometer is used at any temperature other than that considered standard for the

(a)

(b)

Figure 2.5 *Manometers. (a) Well type. (b) U-tube type. Courtesy King Engineering Corp.*

pressure unit. This error is caused by a change in density of the liquids with temperature. If the scale expands or contracts with temperature, an additional error is introduced. The expansion of the scale reduces the reading in the column held at constant pressure, and the expansion of the liquid increases the reading of the column. The two expansions tend to balance each other, but the expansion of the liquid is usually greater. When a scale correction is necessary, temperature corrections are given by the following equations:

$$H_0 = H_t + T \tag{2.3}$$

and

$$T = \frac{S(t - t_s) - a(t - t_0)}{1 + a(t - t_0)} H_t \tag{2.4}$$

where H_0 = height of liquid column at standard temperature of calibration

H_t = indicated height of liquid column at temperature t, including scale error corrections

T = temperature correction

S = scale coefficient of linear expansion

A = liquid coefficient of cubical expansion

t = temperature of liquid column

t_s = temperature at which scale indicates the true height

t_o = standard temperature of calibration at which the height of the liquid column is in terms of the pressure unit used

If the correction is positive in sign, it is added to H_t, as indicated by the equation.

When mercury is used as the liquid, a requires no modification over the temperature range usually encountered during calibration procedures. For water the liquid coefficient of cubical expansion varies considerably with temperature, and is significant for most measurements at all temperatures of interest. The equation for T is not valid for water, and corrections must be applied in two parts, first for the scale error and second for the error of liquid expansion or contraction due to temperature.

When the most exact measurements are needed during the calibration of the pressure instrument, the effect of variation of gravity from the standard value can be evaluated from

$$H_0 = \frac{g}{g_0 H_i} \tag{2.5}$$

where H_0 = height of liquid column under standard gravity conditions

H_i = indicated height of liquid column at standard temperature

g = value of gravity at the location where the calibration is being conducted

g_0 = standard value of gravity, usually designated for sea level

The Smithsonian Meteorological Table can be used to compute the ambient value of gravity at any location, as a function of latitude, when this information is not available from other sources, such as the Geodetic Survey Service. The following relationship is useful when the value of gravity has to be computed as a function of latitude:

$$g = g_0 - 0.000094h \qquad (2.6)$$

where g = value of gravity at the desired location

g_0 = standard value of gravity at sea level

h = height in feet above sea level

While this equation is strictly true only for free air, it is sufficiently accurate for most locations at moderate elevations above sea level.

Deadweight Gages. At pressures above those practical for a liquid column type standard, a deadweight gage tester, shown in Figure 2.6a, is a very useful instrument. These testers employ a piston on which weights are placed to exert a pressure on a hydraulic fluid used to actuate the pressure gage under calibration. These testers can be used for pressures over 5 lb/in.^{2} in the 6 to 2500 lb/in.^{2} range, and over 30 lb/in.^{2} in the 30 to 12,000 lb/in.^{2} range. At the higher maximum pressures the minimum pressure is the minimum weight exerted by the piston and the weight holding fixtures for the particular tester used. This can range from 30 to 250 lb/in.^{2} In all measurements the weight should be free floating and should be revolved slowly to minimize the effects of friction.

Deadweight testers are capable of an accuracy of 0.01% of the reading with a resolution of 5 ppm at full load, decreasing to 50 ppm at no load or empty weight. In the lower ranges these testers have increments as low as 0.1 lb/in.^{2}, and in the higher ranges, 0.5 lb/in.^{2} increments. A hand-operated pump, shown in Figure 2.6b, applies pressure to the liquid system supplied from a reservoir or pressure fluid tank, shown in Figure 2.6c, through a needle valve. Pressure is applied by this hand pump until the entire system is in hydraulic balance with the piston carrying the calibration weights floating in the cylinder.

Deadweight gages can be used when the test gage being calibrated is subjected directly to the pressure of the hydraulic fluid in the deadweight gage, or they can be employed to calibrate gages with an auxiliary

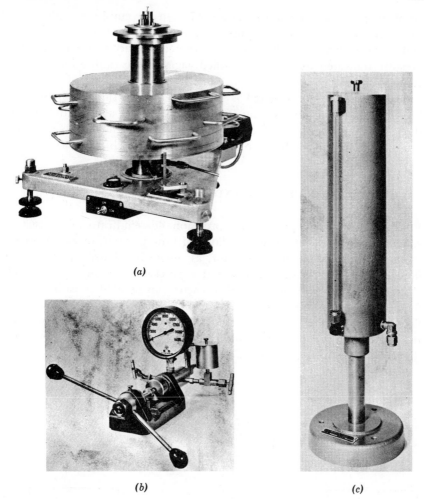

(a)

(b)   (c)

Figure 2.6 *Deadweight gage. (a) Model 2400 master reference standard. (b) Hand pump. (c) Fluid tank. Courtesy Ruska Instrument Co.*

pressure unit in which they are balanced against each other and a differential pressure indicator is being used. The latter method is used when the pressure medium of gas, water, or other liquid used with the gage under calibration is not compatible with the hydraulic fluid used in the deadweight gage.

Standard Gage Hydraulic System. In industrial instrument shops where the cost of a deadweight tester is prohibitive, a hydraulic system and two certified gages, as shown in Figure 2.7, can be employed. When the hydraulic system is used to supply the pressure to the two certified gages and to the gage to be calibrated, it is accepted practice to use the average pressure indicated by the two certified gages as the calibration pressure value.

One precaution to be noted is that oxygen gages should not be calibrated with a hydraulic system in which oil is used as the fluid, unless the gage is thoroughly cleaned to remove all traces of materials that may cause an explosion in the gage or the system in which it is used.

2.4 FLOW CALIBRATION STANDARDS

Precision volumetric pumps and precision flow indicators are used as standards for small and moderate flows of liquid, vapor, or gas. Most instrument shops use a flow indicator, as shown in Figure 2.8, for small flows, and use calibrated orifices and a precision manometer for moderate to large flows. A precision manometer can be used to calibrate differential type flow indicators and controllers when venturi tubes, orifices, flow nozzles, or pitot tubes are used as flow detection devices. These detection devices and other flow indicating and recording instruments are discussed in detail in Chapter 7.

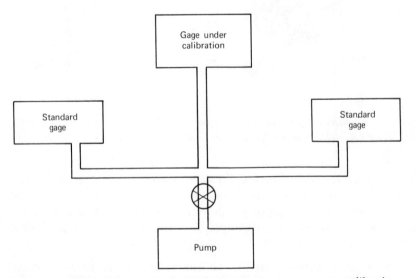

Figure 2.7 *Block diagram of a hydraulic system for pressure gage calibration.*

2.5 WEIGHT CALIBRATION STANDARDS

Every instrument shop should have a set of NBS calibrated and certified weights, similar to those shown in Figure 2.9, for use in calibration of analytical balances and for accurately weighing the mercury needed in manometers. It is also desirable in some instances to have accurate platform scales for weighing liquids for calibration of small flowmeters and for weighing the large volumes of mercury used in differential flowmeters.

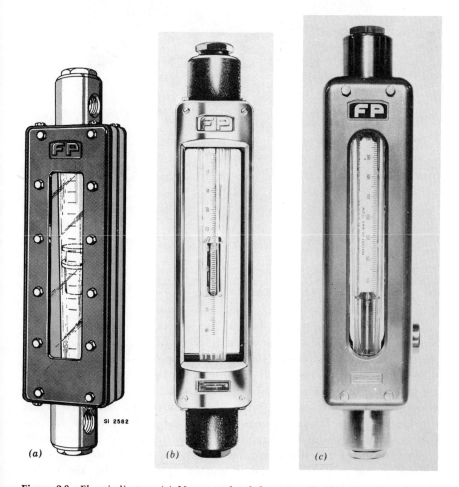

Figure 2.8 *Flow indicators. (a) Master enclosed flowrator. (b) Side plate flowrator, (c) Standard flowrator. Courtesy Fischer and Porter Co.*

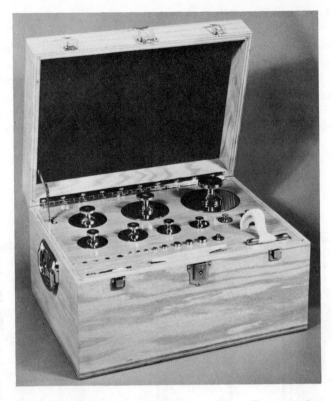

Figure 2.9 *Analytical weights. Courtesy Henry Troemner, Inc.*

2.6 TIME CALIBRATION STANDARDS

A good precision stopwatch, such as the one shown in Figure 2.10, is adequate for calibration of many time response process instruments when the time intervals exceed 1/100 s. Such time measurements as chart speeds, indicator responses, and printing cycles fall into this class. There are many applications in industry and in the laboratory in which millisecond to microsecond responses must be measured and calibrated. Time standards, such as the one shown in Figure 2.11, are available and deliver time intervals of this magnitude. These instruments are normally crystal controlled and temperature compensated. The timing signals delivered by the calibration standard can be compared with the timing pulse of the equipment to be calibrated by projecting them on an oscilloscope. In some cases the time interval can be measured directly by

(a)

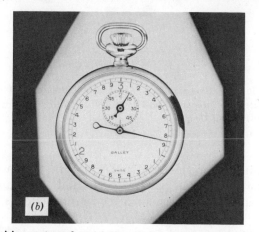

(b)

Figure 2.10 *Precision stopwatches. (a) Two Gallet stopwatches mounted for a permanent installation. (b) A single unit to be used as a portable instrument. Accuracy to $\frac{1}{100}$ s. Courtesy Jules Racine and Co., Inc.*

comparing it with the time base or time delay incorporated in the oscilloscope.

2.7 ELECTRICAL AND ELECTRONIC CALIBRATION STANDARDS

Where large volumes of instruments used for the measurement of current, voltage, and power are serviced and calibrated regularly, a special

Figure 2.11 *Calibration time standard. The Hewlett-Packard 218A digital delay generator can generate time intervals from 1 to 10,000 μs with ±0.1 μs ± 0.001% accuracy. Courtesy Hewlett-Packard Co.*

calibration unit, such as the Cohu and RFL units shown in Figure 2.12, should include ac and dc power sources and precision readouts for microammeters, milliammeters, ammeters, millivoltmeters, and voltmeter calibrations.

The Cohu Model 351 solid state voltage standard shown in Figure 2.12a is used for dc instruments and has a calibration accuracy of ∓0.003% of setting for the voltage to be delivered. This unit covers three major voltage ranges from 10 V in 1 μV steps to 1000 V in 100-μv steps, and currents from 1 to 50 mA at any voltage setting. This unit can be used to certify dc digital voltmeters, dc voltage amplifiers, analog dc voltmeters, voltage controlled oscillators, potentiometers, and transducers.

The Cohu Model 365 shown in Figure 2.12b is a self-contained differential dc voltage standard with an accuracy of ±0.001% for the voltage to be delivered. This unit is useful as a null indicating, potentiometric, dc measuring instrument (differential voltmeter), a direct reading dc voltmeter, or a precision dc voltage source. This unit has nine

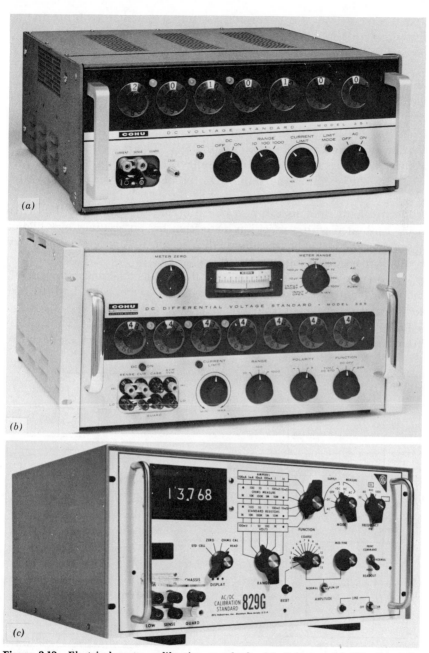

Figure 2.12 *Electrical meter calibration standards. (a) Cohu model 351. (b) Cohu model 365. Courtesy Cohu Electronics, Inc. (c) RFL model 829G. Courtesy RFL Laboratories, Inc.*

decade ranges from 10 μV to 1000 V full scale, high visibility in-line readout with illuminated decimal points, overcurrent and overvoltage protection, and linearity "self-checking."

The RFL Model 829G shown in Figure 2.12c is a self-contained portable standard for calibration of ac and dc electrical measuring instruments with a direct reading, full scale accuracy of ±0.02% dc and ±0.03% ac volts and amperes, ±0.06% resistance to 100 Ω, and ±0.1% from 1 to 10 MΩ. This unit covers a voltage range from 10 mV to 1400 V, and a current range from 10 μA to 14 A. The readout has a repeatability of reading within ±0.01% of range. Automatic protection is provided for the operator and the meters by means of interlocks and high voltage discharge circuits.

Electrical instruments can also be calibrated by comparing the meter being calibrated with a precision meter that has an accuracy of a higher order than the meter being calibrated. A factor of 10 in greater accuracy is usually considered minimal.

Wattmeters can be calibrated by the voltmeter–ammeter method, shown in Figure 2.13, by using dc voltages and currents for both dc and ac instruments of the electrodynometer type.

Potentiometer test sets, like the one shown in Figure 2.14, with a calibrated millivolt output are now available with self-checking features.

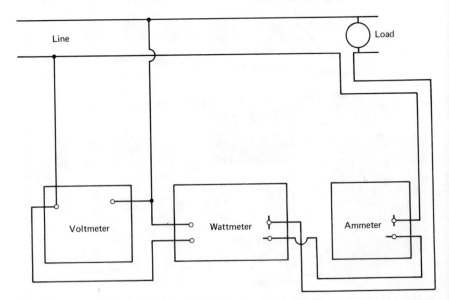

Figure 2.13 *Ammeter-voltmeter block diagram arrangement for calibration of wattmeters.*

Figure 2.14 *Rubicon potentiometer test set. Rubicon series 2920, calibrated in volts and millivolts. Calibrated accuracy 0.1% of reading or one scale division, whichever is larger. Courtesy Penn Airborne Products Co.*

Calibration is checked by means of a certified standard cell. These standards are used for the calibration of pyrometer equipment and other millivolt input indicating, recording, and controlling instruments, such as the Leeds and Northrup Speedomax, the Honeywell Electronik, the General Electric HF, and the Bristol Dynomaster recorders and controllers, which are discussed in Chapter 5.

Voltage standards, such as the one shown in Figure 2.15, are essential for the calibration of oscilloscopes so that the observed signal amplitude can be read directly from the screen. Some of the later oscilloscopes have a self-contained voltage calibration source. Voltage standards are also required for the calibration of the voltage sources used in computer applications.

Precision resistance standards, like those shown in Figures 2.16 and 2.17, are needed to calibrate resistance measuring instruments, such as

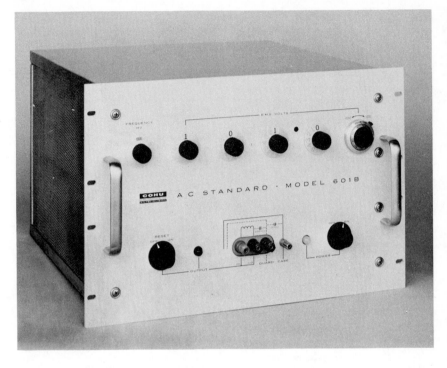

Figure 2.15 *Cohu voltage standard. The model 601B AC voltage standard supplies 1 to 501 V with an accuracy of 0.05% at 60, 400, and 1000 Hz where the frequency accuracy is 1.0%. Courtesy Cohu Electronics, Inc.*

Wheatstone bridges and ohmmeters used in the maintenance and repair of instruments and for measuring the resistance of thermocouples to match them properly to the pyrometer or millivolt meter indicator.

The precision resistance shown in Figure 2.16 are single units. The NBS type, shown in Figure 2.16a, is available in 1, 10, 100, 1000, and 10,000 Ω units with a limit of error of 0.01% up to 0.1 W and 0.04% up to 1.0 W. The Reichsanstalt type, shown in Figure 2.16b, is available in 0.001, 0.01, and 0.1 Ω units with a limit of error of 0.02% up to 2 W and 0.05% up to 10 W.

Precision decade resistances, such as those shown in Figure 2.17, are available with accuracies of 0.05% for 1 to 1,000,000 Ω ranges. The dec-

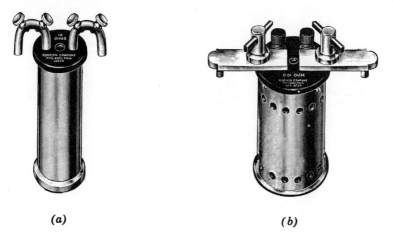

(a)   (b)

Figure 2.16 *Precision resistance standards. (a) NBS type. (b) Reichsanstalt type 1150. Courtesy Honeywell, Inc.*

ade-per-step resistances range from 0.1 Ω for the 1 Ω decade to 100,000 Ω for the 1,000,00 Ω decade.

2.8 RADIOACTIVE SOURCES

In industrial processes in which use is made of radioactive tracers to follow flows, or in which radiation sources are used in nondestructive testing operations or to maintain the thickness of rolled material, a calibrated radiation source is required for calibration of the detecting instruments that control the pumping speed, the flaw detector, or the pressure on the rolls.

2.9 VELOCITY OR SPEED STANDARDS

The speed of a drive mechanism or other moving pieces of equiment is normally measured by means of a tachometer or Strobotac (General Radio Company). A precision drive synchronous motor is adequate as a calibration standard for tachometers if three fixed speeds are available to check the output value of the tachometer shown in Figure 2.18 for range and indicator reading versus speed. The Strobotac, shown in Figure 2.19,

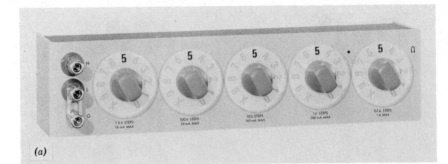

Figure 2.17 *Decade resistors. (a) General Radio 1434-N decade resistor (five decades).*
(b) Decade resistor 1434-N (seven decades). (c) Programmable decade resistor 1435.
Courtesy General Radio Co.

Figure 2.18 *Standco UH hand tachometer. Accuracy 0.5% of full scale deflection on each of six ranges. Ranges cover from 30 to 50,000 r/min normally and 150 to 100,000 r/min when extended. Courtesy Herman H. Sticht Co., Inc.*

is a flashing light source which contains a built-in calibration system which uses the power line frequency for quick, easy checking and adjustment of the flashing rate action.

2.10 FREQUENCY STANDARDS

High sound levels and vibration are not only annoying, but are detrimental to the efficiency of both personnel and equipment. Communication equipment must be kept calibrated to meet Federal Communications Commission (FCC) rules and regulations. Equipment for measuring these parameters is frequency dependent, so that frequency standards such as the one shown in Figure 2.20 are needed for calibration purposes.

Figure 2.19 *Strobotac 1531-AB. Accuracy of ±1% of dial reading after calibration on middle range (670 to 4170). Unit covers three ranges 110 to 690, 670 to 4170, and 4000 to 25,000. Speeds up to 250,000 r/min can be measured. Courtesy General Radio Co.*

These types of standards are normally crystal controlled and temperature compensated.

2.11 SUMMARY

The accuracy of a measuring instrument can not be greater than the accuracy of the standard used in its calibration. The standard should be chosen to have a higher accuracy if possible, so that there is no doubt about the measurement accuracy. The use of calibration standards is discussed in the following chapters, as different standards are employed to calibrate the representative types of instruments used in a particular type of test or process. A general summary is shown in Table 2.2.

Table 2.2 Summary of Standards and Their Applications

Calibration Standard	Range	Application	Medium Generator
Precision thermometer	−80 to 995°F	Mercury-in-glass thermometers, gas bulb thermometers, vapor bulb thermometers, bimetal thermometers, resistance thermal detectors, thermocouples	Water baths, salt baths, sand beds, boiling point metals, freezing point metals.
Noble metal thermocouple	−298 to 2000°F	Thermocouples, thermometers, resistance thermal dectors	Liquid oxygen (bp), liquid nitrogen, Dry Ice, liquid helium, gold fp), palladium (fp), platinum (fp)
Optical pyrometer	1000 to 30,000°F and up	Thermocouples, infrared pyrometers, two-color pyrometers, color-ratio pyrometers, optical pyrometers	Furnaces, blackbodies, plasma jets
Liquid column	0.1 to 100 in. H_2O or Hg	Manometers (U-tube and well), diaphragm pressure gages, bellows pressure gages, pressure transducers	Pumps (pressure and vacuum), gas cylinders, aspirators
Precision manometer	0.1 in. H_2O to 100-lb/in.^{2}	U-tube manometers, well manometers, diaphragm pressure gages, bellows pressure gages, differential pressure gages	Pumps, gas cylinders, aspirators
Deadweight gage	5 to 50,000 lb/in.^{2}	Metal mercury monometers, bellows gages, diaphragm gages, Bourdon tube gages, high pressure transducers, differential pressure gages	Hydraulic pumps, gas cylinders

Table 2.2 (Continued)

Calibration Standard	Range	Application	Medium Generator
Precision pressure gage	−30 to 50,000 lb/in.^{2}	Compound gages, bellows gages, diaphragm gages, Burdon tube gages, differential pressure gages	Vacuum pumps, hydraulic pumps, gas cylinders
McLeod gage	1 to 10^{-6} mmHg	Vacuum gages, compound gages (vacuum half)	Vacuum pumps, diffusion pumps, ion pumps
Volumetric pumps	Cm^3/min to gal/min (small flows)	Liquid flowmeters, gas flowmeters, vapor flowmeters, flowrators	Controlled speed motors, calibrated pumps
Precision flowrator	Cm^3/min to gal/min. (small flows)	Liquid flowmeters, gas flowmeter, vapor flow	Pumps, controlled heads, cylinders
Manometer and orifice	cc/min. to gph	Differential flow indicators, pitot tubes, venturi tubes, flow nozzles	Pumps, hydraulic heads gas cylinders
Analytical weights	Milligrams to kilograms	Scales, analytical balances, small flowmeter outputs	Mercury for manometers
Stopwatch	Time intervals over 0.01 s	Instrument time responses, chart speeds, printing cycles, dead time	Recorders, timers, process operations, sequencers
Time marker standards	1.0 s to 10^{-8} s	Instrument response, recorder frequency markers, oscillograph time markers, time lag responses, feedback responses	Instruments, processes, motors, printers
Voltage standards	10 mV to 5000 V (ac and dc)	Analog voltmeters, digital voltmeters, differential voltmeters	Power supplies (ac and dc), commercial power, batteries
Current standard	10 μA to 50 A (ac and dc)	Analog ammeters, analog milliammeters, analog microammeters, digital	Power supplies, alternators, generators, batteries, ac-to-dc

Table 2.2 (Continued)

Calibration Standard	Range	Application	Medium Generator
		multimeters, digital microammeters, digital milliammeters, digital ammeters	converters
Wattmeter (power standard)	Milliwatts to killowatts	Wattmeters, watt hour meters, circuit consumption	Voltage generators, current generators, power supplies
Potentiometer test set	0.1 mV to 1.6 V	Self-balancing recorders, millivolt indicators, Electronik recorders, Speedomax recorders, Dynomaster recorders	Thermocouples, RTD bridge power supplies, null indicator balance supply
Resistance standard	0-1 Ω 1-100 100 Ω to 1 MΩ	Resistance bridges, resistance multimeters, impedance systems, component readout	Circuits, loads, components
Capacitance standard	1.0 μF to 200 mF (multistep)	Component readout, circuits, bridges	Circuits (networks), components
Tachometer	30,000 to 50,000, 50,000 to 150,000 r/min	Rotating shafts, moving parts	Motors, generators, gear trains transmissions, rotors
Strobotac	110 to 690, 670 to 4170, 4170 to 25,000 r/min		
Frequency generators	5 MHz, 1 MHz, 100 khz	Oscillators, oscilloscopes, television, radar, radio	Cesium, rubidium, quartz
Radioactive sources	1 mCi to 1 Ci	Radiation detectors, level detector calibration, thickness gages, moisture gages	Selenium isotopes, cobalt isotopes, rubidium isotopes

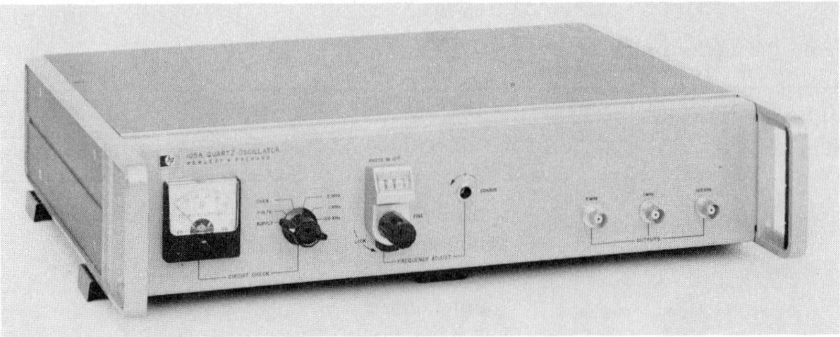

Figure 2.20 *Hewlett-Packard 105A quartz oscillator frequency standard. Output is stable to within ±2 parts in 10^{11} regardless of load changes occurring in any other output frequency. Courtesy Hewlett-Packard Co.*

Review Questions

2.1 What are the recognized types of calibration standards? How do they differ?

2.2 Why are calibration standards needed in the manufacturing and processing industries?

2.3 Under what two conditions are mercury-in-glass standard thermometers calibrated? What are the advantages of each?

2.4 How is the etched scale prepared to ensure the best precision for an etched stem thermometer?

2.5 How is separation or distillation of the mercury in a mercury-in-glass thermometer minimized?

2.6 Under what conditions should noble metal thermocouples be used as standards for calibration?

2.7 Under what conditions should the resistance thermometer be chosen as the calibration standard?

2.8 Under what temperature conditions is the accuracy of a standard not guaranteed? Why?

2.9 If a pressure gage is being used to measure water pressure, explain how to calibrate it using a deadweight tester for the 0 to 500 lb/in² range.

2.10 How would you use a deadweight tester to calibrate a 0 to 3000 lb/in² pressure gage to measure oxygen gas pressure? What accuracy could you guarantee?

2.11 Explain how precision certified standard pressure gages can be used to calibrate other pressure gages.

2.12 Discuss the type of calibration equipment you would use to calibrate a meter with a range from 0 to 5 gal/h liquid flow. Would
you use the same type equipment for a 1% calibration of a meter
covering a 0 to 500 gal/min flow? If not, what would you use?

2.13 What use does a calibration group have for certified standard
weights.

2.14 Where are potentiometer calibration test sets used in industry?
On what features is their accuracy based?

2.15 A precision motor drive requires calibration. Discuss the equipment you would choose for the calibration to cover up to 3600
r/min.

2.16 What features would you choose in a calibration standard to calibrate transmitting equipment for a guarantee to meet FCC specifications?

Problems

2.1 What range thermometer would you choose for a temperature
measurement of 50°F with 0.1°F divisions, 300°F with 0.5°F divisions, and 200°F with 0.2°F divisions?

2.2 If a partial immersion (3 in. immersion) thermometer was calibrated for an ambient temperature of 150°F, what is the immersion stem correction when the temperature of the bath reaches
200°F if the thermometer is immersed to the 50°F mark?

2.3 A pressure system is being calibrated, and the liquid in the column
gage is 6.00 ft high and has a density of 0.491157 lb/in³. What is
the system pressure?

2.4 A manometer has been calibrated in a calibration room at 135 in.
of liquid at 68°F. The manometer is then moved to a process area
where the temperature is 90°F. If the scale coefficient of linear expansion is 0.00005, what height does the manometer read in the
process area for the 135 in. calibration? Is such a correction
necessary?

2.5 A pressure manometer was calibrated at one national laboratory
320 ft above sea level and was shipped to another national laboratory located 7040 ft above sea level for use. What is the indicated
height at the laboratory where the manometer is being used,
assuming no temperature corrections are required?

2.6 A recorder capable of recording 3000 cycles/s has a large range of
chart speeds. What calibration unit would you use for 500, 100, 50,
25, and 10 mm/s chart speed certification?

58

2.7 A triple-range dc instrument with 0.2% accuracy must be calibrated. The ranges are 0 to 5, 0 to 50, and 0 to 500 mA. What type and what accuracy standard would you use? What type would you choose if it were an ac instrument?

2.8 A multirange dc voltmeter with 0.1% accuracy has been given you to calibrate. This meter covers 10 voltage ranges from 0 to 5 mV to 0 to 2000 V. What type or types of calibration standards are you going to use? Justify your choice.

2.9 A Wheatstone bridge with 0.1% accuracy has been brought in for calibration. What standard would you choose to use for the calibration to cover the 0.1 to 10,000 Ω range? Justify your choice.

Bibliography

Cohu Electronics, Inc., Kintel Division, *AC Voltage Standard*, Data Sheet 20-4, San Diego, no date.

M. Ducommun Company, *One Hundred Years of Precision Timing Service*, Catalog No 258, New York, no date.

General Radio Company, *Catalog Q*, West Concord, Mass., 1961.

Herman H. Sticht Co., Inc., *Standco Type UH Universal Hand Tachometer*, Catalog No. 666, Bulletin No. 750, New York, no date.

Hewlett-Packard Company, *Electronic Test Instruments*, Palo Alto, 1959.

Honeywell Rubicon Instruments, Minneapolis-Honeywell Regulator Company, Industrial Division, RS4801-10M, Minneapolis, July, 1959.

Leeds and Northrup Company, *Thermocouples, Assemblies, Parts and Accessories*, Catalog EN-S2, N. Wales, Pa., 1955.

Leeds and Northrup Company, *Thermohm Temperature Detectors*, Catalog EN-S4, N. Wales, Pa., 1957.

Manning, Maxwell and Moore, Inc., *American Industrial Thermometers*, Catalog 100A, Stratford, Conn., 1959.

Manning, Maxwell and Moore, *Ashcroft Gauges*, Catalog 300, Stratford, Conn., 1953.

Radio Frequency Laboratories, Inc., *Product and Application News*, Vol. VI, Boonton, N.J., November 1962.

Radio Frequency Laboratories, Inc., *R. F. L. Test and Service Products*, Boonton, N.J., 1962.

Ruska Instrument Corporation, *Dead Weight Gage, Model 2400*, Houston, 1962.

Ruska Instrument Corporation, *Pressure Calibration Systems and Components*, 12,000 PSI Series, Houston, 1962.

Sensitive Research Instrument Corp., *Electrical Measurements*, Vol. 29, No. 9, Mt. Vernon, N.Y., September 1962.

Soisson, Harold E., *Electronic Measuring Instruments*, McGraw-Hill, New York, 1961.

Trimount Instrument Company, *Manometers and Accessories*, Data Sheets 3, 4, 20, and 50, Chicago, no date.

CHAPTER 3

Pressure and Vacuum

Pressure measurements are some of the more important measurements made in industry, especially in continuous process industries such as chemical processing and manufacturing. The number of pressure measuring instruments may be far greater than that of any other type of instrument used.

The principles used in the measurement of pressure are also applied in the measurement of temperature, flow, and liquid level. Therefore it is essential to know general operating principles, types of instruments, installation principles, how instruments should be maintained to obtain the best possible operation, how they are used for control of a system or operation, and how they are calibrated.

Pressure is a force exerted over a given area and is measured in units of force per unit area. This force may be applied to a point on a surface or distributed over the surface. Every time it is exerted a deflection, distortion, volume, or dimension change occurs. Pressure measurements range from very low values that are considered a vacuum to thousands of tons. See Figure 3.1 for gage types and ranges.

Pressure may be adequately and properly expressed in any of the following units:

Grams per square centimeter (g/cm^2)
Millimeters of mercury (mmHg)
Pounds per square inch (lb/in^2)
Inches of water ($in.H_2O$)
Inches of mercury (in.Hg)
Microns (10^{-3} mmHg)
Torrs (1 mmHg)
Atmospheres (atm)
Tons per square inch or per square foot ($tons/in.^2$ or $tons/ft^2$)

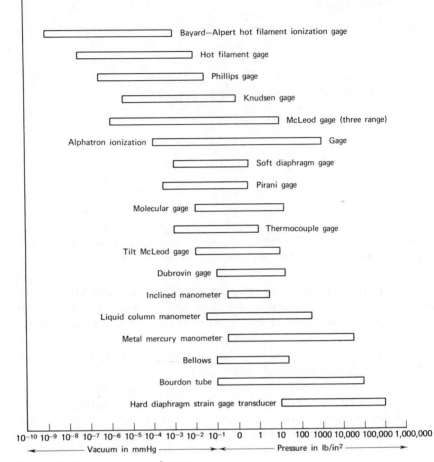

Figure 3.1 *Pressure gages and ranges.*

Pressure is measured either as an absolute value, which is the total force exerted, or as a differential value, which is the algebraic difference between the absolute value and the value due to the surrounding atmosphere at the time and place of the measurement. In equation form the gage and vacuum pressures may be expressed as:

$$P_g = P_a - P_s \tag{3.1}$$
$$P_v = P_s - P_a \tag{3.2}$$

where P_g = gage pressure
P_a = absolute pressure
P_s = atmospheric pressure at the time and place of measurement
P_v = vacuum pressure

In industrial applications pressure is normally measured by means of indicating gages or recorders. These instruments may be mechanical, electromechanical, electrical, or electronic in operation.

Mechanical instruments may be classified into two groups. The first group includes those in which the pressure measurement is made by balancing an unknown force against a known force. The second group includes those employing quantitative deformation of an elastic membrane for the pressure measurement.

Electromechanical pressure instruments usually employ a mechanical means for detecting the pressure, and an electrical means for indicating or recording the detected pressure.

Electronic pressure measuring instruments normally depend on some physical change that can be detected and indicated or recorded electronically.

3.1   PRESSURE AND FORCE BALANCE GAGES

By balancing an unknown force or pressure against a known force, pressure measurements, can be made with liquid column gages, limp diaphragm gages, bell gages, and piston gages.

Liquid Column Gages. The liquid column gage used most in industry is some type of manometer. It may be either the U type or the well type, as shown in Figure 3.2. The U type is made of glass or some other type of transparent tubing with an inner bore of $\frac{1}{4}$ in. or larger diameter and a wall thickness adequate to withstand the pressure for which the manometer was designed. The well type is similar to the U type, however, one leg of the U is replaced by a well. Figure 3.2a and b shows these gages connected into a system for a pressure measurement.

Manometers require two readings to obtain the height of the displaced liquid, which represents the pressure. In equation form,

$$P = Kd(h_1 - h_2) \tag{3.3}$$

where $d =$ density of the liquid

      $K =$ proportionality constant to provide corrections for units and factors

      $h_1 =$ height of liquid in the leg connected to P_1

      $h_2 =$ height of liquid in the leg connected to P_2

      $P =$ pressure in the system

In each case the density of the liquid determines the height of the column representing the pressure. This is illustrated in Figure 3.3, in which

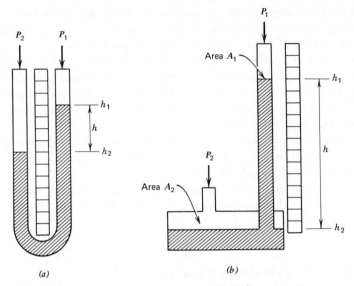

Figure 3.2   *Manometers. (a) U-tube manometer. (b) Well manometer.*

three liquids of different densities are exposed to the same pressure. The manometer in Figure 3.3*a* is filled with water which has a maximum density at 4°C. At this temperature water weighs 62.428 lb/ft³ in the English system of units, or 1 g/cm³ in the metric system. The U tube shown in Figure 3.3*b* is filled with oil, and that in Figure 3.3*c* is filled with mercury. By observation it is quite evident that height *b* is greater than height *a,* and that height *c* is much less than height *a.*

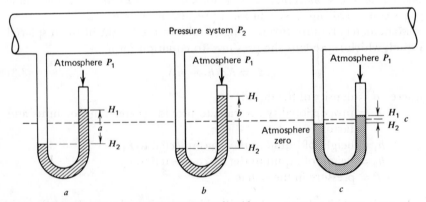

Figure 3.3   *Determining specific gravity of liquids.*

Since the manometer liquid in Figure 3.3*a* is water, whose density is considered unity, the density or specific gravity of the oil in Figure 3.3*b* has a value of a/b, say 0.78, while the mercury in Figure 3.3*c* has a value of 13.56. This merely means that the oil is lighter than water and that the mercury is heavier than water in direct relationship to their densities.

The *inclined manometer* or *draft gage* shown in Figure 3.4*a* is a well manometer whose vertical leg is placed in an almost horizontal position so that a very slight difference or change in the pressure of the gas or air in the well causes a very large change in the measured level of the liquid in the inclined tube. With a large well area A_2, compared to the vertical leg area A_1, there is a very small change in the height of the level of the liquid in the well for large changes in the inclined leg reading. This makes the well unit practically a direct reading manometer. Figure 3.4*b* is an photograph of the schematic representation in Fig. 3.4*a*.

Another industrial well manometer using mercury as the liquid is shown in Figure 3.5. These manometers are metal enclosures capable of

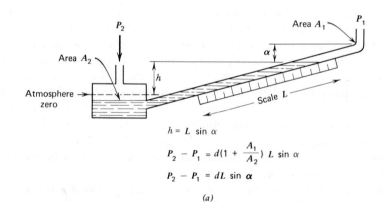

$$h = L \sin \alpha$$

$$P_2 - P_1 = d(1 + \frac{A_1}{A_2}) L \sin \alpha$$

$$P_2 - P_1 = dL \sin \alpha$$

(a)

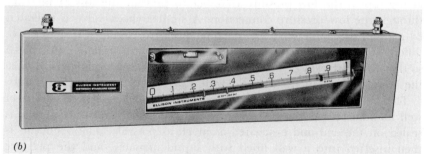

(b)

Figure 3.4 *Well draft gage. (a) Schematic. (b) Photograph. Courtesy Ellison Instrument Division, Dieterich Standard Corporation.*

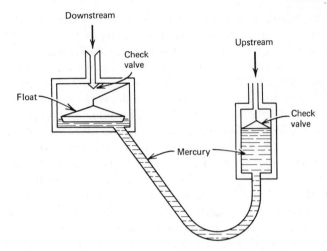

Figure 3.5 *Mercury manometer in cross section. Courtesy Taylor Instrument Co.*

withstanding static pressures as high as 5000 lb/in². The type shown is normally used to measure differential pressures in a high pressure system. The float is shown for zero differential pressure when the mercury is at the same level for both the upstream and the downstream connection. With an increase in pressure, the mercury in the upstream or high pressure chamber falls, and the mercury in the downstream or low pressure chamber rises. This causes the steel float to rise. This float is connected to a lever which rotates a shaft which in turn may drive a recording pen or an indicating pointer, or both, through a suitable linkage. It is also possible to attach a suitable mechanism to transmit a signal to actuate a valve or other control mechanism in the system.

Should the float travel exceed its normal range, it closes a check valve at the top of the low pressure chamber to prevent the loss of mercury through the low pressure connection. A similar check valve is operated by a small auxiliary float on the high pressure chamber if a negative differential pressure occurs. These manometers are usually constructed so that the upstream or high pressure chamber can be changed to make the basic instrument useful for a wide range of pressure measurements.

The *barometer,* used widely in weather forecasts, is a special type of well manometer. As shown in Figure 3.6, the upright measuring tube is sealed on the end and evacuated as much as possible. The open end is then inserted into a well filled with liquid mercury, and the pressure exerted on the surface of the mercury in the well forces the mercury up the evacuated tube. This represents the absolute pressure of the atmos-

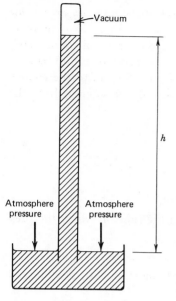

Figure 3.6 *Barometer schematic.*

phere at the time and place of measurement. The standard is 760 mmHg or 29.92 in.Hg for 1 atm, under standard temperature conditions at sea level, and varies for different altitudes and ambient temperatures.

The *ring balance manometer* shown schematically in Figure 3.7 is a special U-tube type. The pressure input connections are made with flexible couplings and tubing, so that the ring can move. As the pressure at

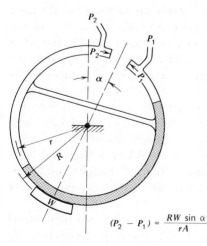

$$(P_2 - P_1) = \frac{RW \sin \alpha}{rA}$$

Figure 3.7 *Ring balance manometer schematic.*

the P_2 connection increases, it lowers the level of the mercury on its side of the ring. This causes the ring to turn about its center, as it becomes out of balance with the counterweight W due to the shift of the mercury in the column. The pressure is measured as a function of the unbalance of the weight W as it moves on its radius R with respect to the change in level of liquid in the ring on its radius r through an angle α from the original vertical position of the ring's Y axis.

$$P_2 - P_1 = \frac{WR}{Ar} \sin \alpha \qquad (3.4)$$

where W = counterbalance weight
     R = radius on which W moves
     A = area of the tube
     r = average radius of the mercury in the tube on which the pressures react
     α = angle of inclination

Many lower pressure manometer gages are made of both flexible and solid clear plastics, as shown in Figure 3.8.

Bell Gages. Bell gages operating on a pressure or force balance principle include balanced lever, beam, spring balanced, and Dubrovin vacuum gages. In each case the bell or bells are immersed in a liquid and measure differential pressures in the range of 1 to 15 in.H_2O, except for the Dubrovin gage which measures less than 1 in.H_2O.

The *balanced lever gage* is shown schematically in Figure 3.9a. As shown, the pressure P_2 is admitted to the bell well above the level of the liquid, and balance is obtained when $P_1 = P_2$ for the zero indication. Any change in pressure under the bell, either an increase or decrease, causes the system to seek a new equilibrium condition, and the pointer indicates the change from the zero balance condition. The indicator is designed and calibrated to give a direct pressure reading for the pressure change occurring. In this type of instrument the lever arm movement is limited to approximately 5°, and the sensitivity of the instrument is a function of the scale beam sensitivity. In turn, the sensitivity of the scale beam depends on the length of the beam, the mass of the system, and the type and condition of the beam pivots. Maximum sensitivity is obtained with a long beam of minimum weight. When a balanced lever gage is used in a system having pulsating or rapidly changing pressures, it is necessary to use a damping mechanism such as a dash pot to slow down the movement of the indicator.

The *beam bell gage* shown in Figure 3.9b consists of two light metal bells supported on a balanced beam. The beam is pivoted on knife

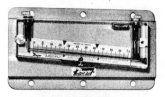

Model 200.5-ST
solid plastic stationary
inclined manometer.

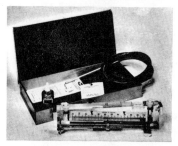

Model 100.5 solid plastic
portable gage kit.

Model 104
well-type manometer.

Model 424 solid plastic
inclined-vertical manometer.

Clear plastic
U-Tube manometer.

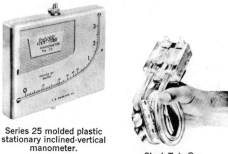

Series 25 molded plastic
stationary inclined-vertical
manometer.

Slack Tube®
roll-up portable manometer.

Figure 3.8 *Different types of manometers. Courtesy Dwyer Instruments, Inc.*

edges, and a counterweight is used to balance the deflection of the system. This counterweight is supported directly below the pivot point on the beam, and a pointer is attached to the pivot point. The pointer in this instrument indicates pressure as a function of the angle of deflection of the beam from the horizontal position as it moves on its Y axis. This is a differential pressure instrument whose action can be expressed as:

$$P = P_2 - P_1 = \frac{wd}{sA} \sin \theta \qquad (3.5)$$

where w = weight of counterweight
$\quad\quad d$ = distance from the pivot to the counterweight
$\quad\quad A$ = areas of the bells
$\quad\quad s$ = distance of each bell support from the pivot point
$\quad\quad \theta$ = angle of deflection of the beam
$\quad\quad P$ = differential pressure
$\quad\quad P_1$ = pressure under bell no. 1
$\quad\quad P_2$ = pressure applied under bell no. 2

The *spring balanced bell gage* has a portion of the bell weight supported by a coil spring, as shown schematically in Figure 3.9c. The bell is sealed with a light oil, and variations in pressure tend to change the position of the bell. Changes in pressure thus move the bell up or down until a balance or equilibrium is established between the weight of the bell, the forces of the coiled spring, and the exerted pressure. The stability and accuracy of this gage is established by the quality of the spring. Any deformations in the spring material, such as permanent set or loss of

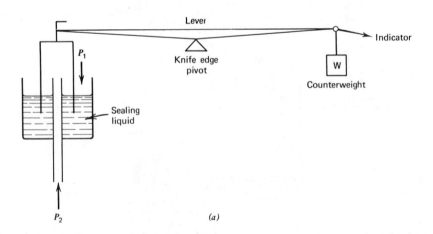

(a)

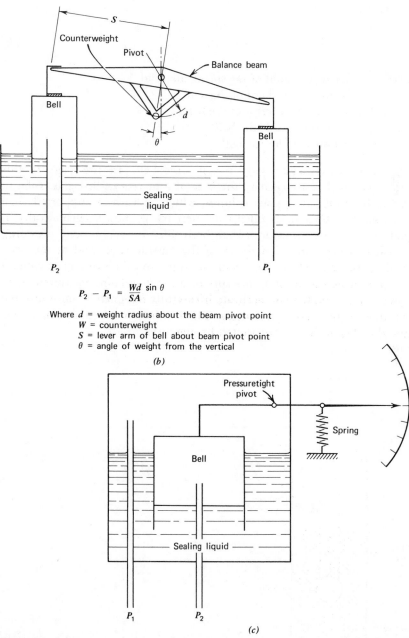

$$P_2 - P_1 = \frac{Wd}{SA} \sin \theta$$

Where d = weight radius about the beam pivot point
     W = counterweight
     S = lever arm of bell about beam pivot point
     θ = angle of weight from the vertical

(b)

(c)

Figure 3.9 *Bell pressure gages. (a) Balanced lever. (b) Double bell. (c) Spring-balanced differential gage.*

elasticity, cause errors in measurements taken. The action of this gage is expressed as:

$$P_2 - P_1 = \frac{F_c h}{A} \qquad (3.6)$$

where F_c = spring constant of the spring material
 h = vertical movement of the bell
 A = area of the interior of the bell
 P_1 = pressure outside the bell
 P_2 = pressure under the bell

The Dubrovin vacuum gage is a floating bell pressure gage. In this gage a glass bell is attached to the base of a mercury filled glass tube to provide buoyancy to a glass column, which in turn is supported in a second glass tube partially filled with mercury, as shown schematically in Figure 3.10. At zero pressure (when both sides are at the same pressure), the bell is deepest in the mercury of the outside tube, and as the pressure decreases the bell becomes more buoyant and the inner glass column rises to indicate the absolute pressure of the system being measured. Such gages are very reliable and accurate if carefully designed to minimize the error due to the mercury meniscus. The mercury meniscus is convex because the glass walls are not wetted by mercury.

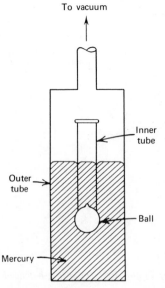

Figure 3.10 *Dubrovin vacuum gage schematic.*

3.2 PISTON PRESSURE GAGES

Piston pressure gages operate on the same principle as deadweight gages used for the calibration of pressure gages of all types. In these gages a calibrated weight is balanced against the piston in a free moving hydraulic system. The Bailey Meter Company employs a similar system in which a balancing float in a mercury pool connected to an auxiliary piston is used to oppose the pressure as applied to the measuring piston. In the Bailey gage the piston is rotated by a separate motor to ensure friction-free motion, and the piston is preloaded so that it can be brought to balance at the pressure to be maintained. Pressures higher or lower than the desired pressure cause the piston to move up or down. This motion is transmitted to the balancing float which moves in such a direction that the pressure piston is restored to its central equilibrium position. The auxiliary piston attached to the balancing float then gives the deviation from the desired pressure through an indicating system.

The Ruska air piston pressure gage shown in Figure 3.11 is useful in precision pressure measurements from tare pressure to full pressure. This gage is mounted in a compact, rigid, cast aluminum frame. Only two piston assemblies are required to cover the pressure range from 0.3 to 600 lb/in.^{2} The low range from 0.3 to 15 lb/in.^{2} is covered with an accuracy of 0.01% of reading, and the high range from 2 to 600 lb/in.^{2} with an accuracy of 0.015% of reading. This gage contains a 0.1 A capacitor motor in the base to rotate the cylinder about the piston to ensure continuous and consistent pressure readings. When the gage is used for

Figure 3.11 *Ruska air piston pressure gage. Courtesy Ruska Instrument Corp.*

absolute pressure measurements and operated at atmospheric reference pressure, a bell jar is placed over the weights to shield against air currents. Weights can be changed on the instrument while the motor is running and pressure is in the gage system. The piston retains its rotational freedom under a full complement of weights at zero pressure or when unloaded and under gas pressure.

3.3 DIAPHRAGM PRESSURE GAGES, NONMETALLIC

The diaphragm gage is probably the best example of a true force balance pressure measuring unit. These gages are designed with a relatively large area of flexible material which has good sealing qualities and is easily deformed, attached to a piston or spring restrained surface. The spring is calibrated to cover a given range of pressure measurements, normally quite low, and a pointer is mechanically coupled to the spring to indicate the pressure for any deformations that occur. Control can also be initiated by mounting pressure switches on these diaphragms to give high and low level signals.

The *limp diaphragm* gage shown in Figure 3.12 is designed for measuring draft pressures of combustion gases. These gages can stand sudden overloads, because the diaphraghm housing is designed to prevent excessive diaphragm travel caused by sudden overloads generated by combustion bursts in furnace firing chambers.

The diaphragm in the Hays unit is made of good pliable leather and is balanced by a heat treated beryllium copper spring. The spring de-

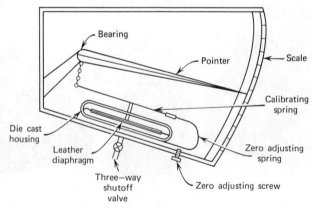

Figure 3.12 *Limp diaphragm gage. Courtesy Hays Corp.*

flects in direct proportion to the magnitude of the pressure, and is adjusted to give accurate readings for zero pressure by means of a zero adjusting screw. If the diaphragm is used in a corrosive or drying atmosphere, it should be inspected periodically, and replaced to prevent the occurrence of serious errors in readings.

The Dwyer gage shown in cross section in Figure 3.13 uses a unique design, called Magnehelic, for converting linear motion into rotary motion to drive an indicating pointer without a physical coupling. This gage uses a silicon rubber diaphragm, which is supported to restrict its motion, preventing damage due to overpressures. The diaphragm separates two pressure-tight compartments in the gage. The interior of the

BEZEL provides flange for flush mounting in panel when desired, simplifies opening gage if required.

CLEAR PLASTIC FACE is highly resistant to breakage, provides undistorted viewing of pointer and scale.

PRECISE PRINTED SCALE is easy to read. Actual size illustrated on page 7.

RED TIPPED POINTER of heat treated aluminum is easy to read and is rigidly mounted to helix shaft.

MOLDED RUBBER POINTER STOPS effectively restrain over-travel of pointer but cannot damage the pointer.

SAPPHIRE BEARINGS are anti-shock mounted, provide virtually friction-free motion for helix. Motion dampened with high viscosity silicone fluid.

ZERO ADJUSTMENT SCREW is conveniently located in plastic cover and is accessible without removing plastic cover. "O" ring seal is used for pressure tightness.

"O" RING SEAL for cover assures pressure integrity of case.

DIE CAST ALUMINUM CASE is precision made, Iridite dipped to withstand 168 hour salt spray test. Exterior finished in baked dark gray hammerloid. One case used for all standard pressure ranges, and for both surface and flush mounting.

SILICONE RUBBER DIAPHRAGM with integrally molded "O" ring is supported by front and rear plates and locked and sealed in position with a sealing plate and retaining ring. Diaphragm motion is restricted to prevent damage due to overpressures.

CALIBRATED RANGE SPRING is a flat leaf of temperature - compensated Ni Span C linked to front diaphragm plate. Small amplitude of motion assures consistency and long life. It reacts to pressure on diaphragm. Live length adjustable for calibration purposes.

"WISHBONE" ASSEMBLY provides mounting for helix, helix bearings and pointer shaft.

ALNICO V MAGNET mounted at end of range spring actuates helix without mechanical linkages.

HELIX is precision milled from an alloy of high magnetic permeability, deburred and annealed in a hydrogen atmosphere for best magnetic qualities. It is mounted in jeweled bearings and turns freely to align with magnetic field of the magnet to transmit pressure indication to dial.

Figure 3.13 *Magnehelic pressure gage in cross section. Courtesy Dwyer Instruments, Inc.*

gage case acts as the high pressure compartment, and a sealed chamber behind the diaphragm acts as the low pressure compartment.

Differences in pressure between the high and low sides of the diaphragm cause the diaphragm to assume a balanced position between the two pressures. The front support plate of the diaphragm is linked to a leaf spring which is anchored at one end and provides calibrated resistance to the diaphragm motion. The motion of the spring is transmitted through an exclusive magnetic linkage to the indicating pointer, as shown and explained in Figure 3.14a and b.

Typical applications are as an air filter gage to measure the pressure drop across the filter, to sense static pressure, measurement of air velocity, air flow measurements, and for liquid level measurements. Except for static pressure measurements, all these applications are for differential pressure measurements. The installations of these applications are shown in Figure 14c–g.

Later in the chapter we discuss the use of metallic diaphragms and the use of strain gages in transducers for pressure measurements.

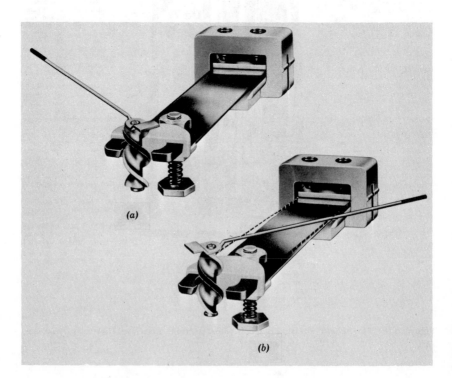

(a)

(b)

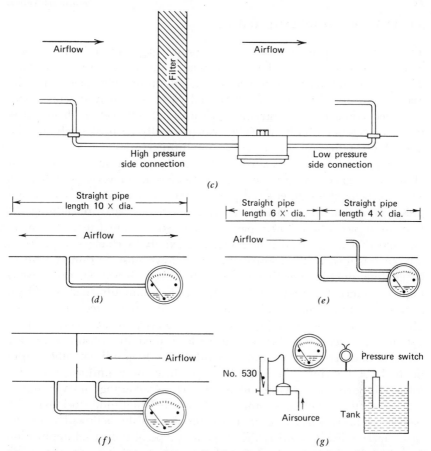

Figure 3.14 *Magnetic linkage and applications of gages using this principle. (a) At zero position, pressures on both sides of the diaphragm are equal. The support plates of the diaphragm are connected to the leaf spring which is anchored at one end. The horseshoe magnet attached to the free end of the spring straddles the axis of a helix but does not touch the helix. The indicating pointer is attached to one end of the helix.*

The helix, being of high magnetic permeability, aligns itself in the field of the magnet to maintain the minimum air gap between the magnet's poles and the outer edge of the helix. (b) When pressure on the high side of the diaphragm increases or pressure on the low side of the diaphragm decreases, the diaphragm moves toward the back of the case. Through the linkage, the diaphragm moves the spring and the magnet. As the magnet moves parallel to the axis of the helix, the helix turns to maintain the minimum air gap.

Movement of the diaphragm is resisted by the flat spring which determines the range of the instrument. Precise calibration is achieved by varying the live length of the spring through adjustment of the spring clamp. (c) As an air filter gauge. (d) To sense static pressure. (e) To measure air velocity. (f) To measure air flow across a sharp orfice plate. (g) As liquid level measurement. Courtesy Dwyer Instruments, Inc.

75

3.4 ELASTIC MEMBRANE GAGES

The most widely used mechanical pressure gages in industry employ quantitative deformation of an elastic membrane to measure the pressure. These are primarily metallic bellows and Bourdon tube gages. Metallic diaphragms are also used in differential gages and relay systems where their deflections can be restrained to withstand relatively high pressures under emergency conditions. They can normally be used safely at relatively high static pressures to detect small differentials.

The *bellows type* is usually limited to lower pressure ranges when absolute or gage pressures are being measured. However, the bellows gage can be used for differential pressure measurements at relatively high pressures. A metallic bellows is a series of circular parts, resembling the folds in an accordion. These parts are formed or joined in such a manner that they are expanded or contracted axially by changes in pressure. The metals used in the construction of bellows must be thin enough to be flexible, ductile enough for reasonably easy fabrication, and have a high resistance to fatigue failure. Materials commonly used are brass, bronze, beryllium copper, alloys of nickel and copper, steel, and monel. Hard-to-work metals or alloys are used primarily to meet corrosion resistance requirements. Most of the bellows used in pressure gages are seamless and are made from drawn tubing by hydraulic or other rapid methods of forming. These methods produce more uniform walls for longer life expectancy. Other methods such as soldering and welding of annular sections, rolling, spinning, and turning from solid stock may also be used to form bellows. A manufacturer chooses the bellows having the best characteristics to satisfy the range and type of industrial application for which his equipment is designed.

Normally a bellows has the ability to move over a greater distance than required in a pressure application, so a range spring which can be calibrated for a particular pressure range is used to oppose the motion. As a general rule, the smaller the deflection the longer the life cycle of the bellows. Nomographs have been developed by bellows manufacturers relating the probable life cycle to the pressure and the length of stroke or deflection length of the bellows material. A separate nomograph is required for each type of material.

The three main configurations for the use of bellows in gage applications are shown in Figure 3.15; *a* is the schematic arrangement for an absolute pressure measurement, *b* is for a gage pressure measurement, and *c* is for a differential pressure measurement. The stroke of the bellows can be increased by using a larger number of convolutions or segments, and its force can be increased by using a larger diameter bellows

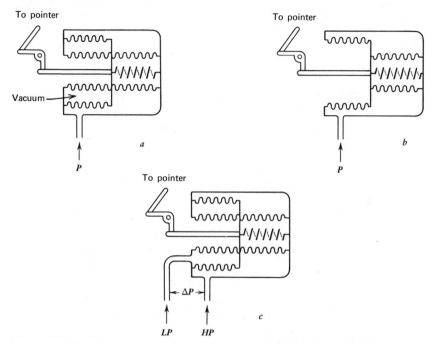

Figure 3.15 *Bellows pressure gage arrangements. (a) Absolute pressure. (b) Gage pressure. (c) Differential pressure. Courtesy Taylor Instrument Co.*

so the pressure has a larger area on which to act. Manufacturers recommend that the stroke not exceed 10% of the length of the bellows. For long life expectancy the stroke should be limited to 5% of the length and 25% of the pressure the bellows material and configuration can withstand.

Bellows are also used in pressure transmitters. Two typical setups are shown schematically in Figure 3.16; *a* is a normal gage arrangement, and *b* is a differential arrangement. There are many elaborations on the basic principles illustrated in the schematics.

Pressure transmitters are used when it is desirable to read pressure values at some remote location nonelectrically. In Figure 3.16*a* the pneumatic pressure transmitter is operated by the pressure to be measured expanding the pressure bellows *a* against the beam and the range spring. This movement causes the beam to reduce the spacing between the orifice and the beam at point *C*. This change in orifice spacing generates a change in pressure in the balancing bellows *B* by causing it to elongate and exert a force on the beam, which in turn opens the orifice spacing a

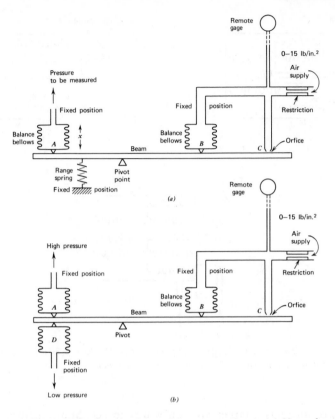

Figure 3.16 *Bellows operated pneumatic pressure transmitters. (a) Normal gage reading arrangement. (b) Differential pressure reading arrangement. Courtesy Honeywell, Inc.*

little. A torque balance is set up between the range spring and the balancing bellows, and the pressure of the bellows is measured. This torque balance can be obtained with a very small movement of the beam with respect to the orifice. Only a small movement is necessary to change the pressure to the maximum supplied at the restriction, which is usually restricted to a nominal 15 lb/in.² For best results pneumatic lines should not be over 30 ft long. Where longer transmission lines are needed, the signal should be converted to an electrical signal by means of pneumatic-to-electrical transducers with electrical outputs. A typical electrical and solid state force balance transmitter is shown schematically in Figure 3.17, in which a diaphragm is used as the pressure transducer. In Figure 3.16*b* the range spring shown in Figure 3.16*a* is replaced by a low pressure bellows *D*. The differential range can be adjusted by changing the position

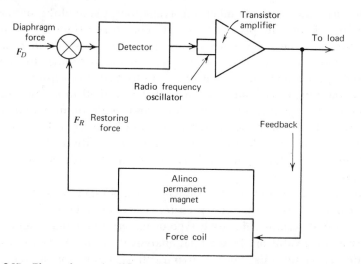

Figure 3.17 *Electroelectronic differential pressure force balance transmitter using a diaphragm as a movable plate air capacitor detector. Courtesy Leeds and Northrup.*

of the pivot point or fulcrum. The operation is the same as that for Figure 3.16a, except that the low pressure bellows, controlled by the low pressure side of the differential, is used in obtaining the torque balance with the balancing bellows *B* in setting the orifice-to-beam spacing against the force exerted by the high pressure bellows *A*. In most industrial units using this principle the orifice is usually referred to as the nozzle, and the bellows are replaced by a diaphragm which can be restrained in the event either of the pneumatic lines becomes clogged and causes full pressure to be exerted on one side of the diaphragm. Diaphragm units can be used in the 3000 to 5000 lb/in.² static range, while the safe limit for most bellows is less than 100 lb/in.² Sudden pressures exerted to only one side of a bellows usually results in a rupture of the seal or the bellows material, requiring replacement of the unit.

Diaphragm Gages. Metallic diaphragms are used in differential pressure gages, pneumatic pressure transmitters, and electrical pressure transmitters in which the static pressure may be well above the rupture strength of the material used in bellows. These diaphragms are constructed as circular discs, and quite often the discs have corrugated surfaces to increase the surface area and the deflection capability of the surface. The deflection of the diaphragm depends on the type of material, the thickness of the material, the diameter of the disc or shell, the shape of the corrugations, the number of corrugations, the modulus of elasticity of the metal, and the pressure applied. The corrugations are formed by hydraulic forming or pressing of material discs or shells in much the same manner as bellows are formed from seamless tubing. In many industrial applications diaphragm shells are formed into capsules by welding or soldering of two shells to each other or a shell to a rigid plate.

The depth, number of corrugations, and angle of formation of the diaphragm face determine the sensitivity and linearity of the diaphragm for use as a pressure detector. The maximum sensitivity for very small deflections or diaphragm motions is obtained with a smooth, flat diaphragm.

The main drawback in the design of diaphragm pressure elements with corrugations is that the pressure–deflection relationship must be empirically determined for each type of material and the number, type, and size of convolution. In the final analysis it has been determined that the deflection for this type of diaphragm is a function of the fourth power of the diameter. This means that, if the diameter is doubled, the deflection is increased 16 times for the same pressure change. These figures are given to emphasize that there are design and calibration problems when a diaphragm is used, and that care must be exercised in the selection of a diaphragm for a given application.

Diaphragms have traditionally been used in gages for relatively low pressure and low vacuum measurements. Aircraft applications requiring high constant forces which are natural characteristics of diaphragm pressure elements, and use in transducers, transmitters, relays, and switches have increased the demand for good diaphragm elements.

A metallic diaphragm arrangement for a differential pressure meter body and a pneumatic-to-electrical transmitter are shown in Figure 3.18. Figure 3.18a is a pressure meter body for sensing differential pressure. The high pressure is admitted through a port (1) to exert its force against a barrier diaphragm (3), and the low pressure is admitted through a port (2) to exert its force against a barrier diaphragm (4). The center section of this meter is filled with a silicon fill fluid which transfers the high and low pressure forces. The high pressure acts on the inner cavity (5), and the low pressure on the outside cavity (6) of the measuring element. When pressure is exerted on the system, the pressure increases on (5) and decreases on (6), so that the high pressure fill is forced through the damping restriction (7). This causes the measuring element to move and exert a proportional torque through a linkage on the force shaft (8). The force shaft extends outside the meter case through a seal tube and can operate an indicating pointer or pneumatic or electric

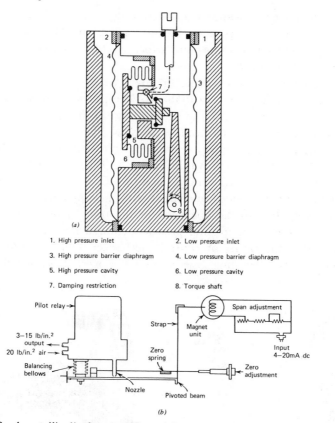

1. High pressure inlet 2. Low pressure inlet

3. High pressure barrier diaphragm 4. Low pressure barrier diaphragm

5. High pressure cavity 6. Low pressure cavity

7. Damping restriction 8. Torque shaft

Figure 3.18 *A metallic diaphragm differential pressure arrangement (a) and a pneumatic to electrical pressure transmitter (b). Courtesy Honeywell, Inc.*

transmitter. Figure 3.18*b* is a pneumatic transmitter, for a 3 to 15 lb/in.² output using a 20 lb/in.² air input to operate a current transducer (pressure-to-current) for long distance, high speed transmission. This eliminates the lags experienced when long runs of pneumatic tubing are used. As stated earlier, runs over 30 ft are to be avoided, if possible, for pneumatic signal transmission.

Bourdon Tube Gages. The Bourdon tube gage is probably the most widely used industrial pressure gage applied to both pressure and vacuum. It, like a bellows or diaphragm gage can be used for both vacuum and pressures, either separately or in a compound gage. The Bourdon tube is usually used whenever (1) the maximum of the required range exceeds 25 lb/in² for measuring combined pressure and vacuum (2) for continuous pressure measurements exceeding 80 lb/in² and up to 50,000 lb/in², or more direct pressure measurements, and (3) especially where sudden pressure fluctuations occur which could cause bellows or normal diaphragms to rupture.

Bourdon tubes may be made of any type of material that has the proper elastic characteristics suitable for the pressure range and the corrosive resistance of the media to be measured in the application. Some of the materials used include brass, alloy steel, stainless steels, bronze, K-Monel, and beryllium copper. A Bourdon tube may be in the shape of a C, a spiral, or a helix, as shown in Figure 3.19. It is shaped by flattening a rounded tube and then bending it into a C, a spiral, or a helix. One end of the tubing is sealed and fitted with a pointer mechanism. When pressure is applied to the open end of the tube, it tends to straighten out into its original shape and produces enough force to move a sector gear or other indicating or control mechanism. The Bourdon tube is anchored to its restraining base, so that the exerted pressure is propor-

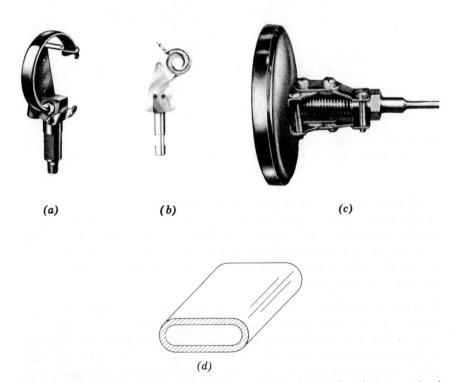

(a)   (b)   (c)

(d)

Figure 3.19 *Bourdon tube elements. (a) C type. (b) Spiral. (c) Helix. (d) Cross-sectional area. Courtesy Dresser Industries.*

tional to its movement. Sector gears or other mechanisms such as a taut band are used to multiply the magnitude of the tube movement to make the reading of the measurement easier and more accurate. Each arrangement requires careful workmanship to produce a linear pointer movement on a calibrated scale or recording mechanism.

Sector Gear Arrangement.   The usual sector gear arrangement has a large sector gear mounted at right angles to a link connecting the sector gear arm and the Bourdon tube tip. This connecting link is mounted parallel to the tip movement. A small pinion gear, to which the pointer is attached, is then matched to the sector gear, as shown in Figure 3.20. It is easy to see that any mismatch in the teeth of the sector or pinion gear will cause errors in measurement and nonlinearity of pointer movement. It is also evident why careful workmanship is required to obtain the needed multiplication to give accurate pointer movement on the

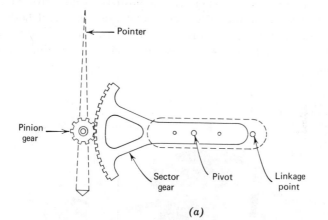

(a)

(b)

Figure 3.20   *Sector and pinion gear arrangement. (a) Sketch. (b) Photograph. Courtesy Dresser Industries.*

gage dial or pen movement on a recorder. Wear on the sector and pinion gear teeth causes backlash and other inaccuracies during the life cycle of the gage. These gears are most commonly made of bronze and may be machined, broached, or stamped, depending on the quality and accuracy required of the gage. Steel sector and pinion gears are also used and, while they are more expensive, they will outwear several sets of bronze movements.

Cam and Roller Arrangement.   Systems developed to overcome most of the disadvantages of sector and pinion gear arrangements are the Helicoid movement and the taut band drive mechanism. The Helicoid movement is shown in Figure 3.21. It employs a cam sector and a Helicoid roller to which a pointer is attached. The Helicoid stainless steel roller is long wearing and is used especially in services on engines, turbines, blowers, hydraulic presses, pumps, and compressors where violent pressure pulsations or severe mechanical vibrations occur. The band drive mechanism uses a sector cam with a positive band drive to the roller which carries the pointer. Neither system uses gears and both are more stable under pulsating pressure conditions. In each case the mechanism is used to translate linear motion into circular motion to move an indicator.

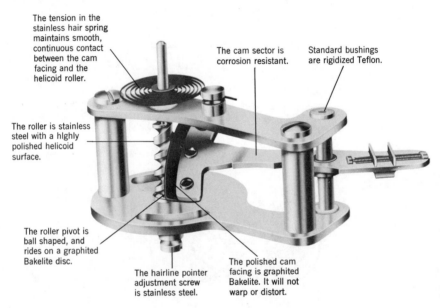

Figure 3.21 *Helicoid movement. Courtesy Helicoid Gage Division, American Chain and Cable, Inc.*

Special Pressure Gage Applications.   In applications in which fluids are used that attack the materials available for use in Bourdon tubes or in which pulsations occur that reduce the accuracy of the measurement or cause excessive wear on the moving parts of a Bourdon gage, certain precautions or special equipment may be required. When Bourdon tubes are used with corrosive chemical liquids or liquids that solidify at normal room temperatures, a diaphragm may be placed in the line, as shown in the breakdown assembly of Figure 3.22, and the gage line filled with water or oil and sealed. The sealed system then senses the diaphragm movement and indicates the pressure. When Bourdon tubes are used to measure steam pressure, a loop is placed in the gage line so that the liquid condensate is trapped and used to transmit the live steam pressure to the gage. This prevents gage error and damage caused by the elevated temperature of live steam. In applications involving rapid fluctuations or pulsations in pressure, some type of throttling action is needed to reduce the wear on gage parts. So-called snubbers shown in Figure 3.23, are available and do a good job of stabilizing the pressure for the gage measurement. Care must be used so that the throttling orifice is not too small because, if the liquid or gas contains any dirt, the orifice may clog and block the line to the gage.

Another caution to be observed with most pressure gages is that they should not be mounted on equipment subject to excessive vibration. External vibrations cause excessive wear and inaccuracies in gage indications. Use only those gages least affected by vibration.

When amplitude and frequency cause an acceleration of less than 1 g, usually little or no harm is done to a good movement but, if the acceleration is greater than 1 g, there will be wear on pivots and bearings, which will mean higher repair or replacement rates and additional calibrations.

3.5   ELECTROMECHANICAL PRESSURE GAGES

Electromechanical pressure gages are combinations of mechanical bellows, metallic diaphragms, or Bourdon tubes with electrical sensing, indicating, recording, or transmitting devices.

Pressure Transducers.   In a true interpretation of the word transducer, it could be said that any device that converts one type of motion or signal into another is a transducer. Here it is assumed that some type of mechanical motion generated by pressure forces is converted into an electrical or electronic signal for use in measurement and control. The most used transducers for pressure detection are those operating on strain

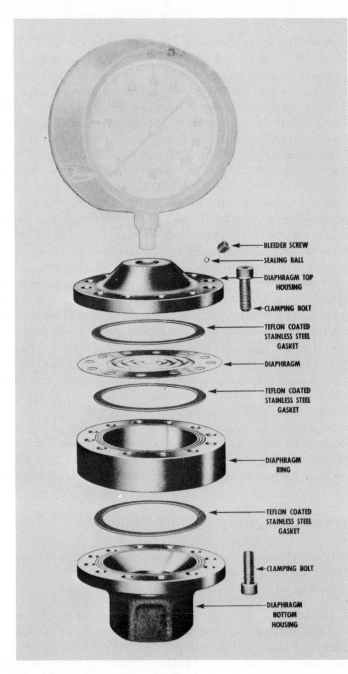

BLEEDER SCREW

SEALING BALL

DIAPHRAGM TOP HOUSING

CLAMPING BOLT

TEFLON COATED STAINLESS STEEL GASKET

DIAPHRAGM

TEFLON COATED STAINLESS STEEL GASKET

DIAPHRAGM RING

TEFLON COATED STAINLESS STEEL GASKET

CLAMPING BOLT

DIAPHRAGM BOTTOM HOUSING

Figure 3.22 *Ashcroft gage arrangement for use on a pressure system containing corrosive fluids. Courtesy Dresser Industrial Valve and Instrument Division, Dresser Industries, Inc.*

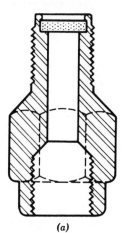

(a)

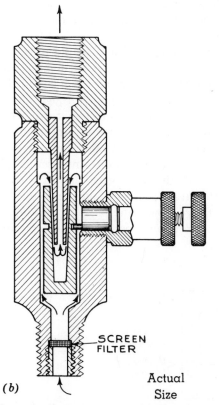

SCREEN
FILTER

(b)

Actual
Size

Figure 3.23 *Stabilizing units for pressure gages. (a) Chemquip snubber for pulsations and surges in line pressures. Courtesy Chemquip Co. (b) Throttler with screen filter and adjustable orifice for pressure gage protection. Courtesy Marshalltown Manufacturing Co.*

gage, inductive, piezoelectric, capacitative, oscillating, and similar principles.

Strain Gage Pressure Measuring Units.   A strain gage (discussed in detail in Chapter 9) can be used to produce an electrical signal in proportion to the resistance change caused by the distortion of a flexible membrane to which it is attached. Such gages are designed so that the electrical output is directly measured in pounds per square inch or other suitable pressure units. Strain gages can be used throughout the pressure range and can be applied as either unbonded or bonded units. Such transducer units can be designed for absolute, gage, and differential pressure applications.

The gage schematic shown in Figure 3.24 uses a bellows in which part *A* moves in respect to part *B*. Strain gages are placed between parts *A* and *B,* and they are conditioned to produce a signal that can be calibrated directly in pounds per square inch and shown on an indicating or recording mechanism.

Basically, the conditioning system is simple in concept and operating principle. The unstrained gage has a fixed nominal electrical resistance. This resistance is formed into part of a half or full bridge. A nominal voltage is applied so that a current flows through the resistance to create a voltage drop for a constant current. This current must be kept small enough so that it does not cause gage heating and change the gage resistance through thermal changes. Under these conditions the bridge is balanced for a zero pressure condition. Any pressure changes that stress the membrane to which the strain gages are attached cause a change in the resistance of the strain gage and unbalance the bridge circuit. The bridge unbalance thus becomes the pressure measurement, and this unbalance is calibrated in pounds per square inch. Strain gage transducers

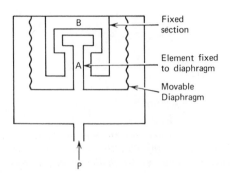

Fixed section

Element fixed to diaphragm

Movable Diaphragm

Figure 3.24 *Bellows gage using strain gage principles.*

have very good frequency response, can be used for static or dynamic measurements, and can be excited with either ac or dc power. They provide continuous resolution with a small displacement and require only simple resistive balance. In general these transducers are not affected by electromagnetic fields, and temperature compensation is relatively easy to accomplish. The disadvantages are that they are relatively complex to fabricate and have a low output, which requires conditioning accessories such as bridges and power supplies.

Strain gage pressure transducers are also very adaptable to analog-to-digital conversion, so that the pressure units can be readily displayed on the latest LED and numerical or alphabetical readout devices. The analog-to-digital conversion output is also compatible with computer and magnetic binary coded decimal storage and display.

Another pressure transducer, using flexible membranes and variable resistance to measure pressure, is the potentiometric type. This type is especially well suited to compound gage applications in which there is either slowly changing or steady state pressure. A potentiometer with good linear output characteristics is attached to a diaphragm or bellows and set for zero output in a balanced system for zero pressure. As the pressure expands the bellows or diaphragm the resistance of the pot increases, and as the bellows or diaphragm contracts the resistance decreases. These resistance changes are then calibrated as pounds per square inch and establish the range and scale for the unit. As in a strain gage, a voltage can be impressed across the resistance with a constant current flow and the output converted to binary decimal codes for numerical readout, tape storage, or computer input.

Inductive Transducers. Inductive pressure transducers are magnetic coupled units which can be used for both differential and gage pressure measurements. The basic transducer consists of a diaphragm or other similar type of driver magnetically coupled to a balanced electric pickoff system which delivers a full sized output in millivolts per volt in an ac bridge circut. There is no mechanical coupling between the diaphragm or driver and the pickoffs imbedded in the rigid body of the transducer. The action of the transducer is quite similar to that of the float shown in the inductive manometer gage in Figure 3.25. This inductive pressure gage is a metal mercury manometer using a floating bell to move an iron rod up and down inside a divided coil designed to give an electrical output proportional to the movement of the float caused by pressure changes. The signal generated by the internal coil is transmitted to an external coil from which the signal is taken either mechanically or electrically. Accuracies of ±1% of full scale are available for maximum pressures up to 1000 lb/in^2 where the maximum pressure across a diaphragm can be as much as 500% of the range. The natural frequencies of the diaphragm units normally lie between 3000 and 25,000 Hz and increase with range. The ac excitaton frequency lies between 400 and 20,000 Hz and depends on the inductance coil used in the transducer construction. Ambient temperatures for application are between −65 and 225°F.

Piezoelectric Transducers. Piezoelectric transducers are composed of crystalline materials which produce an electrical signal when they are deformed physically by pressure. Conversely, when they are charged electrically they become physically deformed. Two of the important crystalline materials used in these pressure transducers are quartz and barium titanate. Both withstand temperatures up to 300°F continuously and 450°F in intermittent service. This type of transducer is extremely useful in hydrophones and other underwater detection applications. The advantages of piezoelectric transducers are high output, high frequency response, self-generation, negligible phase shift, small size, and rugged construction. Disadvantages are sensitivity to temperature changes, unsuitability for static applications, high impedance output, sensitivity to cross accelerations, need for impedance matching, and zero shift after extreme shock. See Figure 3.26.

Capacitive Transducers. Capacitive transducers operate on the principle that, when one plate of a simple capacitor is displaced, the capacitance is changed. This is accomplished by applying pressure to a diaphragm which acts as one plate of a plate capacitor. The transducer consists of a deflectable diaphragm as the movable plate, separated from the fixed plate by a compressible dielectric material. The components are mounted in a pressure housing rated for the application. This type of transducer

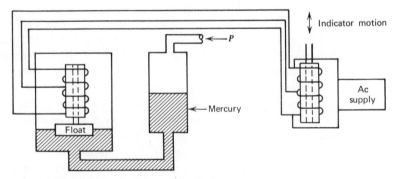

Figure 3.25 *Inductance electromechanical pressure meter.*

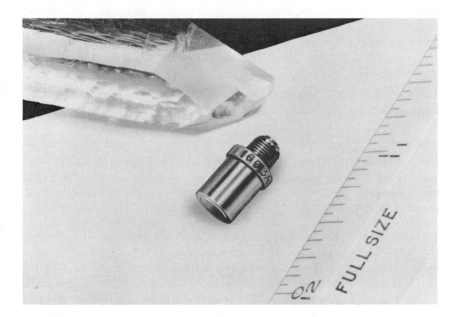

Figure 3.26 *Quartz crystal transducer. Courtesy Kistler Instrument Corp.*

is useful as part of a resistive-capacitive or inductive-capacitive network in oscillator circuits, and as a reactive element in ac bridges. Advantages of capacitive pressure transducers are very good frequency response, relative simplicity in construction, good G response, relative low cost, suitability for static and dynamic measurements, small volumetric displacement for minimum diaphragm mass, and continuous resolution. Disadvantages are the same as those for piezoelectric units, plus the requirement for reactive and resistive balancing.

Oscillating-Type Transducers. Oscillating transducers make use of semiconductor materials and operate on the principle that the frequency of a transistor oscillator is changed as a function of inductance or capacitive change in a force summing member of a transducing element. The stability of the oscillator is of prime importance for the stability of such a pressure gage unit. Advantages are small size, high output, ability to measure both static and dynamic pressures, and convenience for telemetering purposes. The output is adaptable for numerical display in pressure units, magnetic storage, and binary decimal codes for computer input. Disadvantages are limited temperature range determined by the semiconductor materials, expense, low accuracy, and need for special equipment for analog conversion. They also have poor thermal zero shift and poor sensitivity.

3.6 VACUUM GAGES—MECHANICAL, ELECTRICAL, AND ELECTRONIC

The pressure gages used primarily for measuring pressure below atmospheric pressure, which is most often referred to as a vacuum, are McLeod gages, Pirani gages, Knudsen gages, thermocouple gages, Phillips gages, and ionization gages. The different types, except for the Knudsen gage, are shown in Figure 3.27.

McLeod Gage. The McLeod gage is a mercury gage for the measurement of absolute pressure. It is one of the most basic types and has a measurement range from 2 μm to 100 mmHg. There are three types of McLeod gages. The swivel McLeod gage has an accuracy of 3% of reading or 1 mm of scale reading. The standard tilt high precision McLeod gage has been modified to simplify its operation, use less mercury, be more rugged and compact, and still retain its precision. The newer modified gage is known as the adjustable closed end improved McLeod gage and is shown schematically in Figure 3.27a.

The advantageous properties of the standard gage preserved on the improved gage are:

1. High precision single linear scale, 100 mm long
2. Multiple ranges achieved with an adjustable closed end: 0 to 1.00 mm in 0.01 mm divisions; 0 to 0.1 mm in 0.001 mm divisions, 0 to 0.01 mm in 0.0001 mm divisions
3. Closed end easily opened for cleaning
4. Micrometer adjustment of the meniscus
5. High vacuum O-ring seals throughout
6. High precision maintained by use of precision bore capillaries and exact adjustment of volume
7. Permanent ceramic scales and white background for accurate and easy reading

New advantages include:

1. Rapid adjustment of the meniscus made by means of a positive acting multiple thread

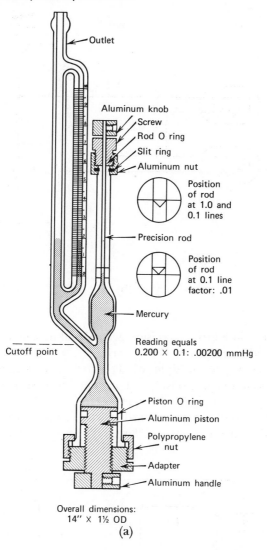

Outlet

Aluminum knob
Screw
Rod O ring
Slit ring
Aluminum nut

Position
of rod
at 1.0 and
0.1 lines

Precision rod

Position
of rod
at 0.1 line
factor: .01

Mercury

Cutoff point

Reading equals
0.200 × 0.1: .00200 mmHg

Piston O ring
Aluminum piston
Polypropylene
   nut
Adapter
Aluminum handle

Overall dimensions:
14″ × 1½ OD
(a)

2. Conical point on precision ground rod to compensate for mercury meniscus to give better accuracy

3. Mercury volume reduced one-half

4. Overall length reduced considerably

5. Glass parts much sturdier

The principle of operation is as follows. The fixed closed end capillary of the standard McLeod gage has been replaced with an adjustable unit

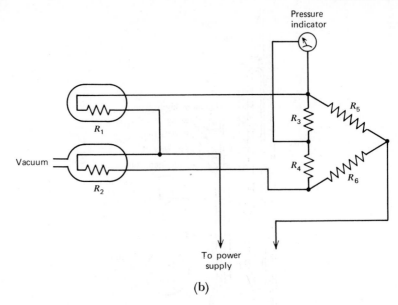

(b)

consisting of a close fitting precision ground rod in a precision bore capillary. Sliding this rod up or down varies the volume of the compressed attenuated gas. Using multiples of 10, a threefold change in the magnitude of the scale range is possible. The same linear scale can be used for all three ranges by applying the appropriate range factor.

The conical tip of the ground rod exactly compensates for the volume created by the curved meniscus. For this reason it is possible to work

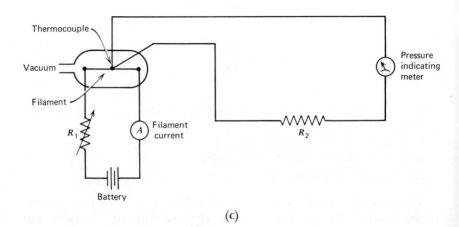

(c)

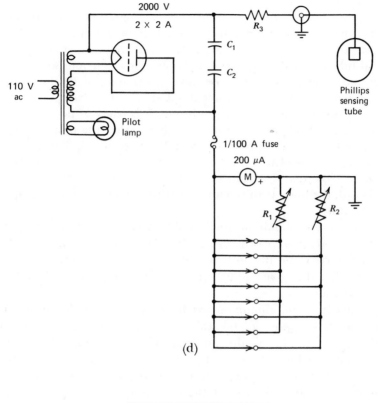

(d)

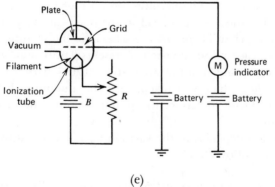

(e)

Figure 3.27 *Vacuum gages. (a) McLeod gage Schematic. Courtesy Roger Gilmont Instruments, Inc. (b) Schematic of a Pirani gage. (c) Thermocouple vacuum gage schematic. (d) Phillips gage schematic. (e) Ionization tube vacuum gage schematic. Courtesy Consolidated Vacuum Corp.*

with very small volumes of compressed gas and still maintain relatively high accuracy.

Directions for use—Filling with Mercury.   Referring to the schematic drawing, unscrew the aluminum piston all the way down and add clean instrument grade mercury (10 to 12 ml) until the meniscus reaches the vertical tube below the cutoff point. Apply a good vacuum source to the outlet and remove the excess gases and vapors trapped by the mercury. Tilting and tapping the instrument speeds up the process. When the evacuated instrument reaches equilibrium, it is ready to be read.

Reading the Modified McLeod gage.   First, select the proper range by raising or lowering the precision ground rod in the capillary until the conical end is coincident with the desired range zero line as shown in the schematic. Note that at factors of 1.0 and 0.1 the base of the cone corresponds with the line. At the factor of 0.01 the tip of the cone corresponds with the line.

The screw and aluminum nut should be loosened slightly to permit the rod to slide more easily. When the rod is set at the desired range, tighten the screw and nut to lock the rod in place. As shown, the range has been selected to give 0 to 0.01 in 0.0001 mm divisions on the 0 to 100 mm linear scale.

Carefully turn the aluminum handle at the base of the meter. This causes the mercury to rise and seal off the attenuated gas in the bulb. Continue to turn the handle until the top of the mercury meniscus reaches the zero line. With the tip of the rod set to the zero line as shown for the 0 to 0.01 range the positioning of the meniscus can be precisely made, since the top of the cone just contacts the top of the hemispherical meniscus. The absolute pressure of the system is now obtained by reading the meniscus level in the open capillary and multiplying by the appropriate factor. In the schematic the reading is at 0.2 on the scale, where each minor division represents 0.0001 mm. The absolute pressure reading as shown in the schematic is $0.20 \times 0.01 = 0.002$ mmHg.

When the reading has been taken, turn the aluminum handle on the base counterclockwise to retract the piston to lower the mercury below the cutoff point.

Range selection depends on the vacuum to be measured by the system. Accuracy is as follows:

1.0 range: 1% of reading or 0.5 divisions on the linear scale
0.1 range: 2% of reading or 1.0 divisions on the linear scale
0.01 range: 10% of reading or 2.0 divisions on the linear scale

When the meniscus in either the open or closed end capillary deviates significantly from a hemisphere, the instrument should be cleaned and the mercury replaced.

Pirani Gage. The Pirani gage shown in Figure 3.27b is a hot wire vacuum gage. These gages employ a Wheatsone bridge circuit to balance the resistance of a tungsten filament or resistor sealed off in a high vacuum against that of a tungsten filament which can lose heat by conduction to the gas whose pressure is being measured. In this circuit the zero drifts caused by slight deviations of the bridge voltage are compensated for by the resistor sealed in the high vacuum. A change in pressure causes a change in the filament temperature. This causes a change in the filament resistance and unbalances the bridge. The bridge unbalance is then read across R_3 as the dry air pressure, by means of a microammeter, calibrated in pressure units. The useful range for the Pirani gage is from 1 μm to 100 mmHg. The Pirani gage has the advantage of being compact, simple to operate, and can be opened to the atmosphere without burnout failure. The main disadvantage is that the calibration depends on the type of gas in which the pressure is being measured. These gages are useful for pressure measurements involving acetylene, air, argon, carbon dioxide, helium, hydrogen, and water vapor for the general pressure range of 1 to 200 μm (1 μm $= 1 \times 10^{-6}$ meter which is equal to 1×10^{-3} mm) and is most useful and accurate in the 20 to 200 μm range.

Knudsen Type Vacuum Gage. The Knudsen gage operates on the principle of heated gases rebounding from a heated surface and bombarding a cooled movable surface (vane) spaced less than a mean free path length from the heated surface. The gas particles rebound from the cool vane with less energy than from the heated vane which tends to rotate the cool movable vane away from the heated vane within the restrictions of a suspension system designed to carry a galvanometer mirror for producing a reading on a fixed scale. The particular advantage of the Knudsen gage operating principle is that the gage response is relatively independent of the composition of the gas whose pressure is being measured. In spite of this very desirable feature, the gage is not widely used because the torsion system is rather delicate and sudden inrushes of air cannot be tolerated. New developments are being investigated that may make the gage more acceptable for industrial applications.

Thermocouple Vacuum Gages. Thermocouple vacuum gages as shown in Figure 3.27c employ a thermocouple attached to a filament, a battery, a resistor R_1 for adjusting the filament current, a filament, and a readout meter for measuring the output emf of the thermocouple. The filament is heated by a constant current, and its temperature depends on the amount of heat lost to the surrounding gas by convection and conduction. This heat loss is determined by the pressure in the system, where the filament temperature is a function of the pressure at pressures below approximately 100 μm. The output of the thermocouple is calibrated to read directly in micrometers on the indicating meter. These gages have characteristic nonlinear readout scales and operate at low heater currents. They are subject to burnout if exposed at atmospheric pressure when hot and current is flowing.

Phillips Vacuum Gage.   Phillips gages are cold cathode ionization gages which provide direct measurement for pressure values both above and below 1 μm. These gages cover the 0.05 to 10^{-7} mmHg pressure range. The schematic in Figure 3.27d shows the basic gage circuit. The pressure measurement is a function of the current produced by a high voltage discharge. The electrons drawn from the cold cathode are caused to spiral as they move across a magnetic field to the anode. This spiral motion greatly increases the possibility of collisions with the gas molecules between the cathode and anode, and produces a higher sensitivity by creating a higher ionization current. The output is read out on a microammeter calibrated directly in pressure units. The range is divided into four separate outputs with direct reading for each portion of the total range. The advantages of this gage are the wide range that can be covered, absence of filaments to burn out, rugged metal construction, and ease of cleaning and maintenance. The disadvantages are that cold cathode tubes are slower to outgas than hot filament tubes, they are adversely affected by mercury, and there is a higher breakdown of organic vapors at higher voltages. These factors limit the use of these gages to applications in which oil diffusion pumps are used.

The hot filament *ionization vacuum gage* is shown schematically in Figure 3.27e. The ionization tube is the primary detector and is constructed of glass. It contains an anode, a grid, and a filament cathode. The electrons emitted by the filament are attracted to the grid, pass through the grid, and form ions by collision with the molecules present between the grid and the anode. The positive ions are collected on the anode, and the electrons are collected on the grid. The positive ion current created is proportional to the amount of gas present, if the electron current is kept constant. The electron current is maintained at a constant rate by means of a grid current regulator. The advantage of this type of gage is that very low pressures can be detected and measured in vacuum furnace and mass spectrometer applications. The ionization gage can be used in the 1 micron to 2×10^{-9} mmHg pressure range with relatively good accuracy. The disadvantage of these gages is that the filament can burn out quickly if it is heated before the pressure is at a low enough value. For this reason it is usually necessary to have a Pirani or thermocouple gage in the system to ensure a low enough vacuum, and to have an automatic cutout circuit to protect the ionization tube in case of a system leak or break.

The *Alphatron* gage (National Research Corporation) uses a radium source sealed in a vacuum chamber where it is in equilibrium with its immediate decay products. This provides a constant source of alpha particles for ionizing the gas particles present in the vacuum chamber. The alpha particles collide with the gas molecules in the same manner as the electrons in the hot filament tube just discussed. The advantages of this gage are the same as those of the hot filament gage, but it overcomes the burnout problem, the fragility, and the emission instability. Some of the disadvantages are that at very low pressures a preamplifier is required to give an undistorted output and the currents produced are in the order of 10^{-11} to 10^{-13} A and are directly proportional to the number of ions collected on the grid in a given time. With proper circuitry the response of the gage, within its range, and the indicator or recorder can be made linear with respect to the pressure, regardless of the nature of the gas under measurement.

All electrical and electronic vacuum gages now employ the latest solid state circuitry to maintain constant currents and voltages. This type of circuitry has added to both the stability and accuracy of measurements.

3.7 PRESSURE GAGE CALIBRATION

The pressure calibration standards discussed in Chapter 2 are used to calibrate pressure gages and control devices to ensure product quality and the safety of an industrial process or operation. For the most accurate calibration, the calibration points should be checked from zero to full scale and from full scale to zero. A normal calibration run will show that the majority of gages are the least accurate in the lower one-fifth and upper one-fifth areas. This means that the low and high ends of gage scales should not be relied on for steady process or system readings. A gage with the proper range should be chosen to have the process or system reading fall in the more accurate area of gage operation.

Manometer Standards. Low pressure gages such as manometer and straight bellows types are usually calibrated by comparison methods using precision manometers or liquid columns. Calibration manometers may use a liquid of lower density than that used in the gage being calibrated to produce a greater motion of the liquid in the capillary tube of the calibration unit than in the unit being calibrated. This technique increases the accuracy of the calibration.

Deadweight Standards. When calibrations are made using a deadweight tester, the gage to be calibrated can be calibrated directly by attaching it to an auxiliary gage connector block in the hand pump line. Under these conditions the test gage is subjected directly to the pressure of the deadweight gage fluid which is usually an oil. This setup is shown in Figure 3.28. Any test gages calibrated in this setup must be used with a pressure medium compatible with the oil used during the calibration. Extreme care must also be taken to keep the oil in the calibration unit from becoming contaminated.

For the calibration of pressure gages used in applications not compatible with oil, such as those involving water, gasoline, benzene, or alcohol, direct calibration cannot be made with a deadweight gage. This type of calibration is made by having a separate pump containing the liquid medium for the test gage service. The gage pressure is balanced against the pressure of the deadweight tester through a null balance or differential pressure indicator. A typical setup is shown in Figure 3.29, where the balance between the two gages is read on the differential pressure indicator. To reduce the pressure on both systems, a bleeder valve is required in the system piping to the reservoir.

Test gages for use in gas services are calibrated by the same method as for noncorrosive liquids, except that a compressed gas supply is used in

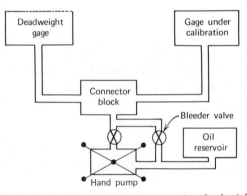

Figure 3.28 *Setup for direct calibration of test gage using deadweight gage fluid.*

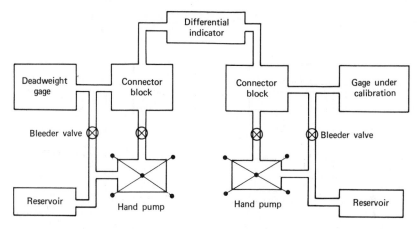

Figure 3.29 *Test gage setup for using fluid whose pressure is measured in service in the test gage system and deadweight fluid in the deadweight gage.*

place of the second liquid hand pump. The two systems are balanced, and the balance is measured on the differential indicator. The gas supply is normally a compressed gas cylinder capable of delivering the required pressure. To make full-scale-to-zero readings at the calibration checkpoints, the bleeder valve is required to vent the compressed gas in the system to the atmosphere, as shown in Figure 3.30.

Standard Test Gages. Calibration of operating gages using one or more standard test gages to evaluate the checkpoints is carried out in the same manner as discussed for deadweight gages. The only difference is that precision calibration gages are used in place of the deadweight gage. Two calibration gages are used for greater accuracy, although one gage is sufficient in many industrial applications.

Vacuum Gage Calibration. The majority of industrial vacuum applications do not require the ultimate in vacuum calibration techniques. To calibrate the most industrial vacuum gages and equipment, a comparison gage that covers the calibration points from the one 1 μm to 10^{-7} mmHg range is sufficient. A precision McLeod gage can be used as the standard. The calibration gage is placed on a manifold system, and the calibration points plotted for the vacuum gage being calibrated as the manifold system is evacuated. Care must be exercised to ensure that a sufficiently low pressure is reached with hot filament vacuum gages to

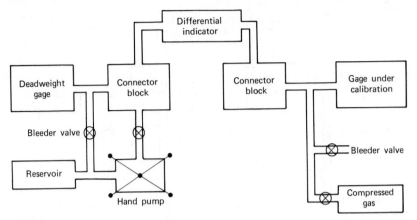

Figure 3.30 *Test setup for calibration of test gages for gas services using a deadweight gage as the standard.*

prevent filament burnout. Care must also be taken to guarantee that filament gages are properly outgassed during the calibration procedure.

Where calibrations are needed for very high vacuum technique measuring gages, a calibrated precision ionization gage should be used as the standard. This type of gage has a range down to pressures of 10^{-11} or 10^{-12} torr. This type of calibration equipment is expensive and finds applications in the industrial laboratory rather than in process or manufacturing systems. There are some exceptions such as mass spectrometer applications for isotopes, and use in special electron welding chambers.

3.8 SUMMARY

The more widely used types of pressure instruments have been discussed to give an understanding of operating principles, how they can be applied in an industrial process or operation, and how the different types are calibrated. The maintenance of each type of gage should be carried out in accordance with the recommendations of the manufacturer, and should be based on the actual type of service in which the gage is used.

A comparison of various types of pressure sensors is shown in Table 3.1, and some conversion factors for various pressure units are listed in Table 3.2.

Table 3.1 Comparison Table for Various Pressure Sensors[a]

Parameter	REC[b] Single Plate	REC[b] Dual Plate	Variable Reluctance Type	Bonded Strain Gage	Unbonded Strain Gage	Potentiometer Type	Differential Transformer Type	Semiconductor Type
Linearity	0.1%	0.5%	0.5%	0.25%	0.5%	0.5%	1.0%	0.75%
Hysteresis	0.015%	0.015%	0.2%	0.25%	—	0.5%	0.25%	0.25%
Repeatability	0.02%	0.02%	0.1%	0.1%	0.1%	0.5%	0.1%	0.1%
Stability	0.1%	0.1%	0.5%	Not specified	Not specified	Not specified	Not specified	Not specified
Resolution[c]	Continuous	Continuous	Continuous	Continuous	Continuous	0.25%	Continuous	Continuous
Over-range or overpressure	150%	150%	150%	150%	200%	150%	80%	150%
Output level	0–5 V	0–5 V	0–5 V	35 mV	35 mV	Resistance change	0.250V/V (input)	0–5 V
Output impedance	5 kΩ	5 kΩ	2 kΩ	350 Ω	350 Ω	0 to full resistance	5000–2000 Hz	100–1300
Temperature effect	0.01%/°F typical	0.01%/°F typical	0.02%/°F	0.01%/°F	0.01%/°F	0.015%/°F	0.03%/°F	0.025%/°F
Vibration acceleration sensitivity	0.02%/g	0.02%/g	Not specified	0.015%/g	0.015%/g	0.1%/g	Not specified	0.02%/g

[a] Courtesy Rosemount Engineering Co.
[b] REC, Rosemount Engineering capacitor.
[c] Does not include signal-to-noise ratio.

Table 3.2 Conversion Factors for Various Pressure Units, Equivalent Value

Pressure Unit	Pounds Per Square Inch	Inches Hg, 0°C	Millimeters Hg, 0°C	Millibars	Inches H₂O, 60°F	Centimeters H₂O, 60°F
Pounds per square inch	1	2.0360	51.715	68.947	27.707	70.376
Inches Hg	0.49116	1	25.400	33.864	13.609	34.566
Millimeters Hg	0.019337	0.03937	1	1.3332	0.53577	1.3609
Millibars	0.014504	0.029530	0.75006	1	0.40186	1.0207
Inches H₂O, 60°F	0.036092	0.073483	1.8665	2.4884	1	2.5400
Centimeters H₂O, 60°F	0.014209	0.028930	0.73483	0.97970	0.3937	1
Inches H₂O, 20°C	0.036063	0.073424	1.8650	2.4864	0.99919	2.5380
Centimeters H₂O, 20°C	0.014198	0.028907	0.73424	0.97891	0.39338	0.99919
Inches H₂O, 25°C	0.036021	0.073339	1.8628	2.4835	0.99803	2.5350
Centimeters H₂O, 25°C	0.014181	0.028873	0.73339	0.97777	0.39292	0.99803
Grams per square centimeter	0.014223	0.028959	0.73556	0.98066	0.39409	1.0010
Pounds per square foot	0.0069444	0.014139	0.35913	0.47880	0.19241	0.48872
Ounces per square inch	0.0625	0.12725	3.2322	4.3092	1.7317	4.3985
Atmospheres	14.6960	29.9212	760	1013.25	407.183	1034.25

[a] Courtesy Rosemount Engineering Co.
[b] 1 Millibar = 1000 dyn/cm²; 1 kq/m² = 0.1 g/cm²; 1 kg/cm² = 1000 g/cm²; 1 g/m² = 0.0001 g²/cm².

Review Questions

3.1   Name the classes of mechanical pressure instruments and give examples of each.

3.2   What is the difference between a U-tube and a well manometer?

3.3   What is a draft gage? What group of pressure instruments does it represent and where is it used?

3.4   What type of manometer would you choose for use at a static pressure of 2500 lb/in.² gage? What would happen if the pressure were to increase to 2600 lb/in.² gage in the unit you choose?

3.5   What so-called manometer can be used without liquids?

3.6   What mechanical method is used to indicate differential pressure?

3.7   What are the advantages and disadvantages of limp diaphragm gages?

3.8   On what principle do bell pressure gages operate?

in Various Units[a,b]

Inches H₂O, 20°C	Centimeters H₂O, 20°C	Inches H₂O, 25°C	Centimeters H₂O, 25°C	Grams per square centimeter	Pounds per square foot	Ounces per square inch	Atmospheres
27.729	70.433	27.762	70.515	70.307	144	16	0.0680457
13.620	34.594	13.635	34.634	34.532	70.727	7.8585	0.0334211
0.53620	1.3620	0.53682	1.3635	1.3595	2.7845	0.30939	1.31579×10^{-3}
0.40218	1.0215	0.40265	1.0227	1.0197	2.0886	0.23206	9.86923×10^{-4}
1.0008	2.5421	1.0020	2.5450	2.5374	5.1972	0.57747	2.45590×10^{-3}
0.39402	1.0008	0.39448	1.0020	0.99901	2.0461	0.22735	9.66887×10^{-4}
1	2.5400	1.0012	2.5430	2.5355	5.1930	0.57700	2.45392×10^{-3}
0.3937	1	0.39410	1.0012	0.99821	2.0445	0.22717	9.66105×10^{-4}
0.99884	2.5371	1	2.5400	2.5325	5.1870	0.57633	2.45106×10^{-3}
0.39324	0.99884	0.3937	1	0.99705	2.0421	0.22690	9.64984×10^{-4}
0.39441	1.0018	0.39487	1.0030	1	2.0482	0.22757	9.67841×10^{-4}
0.19257	0.48912	0.19279	0.48969	0.48824	1	0.11111	4.72540×10^{-4}
1.7331	4.4021	1.7351	4.4072	4.3942	9	1	4.25286×10^{-3}
407.512	1035.08	407.986	1036.29	1033.23	2116.22	235.136	1

3.9   Explain the difference between spring balanced and beam bell gages.

3.10 Explain how friction-free motion is obtained in piston pressure gages.

3.11 Why are metallic diaphragm gages not more widely used in industry?

3.12 What are the advantages of Bourdon gages in making pressure measurements? What is the main disadvantage of this type of gage?

3.13 What are the advantages of Helicoid and band drive arrangements over gear and sector arrangements for pressure gage indicators?

3.14 Under what conditions would you use a diaphragm line loop or a snubber in a Bourdon tube gage input line?

3.15 In what type of service is the wear on indicating mechanisms of Bourdon tube gages the greatest?

3.16 What precautions should be observed in the use of bellows pressure gages? What advantages do these gages offer?

3.17 Give the advantages and disadvantages of a resistance contact pressure gage?

3.18   What advantage do inductance manometer pressure gages offer? What are their disadvantages?

3.19   Describe piezoelectric transducer action. At how high a temperature can these transducers be operated and under what conditions?

3.20   What are the advantages and disadvantages of piezoelectric transducers?

3.21   Explain how a capacitive transducer operates.

3.22   Give the advantages and disadvantages of capacitive transducers.

3.23   What features of an oscillating transducer require the closest control?

3.24   Why are oscillating transducers used although they have a limited temperature range and are expensive?

3.25   Describe the operating principle of inductive transducers.

3.26   How is the output of an inductive transducer measured?

3.27   What are the two main types of strain gage transducers? How do they differ?

3.28   Give the advantages and disadvantages of strain gage transducers.

3.29   What is a McLeod gage? Where does it find its major applications?

3.30   What is meant by a hot wire filament vacuum gage? Name the types available.

3.31   Give the advantages and disadvantages of the Pirani gage compared to the thermocouple gage.

3.32   What is an ionization gage and how does it work? Where is it useful?

3.33   How does an Alphatron gage function? What makes it more useful than other types of vacuum gages? Why is it not more widely used?

3.34   In what scale areas does one find the least accuracy in the majority of pressure gages? What does this tell you if you plan to take a pressure measurement?

Problems

3.1   Show that the dimensions in the equation $P = F/A$ are an expression for pressure in basic physical units of mass, length, and time.

3.2   In what types of values can pressure be measured? Give an equation to find each type of value.

3.3   What pressure is represented in a manometer 72 in. high using a liquid with a density of 0.32755 lb/in.^{3}

3.4   You have a beam gage available to make pressure measurements. What is the range of pressures for which this instrument is useful if a counterweight of 8 oz is placed 3 in from the pivot which is attached to two 4-in. diameter bells each 6 in. from the pivot? Assume the indicator to have a 90° range and to be linear.

3.5 Explain how the Phillips gage operates and what advantages this type of operation offers in vacuum measurements.

3.6 Explain how to calibrate a manometer calibrated in both inches and centimeters.

3.7 Sketch the equipment and explain how you would calibrate a pressure gage used on alcohol or gasoline lines using a deadweight tester as the calibration instrument.

3.8 Explain how you would make a gas gage calibration and sketch the calibrating setup using (a) a deadweight tester, and (b) standard test gages.

3.9 Outline how, and sketch the setup of equipment, to calibrate (a) an ionization gage, and (b) a filament vacuum gage.

Bibliography

Astromics, A Division of Mitchell Camera Corporation, San Jose, Calif., *Variable Reluctance Pressure Transducers*, October 1961.

Bourns Inc., Instrument Division, *Bourns Instruments*, Instrument Summary Brochure No. 2, Riverside, Calif., August, 1960.

Consolidated Electrodynamics Corporation, *Transducers*, Pressure Pickup Bulletins CEC-1539B, CEC-1540B, CEC-1541A, CEC-1553A, CEC-1556A, 1568 and 1558, Pasadena, Calif., no date.

Dwyer Instruments, Inc., *Dwyer Controls and Gages*, Michigan City, Ind., 1971.

Dynisco, Division of American Brake Shoe Company, Westwood, Mass., *Strain Gage Transducers*, Short Form Catalog, 5/62/5M.

Eckman, D. P., *Industrial Instrumentation*, John Wiley, New York, 1950.

Foxboro Instrument Company, *Foxboro Dynaformer Cell*, Technical Information Bulletin 27A-12A, Foxboro, Ma. September 1953.

Helicoid Gage Division, American Chain and Cable Company, *The Exclusive Helicoid Movement, Helicoid Gage*, Bridgeport, Conn., DH-226, no date.

Hernandez, J. S., *Introduction to Transducers for Instrumentation*, Statham Instruments, Inc., Oxnard, Calif., no date.

Manning, Maxwell & Moore, *Ashcroft Gauges*, Catalog 300, Stratford, Conn., 1953.

Norden-Ketay Corporation, *Solid Front Bandrive Gages*, Bulletin 403, Norwalk, Conn., no date.

Pace Engineering Company, *Pace Model P-1 Pressure Transducer*, Los Angeles, no date.

Pace Engineering Company, *Pressure Transducers and Instrumentation Systems*, Los Angeles, no date.

Rhodes, Thomas J., *Industrial Instruments for Measurement and Control*, McGraw-Hill, New York, 1941.

Roger Gilmont Instruments, Inc., *Instruments by Gilmont*, Catalog G68, Great Neck, N.Y., no date.

Rosemont Engineering Company, *Capacitive Pressure Sensors*, Bulletin 1672, Minneapolis, 1968.

Ruska Instrument Corporation, *Low Pressure Standards, Tilting Piston Gage-Air Piston Gage*, Houston, no date.

Ruska Instrument Corporation, *Pressure Calibration Systems and Components*, Houston.

Studier, Walter, Quartz Pressure Sensors, *Instruments and Control Systems*, Vol. 35, No. 12, December 1962, pp. 94–95.

Taber Instrument Corporation, *Teledyne Pressure Transducers*, Models 206, 206-1, 206-2, 10M-2-62, North Tonawanda, N.Y.

Taylor Instrument Companies, *Taylor Instruments for Pressure*, Catalog 76JF, Rochester, N.Y., December 1952.

CHAPTER 4

Thermometers

Temperature is an important industrial measurement, and a temperature measurement is required in every case in which the application of heat or cold is needed for the control of a process or manufacturing operation. The accuracy of the measurement and the speed at which it can be made depend on the application. The application also dictates whether a simple indicator or a more complex recorder or controller is necessary. Use can be made of relatively simple thermometers where only an indicator is needed and adequate installation space is available for temperatures below 975°F, and where the system can normally reach equilibrium with the measuring instrument. Some thermometers can also be used with recorders and control devices.

A thermometer is one of the more familiar instruments used for the measurement of temperature. The liquid-in-glass type of thermometer is probably the most easily recognized, since it is used to indicate human body temperatures and to indicate indoor and outdoor temperatures. A thermometer can be constructed from several materials that expand when they are exposed to heat and contract when they are exposed to cold. The materials used industrially are fluids such as oils and other nonfreezing liquids, liquid metals such as mercury, gases, and vapors. Thermometers are also constructed of metals that have the proper coefficient of expansion or contraction as bimetallic devices, or exhibit a change in their resistance in proportion to a temperature difference.

4.1 TEMPERATURE SCALES

One of the first requirements for the measurement of temperature is to establish a temperature scale to be used on the indicating, recording, or controlling instrument. As discussed in Chapter 2, the Celsius and the

Fahrenheit scales are the two temperature scales primarily used in industrial applications. In the case of recorders, it is more essential that the relationship between the detecting medium and the temperature change have a definite proportionality, because the charts have to be changed at fixed intervals and are standard for certain types of instruments. It is highly desirable that these records be interchangeable. Otherwise, every recorder would have to have special charts to match the recording mechanism. This definitely does not mean that the relationship has to be one to one. The relationship could be based on any physical expansion of a liquid, vapor, gas, or metal. The only specific requirement is that there be a thermodynamic relationship between the two ends of the indicating or controlling span, with a known temperature range, which can be inscribed on the measuring or controlling instrument scale. The reading is just as reliable and indicates the same temperature change if the scale is marked from 0 to 100°C or 32 to 212°F. In fact, many scales are conveniently marked in both measurement systems.

Fundamentally, the Celsius and Fahrenheit temperature scales measure the same temperature differences, but different values have been arbitrarily chosen for the fixed points upon which each of the systems is based. There is a simple relationship between the two systems, as shown by the two equations:

$$°C = \tfrac{5}{9} (°F - 32) \tag{4.1}$$

$$°F = \tfrac{9}{5} (°C) + 32 \tag{4.2}$$

This means that at the ice point or freezing point of 0°C the Fahrenheit scale reads 32°. For the 100°C boiling point, the conversion is:

$$°F = (\tfrac{9}{5} \times 100) + 32 = (9 \times 20) + 32 = 180 + 32 = 212$$

To permit ready reference to the Kelvin and the Rankine scales, which are, respectively, used in scientific and engineering literature, the following relationships may be helpful. The Kelvin scale (indicated as K) has an absolute zero equivalent to −273.2°C, and 1 K is the same as 1°C for all practical applications. The Rankine scale (indicated as °R has an absolute zero equivalent to −459.7°F and has been assigned a value of 491.7°R for the ice point and a value of 671.7°R for the boiling point. It is often referred to as the absolute Fahrenheit scale.

The types of thermometers to be discussed are liquid-in-glass, liquid-in-metal, vapor tension, gas filled, bimetallic, and changing resistance. Each type is based on certain theory and has certain advantages and certain disadvantages. Each has its place in industry, and this place is based on the process application and the accuracy required for product quality, safety of operation, or necessary control.

4.2 LIQUID-IN-GLASS THERMOMETERS

The liquid-in-glass thermometer is one of the simplest temperature measuring instruments and is widely used in industrial applications. These temperature indicating devices operate on an expanding volumetric principle. A typical industrial thermometer is shown in Figure 4.1. This is strictly an indicating instrument. In this instrument, as the temperature rises, the liquid in the well absorbs heat and expands. This expansion

Figure 4.1 *Typical industrial thermometer. Courtesy Taylor Instrument, Consumer Product Division, Sybron Corp.*

causes the liquid to rise in the capillary tube in proportion to the temperature applied. The scales on these thermometers are not perfectly linear, because as the heat is absorbed by the liquid, the well, the glass tube, and the mounting are also affected by the temperature rise. At higher temperatures the metal well and glass tube expand. This causes a change in the total volume. In addition, the coefficient of expansion of the liquid may vary slightly at different temperatures. As discussed in Chapter 2, the glass tube may be filled with an inert gas and, as the column rises, more pressure is exerted. While most good industrial thermometers are constructed of well-aged glass, there may be some change after an extended period of time. All these variations contribute some error in the measurement. Good industrial thermometers are normally accurate to within ±1%, and may be made as accurate as ±0.5% of full span.

The Taylor instrument shown in Figure 4.2*a* shows the construction features of an industrial liquid-in-glass thermometer. The case (1) is a shallow one-piece unit finished in dull black enamel, practically fume-proof, and designed for easier reading through a wide angle of vision. The bezel (2) is chromium plated to resist corrosion and is constructed of two interlocking pieces which hold the clear thick protective glass front (4) against two wavy tension springs. The case is grooved to hold the bezel securely in place. The tension springs (3) are permanently attached to the case to provide rigid rattleproof construction. The easily read binocular tube (5) is a feature of all Taylor thermometers with ranges up to 750°F. The liquid column stands out vividly through a wide range of vision and, for the higher ranges where mercury is used, a lens front borosilicate glass is used to provide magnification for easier reading. These tubes are provided with an expansion chamber at the top to prevent breakage due to overheating. The amount of overheating that can occur without damage depends on the range of the thermometer and the liquid used. In higher range instruments, triple-distilled mercury is used, and an inert gas (usually nitrogen) is used above the mercury column to prevent oxidation and minimize distillation of the mercury. Each scale (6) is individually graduated for Taylor thermometer tubes, and the open faced numerals stand out clearly against a sharply contrasting background.

Figure 4.2*b* shows an actual cross sectional view of the stem construction features. The swivel nut (1) turns freely on the thermometer stem to permit rapid installation in separate well or union hub connection without thermometer rotation. The well (2) can be separately and permanently installed in the process system or apparatus. These wells are nor-

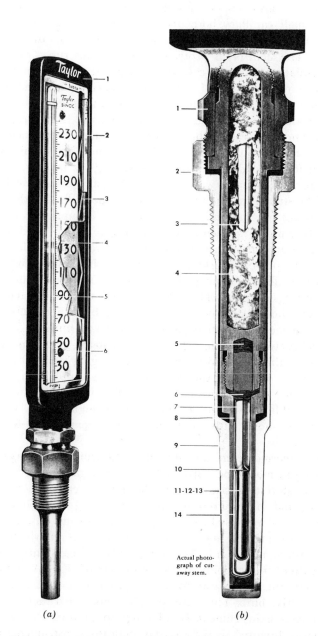

(a)

(b)

Actual photo-
graph of cut-
away stem.

Figure 4.2 *Construction features of a liquid-in-glass thermometer. (a) External. (b) Internal. Courtesy Taylor Instrument, Consumer Products Division, Sybron Corp.*

mally made of brass or steel for standard applications, and can be supplied in such metals as copper, stainless steel, monel, aluminum, and cast iron. The small capillary bore diameter (3) minimizes the volume of liquid in the tube and aids in the elimination of errors due to variable immersion. The stem (4) is protected with asbestos rope against breakage, and insulates it from the case to minimize the conduction of external heat to the temperature sensitive bulb. The packing (5), forced in around the tube under pressure, prevents error due to a slipping tube. The packing (6) provides a positive seal to prevent mercury from escaping from the bulb chamber in case of bulb breakage. Tube (7) breakage is minimized by thorough annealing of the glass to remove all strains. Space (8) is provided for mercury expansion. The stem (9) is taper ground to fit the taper reamed separate well precisely to ensure maximum heat conduction. The streamlined joint (10) between the bulb and the capillary tubing provides free movement of the mercury column. For the highest accuracy, the temperature sensitive bulbs (11) to (13) are aged to prevent changes in volume after installation. They have uniform thin walls of special glass to prevent cracking with changes in temperature, and are shaped to provide maximum speed of response to temperature changes. The conducting medium (14) provides rapid transfer of temperature changes from the well to the thermometer bulb. This chamber is completely free of moisture, which prevents a pressure buildup.

Industrial thermometers may have straight stems, as shown in Figures 4.1 and 4.2, or they may have angled stems ranging from 90 to 225°, as shown in Figure 4.3. Angle thermometers can be supplied in rear, right, or left side 90° and oblique angles. Handled industrial thermometers are also available, as shown in Figure 4.4. In Figure 4.5 a few typical applications are shown with the thermometers recommended for the type of application.

Industrial thermometers are used for measuring the temperature of molten metal in monotype casting machines, flue gas, ovens, kilns, air in air ducts, dough testing, cruller frying, hard candy, cream cooking, chocolate melting and mixing, refrigerators and cooling units, hot and cold water, steam, cooking vessels, brewing vats, lubricating oils, air compressors, and diesel engines, and for other applications in which the temperature sensitive bulb can be kept completely and constantly submerged in the medium at the point of maximum circulation. A thermometer measures only the temperature of the medium in contact with the temperature sensitive bulb. There are specially constructed thermometers for use in which the instrument is subjected to unusual vibrations. The thermometers discussed so far are only indicating instruments and provide no control.

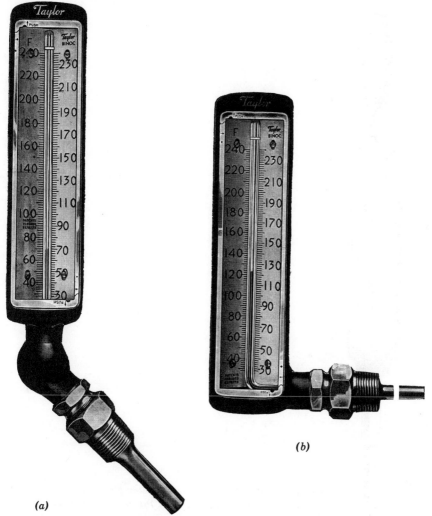

(a)

(b)

Figure 4.3 *Angle thermometers (a) 135° Right oblique side angle. (b) 90° Right side angle. Courtesy Taylor Instrument, Consumer Products Division, Sybron Corp.*

To obtain the best accuracy with an industrial thermometer, which is approximately ±1% of the span, as stated earlier, it must be properly installed so that the temperature sensitive bulb can reach temperature equilibrium with the surrounding medium. The temperature sensitive bulb should also be properly immersed in the medium to be measured to eliminate the immersion error discussed in Chapter 2. It should also

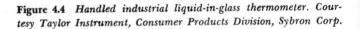

Figure 4.4 *Handled industrial liquid-in-glass thermometer. Courtesy Taylor Instrument, Consumer Products Division, Sybron Corp.*

be installed at the point of maximum flow to provide the most rapid heat transfer from the medium under measurement to the bulb.

The speed of response ranges from a fraction of a second to minutes, depending on the construction of the thermometer. The construction features affecting response to a temperature change are the size of the thermometer, the size and type of metal of the thermal well, the wall thickness of the temperature sensitive bulb, the heat transfer medium between the bulb and the well, and the type of medium in which the

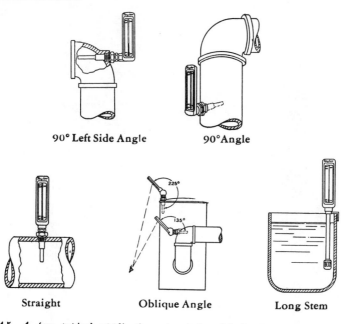

90° Left Side Angle   90°Angle

Straight   Oblique Angle   Long Stem

Figure 4.5 *A few typical applications for industrial liquid-in-glass thermometers. Courtesy Taylor Instrument, Consumer Products Division, Sybron Corp.*

thermometer is immersed. Thermal wells add a time lag which may extend into minutes, especially if the well has relatively thick walls to withstand high pressure. Very thin bulbs directly exposed to the medium to be measured can respond in less than 0.1 s. and reach full equilibrium in 0.3 to 0.5 s. It should be remembered that thin bulbs cannot withstand much pressure and should always be enclosed in wells if exposed to pressure or corrosive materials.

4.3 LIQUID-IN-METAL THERMOMETERS

The heart of a liquid-in-metal thermometer, which may contain mercury or organic liquids, is shown in Figure 4.6. The Taylor unit shown contains mercury and is used to actuate a recorder. This system contains a bulb A, mercury B, stainless steel Accuratus tubing D, a special alloy filler, E, and a Bourdon spring G. The special tubing is welded into the system at C and F.

In the type of thermometer shown in Figure 4.6, the temperature coefficient of the wire is selected so that the effective volume of the capil-

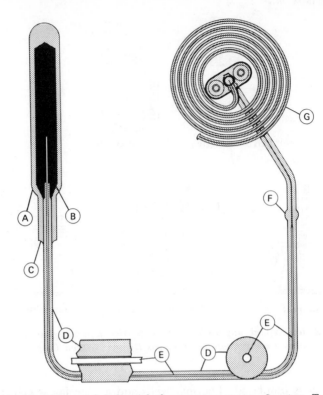

Figure 4.6 *Liquid-in-metal industrial thermometer system. Courtesy Taylor Instrument, Consumer Products Division, Sybron Corp.*

lary increases with temperature just enough to hold the expanded volume of the mercury in the capillary. In the Taylor unit the stainless steel Accuratus tubing can be used in lengths exceeding 200 ft. The accuracy of the measurement is not affected by variations in temperature along the tubing, because these variations in temperature are accurately counterbalanced at the point of fluctuation. However, there is a difference in the speed of response with different lengths of connecting tubing. The temperature limits range from −38 to 1200°F, or the equivalent Celsius range.

The use of the filler wire in a hermetically sealed system permits the bulb to be small enough for use in limiated space applications, such as pipelines, and to be mounted in a variety of positions. The dimensions of the mercury bulbs and their separate wells are determined by the length of the temperature range. With mercury actuation, charts with uniformly spaced graduations can be used. These records can be read with equal accuracy over the entire scale, although the actual response is not linear.

Actually, the liquid-in-metal thermometer operates on a pressure principle and the volume is relatively well fixed. According to the fundamentals of basic physics, this means that the temperature ideally follows equation 4.3:

$$\frac{P_1 V_1}{T_1} = \frac{P_2 V_2}{T_2} \tag{4.3}$$

$$\text{with } V_1 = V_2 \quad \text{and} \quad P_1 T_2 = P_2 T_1 \tag{4.4}$$

where P_1 = lower pressure
P_2 = higher pressure
T_1 = lower temperature
T_2 = higher temperature
V_1 = volume at lower temperature and pressure
V_2 = volume at higher temperature and pressure

Equation 4.4 makes it evident that the pressure varies as a function of the temperature. Therefore, when the temperature increases, the pressure increase causes the indicating helix, spiral, or C-shaped Bourdon tube to move in proportion to the pressure increase and indicate the temperature at that time.

Unfortunately, no liquid behaves ideally, nor are V_1 and V_2 exactly alike. However, these systems are accurate enough to find wide usage in industrial applications. They can be used as self-actuated control systems in which the control for an on-off operation is in the recording or indicating device.

In systems in which Accuratus tubing or its equivalent is not used, the length of the capillary tubing is limited to approximately 30 ft, because of the volume involved and the time required for the temperature change to be indicated or recorded. Such systems are not adequate when rapid responses and controls are needed. They are useful and economical for use in close coupled systems having relatively slow changing variables. These systems must be compensated for temperature changes occurring along the capillary.

Mercury is usually used in most recording thermometer systems, but for cases in which it cannot fulfill the requirements organic liquid actuated recording thermometers are available. These systems are particularly adaptable to applications in which a small bulb is required, where short range intervals are desirable, for low temperature measurements, and where a manufacturing process is sensitive to mercury. Such systems are normally limited to capillary lengths of 15 ft or less, because longer capillary lengths require careful compensation with standard types of capillary tubing. The length of tubing can be extended to 200 ft by the use of Accuratus tubing or its equivalent. Where the range interval is 280° or less, the systems are normally calibrated to use uniformly graduated charts. Liquids used include monochlorobenzene, methyl alcohol, and ethyl benzene.

The operations of the mercury (class V) thermal system and the organic liquid (class I) thermal system are identical. Class V and class I classifications have been established by the Scientific Apparatus Manufacturer's Association to identify the type of fill used in filled system thermometers. We also discuss class II vapor pressure and class III gas fill systems in Sections 4.4 and 4.5.

4.4   VAPOR ACTUATED THERMOMETERS

A class II vapor pressure system has two possible modes of operation, as shown in Figure 4.7a and 4.7b. A volatile liquid is used as the fill medium, and at any particular temperature an equilibrium condition exists. This means that both vapor and liquid states are present and the

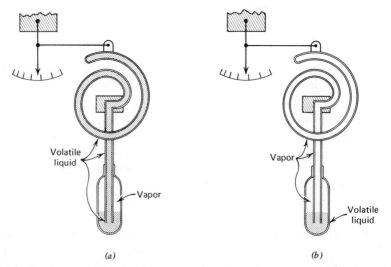

(a)   (b)

Figure 4.7   *Two modes for a class II vapor pressure system. (a) Liquid-filled. (b) Vapor filled. Courtesy Honeywell, Inc.*

vapor pressure is just adequate to prevent further change of liquid to vapor. As the temperature rises, more of the liquid becomes a vapor and exerts more pressure on the Bourdon tube and causes it to deflect. Conversely, when the temperature drops, the vapor condenses to the liquid state and the pressure is reduced, with a resultant contraction of the Bourdon tube and capillary with liquid.

In Figure 4.7b the bulb is at a temperature lower than that of the Bourdon tube and the capillary. Under these conditions the vapor condenses in the bulb and the liquid collects in the bulb.

These modes of operation demonstrate that vapor actuated systems are not suitable for use where the ambient or surrounding temperature is the same as or is close to the measured temperature, or where the bulb temperature goes back and forth past the capillary temperature. In all cases the bulb must be large enough to contain enough liquid so that a liquid-vapor interface is always maintained in the bulb. When the measured temperature cycles around the ambient temperature, the liquid shifts from the Bourdon tube and capillary to the bulb, or from the bulb to the capillary and the Bourdon tube. This generates a condition in which there may actually be several different surfaces within the system. This type of instrument can read accurately only when the liquid is either in the bulb or in the capillary and Bourdon tube. When the bulb temperature oscillates above and below the capillary and Bourdon tube temperature, the liquid must transfer from the capillary to the bulb and back. This transfer action involves considerable time delay which is highly undesirable.

In vapor actuated thermometers the scale graduations expand as the temperature increases, and they are more accurate at higher temperatures. These systems have temperature ranges from −40 to 600°F, and the minimum span is approximately 60°F. The maximum tubing length is 200 ft. It is not necessary for the bulb to be installed at the same elevation as the capillary and Bourdon tube if the difference in elevation can be measured and included in the calibration of the system for the particular application. Class II thermal systems do not require ambient temperature compensation, because vapor pressure depends only on the temperature at the surface between the liquid and the vapor.

It should be noted that the pressure of saturated vapor is independent of the volume occupied by the vapor. The only difference this causes is in the ratio between the amount of liquid and the amount of vapor present in the volume at a constant temperature. The pressure depends only on the volatile liquid used as the fill material and the temperature.

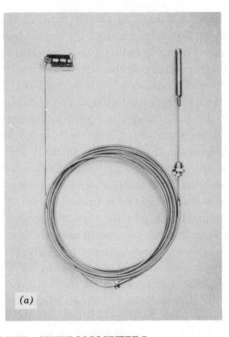

(a)

4.5 GAS ACTUATED THERMOMETERS

Gas filled temperature systems operate fundamentally in accordance with Charles's law. This law may be interpreted by saying that an ideal gas of a given weight at a constant volume produces an absolute pressure in direct proportion to the absolute temperature of the gas. With such an ideal gas, under ideal conditions, the following equation is perfectly valid, and a gas filled thermometer has equal scale divisions:

$$\frac{P_1}{P_2} = \frac{T_1}{T_2} \tag{4.5}$$

where P_1 = initial absolute pressure
P_2 = final absolut pressure
T_1 = initial absolute temperature
T_2 = final absolute temperature

Unfortunately, we do not have perfect gases and perfect conditions, so we have deviations from the ideal. However, the deviations for some gases are so small that no corrections are necessary for the accuracies normally encountered in industrial gas filled thermometer systems designated class III.

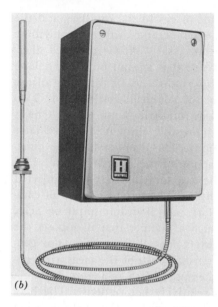

(b)

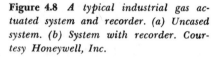

Figure 4.8 *A typical industrial gas actuated system and recorder. (a) Uncased system. (b) System with recorder. Courtesy Honeywell, Inc.*

Gas actuated thermometers can be very accurate temperature measuring devices, and this is emphasized by the fact that the NBS uses a constant volume gas thermometer as a primary standard for temperature measurement.

A typical industrial gas filled thermometer is shown in Figure 4.8. The basic units consisting of the bulb, capillary, and helix are shown in Figure 4.8a, and the bulb, capillary, and a circular chart recorder are shown in Figure 4.8b. A gas actuated industrial thermometer is available in minimum spans of 150°F for the −125 to 1000°F range. The actuating gas is normally nitrogen, although other gases can be used. The gas used should be chemically inert, have a high coefficient of expansion at a constant volume, have low viscosity, have low specific heat, and be readily available as a highly pure commercial product. The use of gas actuated thermometers should be restricted to temperatures under 1000°F and far enough above the condensation temperature of the gas to permit action obeying Charles's law.

Installation of gas actuated thermal systems may require compensation if the application prohibits the use of a large enough bulb to contain the majority of the gas used in the system, or if the capillary system has to pass through more than one ambient temperature region. Where a large temperature sensing bulb and a short capillary leading to the Bourdon spiral or helix can be used, the amount of gas in the capillary and the Bourdon spiral or helix at the ambient temperature will not contribute enough error to make compensation necessary. A gas actuated thermometer gives the average temperature, so that a small bulb and a long capillary will not give a true temperature indication or recording of the bulb temperature unless some method of compensation is used. Two methods of compensation can be used in industrial applications. One is called *case compensation,* in which a bimetallic element, discussed later in this chapter, is used as shown in Figure 4.9. The action of the bimetallic element is in opposition to the action of the Bourdon spiral. This type of compensation is used when the case and capillary are subjected to only one ambient temperature condition. The second method of compensation involves an identical capillary and Bourdon spiral or helix without a sensing bulb, installed in such a manner that it opposes the action of the measuring system to correct for all the motion not generated by the gas in the temperature sensing bulb of the measuring system. Such a system is shown by the schematic sketch in Figure 4.10.

Gas actuated thermometers have a relatively fast response. The bare bulb reacts in seconds when completely immersed in the medium being measured. The reaction time to reach the equilibrium condition, usually considered to be 90% of the total response time may be extended to minutes by the use of protective wells.

Gas actuated thermometer systems withstand an overrange of from 150 to 300% of range above the nominal maximum for short periods of time. What overrange the system stands depends on the actual range of the instrument system. This compares to the overrange for mercury actuated systems which withstand 150% of the scale range above the nominal

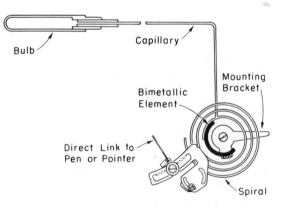

Figure 4.9 *Case compensation using a bimetallic element. Courtesy Honeywell, Inc.*

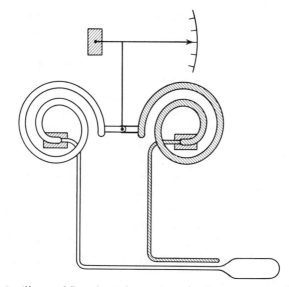

Figure 4.10 *Capillary and Bourdon tube compensation for a gas actuated thermometer.*

maximum temperature for short periods of time. Vapor actuated systems are not recommended for applications where sizable overranges are likely to occur.

Calibration of gas actuated thermometers should be performed using three separate temperature baths. A low temperature bath should be used to check the zero adjustment, a medium temperature bath is needed to check approximately the one-third full scale value, and a high temperature bath is needed to check the three-fourths full scale value. If the calibrations do not check properly, it is then necessary to adjust the gas pressure in the system to bring the system into calibration. This is not a simple one-adjustment operation, since there are three successive steps involved because full scale adjustments require a readadjustment of the zero setting. This means that any pressure change creates a need for both full scale and zero value adjustments.

4.6 SPEED OF RESPONSE IN BULB-THERMOMETER SYSTEMS

Many times the rate of response of temperature bulb system is over-emphasized. Occasionally, this factor is considered at the expense or ex-clusion of other equally important qualities vital to a good thermometer or thermometer system. However, the rate or speed of response should be rapid enough to follow accurately the temperature changes that are to be measured.

In industrial applications the ideal bulb thermometer is not always the fastest response system. It is usually the one specifically designed or engineered for the application. This application engineering involves serious consideration of the process, the apparatus, the minimum and maximum operating temperatures, the process time lag, the rate at which temperature changes occur, the amount of space available for mounting the bulb, the rate at which the medium circulates at the proposed bulb location, the type of medium to which the bulb is to be exposed, the length of the connecting capillary and, last but not least, the possible abuse to which the bulb, capillary, or instrument may be subjected in the specific application. Figure 4.11a illustrates observed response times for bare bulb type systems, and Figure 4.11b (Taylor Instrument Companies) shows a comparison of rates of response of various temperature measuring devices.

4.7 BIMETALLIC THERMOMETERS

Bimetallic thermometers are usually constructed of two thin strips of dissimilar metals which are bonded together for their entire length. In industrial thermometers these bonded strips are often formed into a helical coil. One end of the coil is welded to the thermometer stem, and the other end to the pointer staff. A typical unit is shown in Figure 4.12.

The operating principle of a bimetallic thermometer is relatively simple. The metals have different coefficients of expansion and, when heat is applied, expand at different rates and in different amounts. A helix coil is formed by placing the metal with the greater coefficient of expansion on the outside. When heat is applied to the thermometer stem, the coil actually winds up, and this winding motion rotates the shaft and the pointer. With proper design of the helically wound coil, adequate angular motion can be obtained with negligible helix elongation.

Bimetallic thermometers are not recommended for use at temperatures above 800°F on continuous duty or above 1000°F in intermittent duty. All metals have physical limitations and are subject to creep and permanent warp distortion. This means that the metals do not return to their normal condition and the temperatures indicated are not accurate. Bimetallic thermometers can be used between the limits of −40 and 1000°F and can be expected to reproduce with an accuracy of ±1% when not used above 800°F in continuous service.

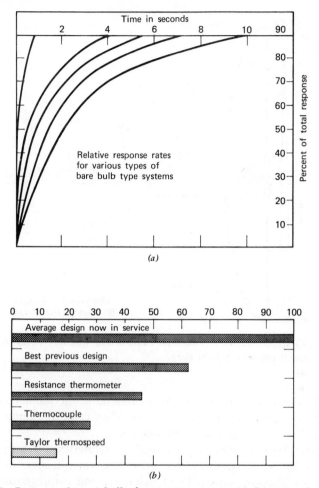

Figure 4.11 *Response time of bulb thermometer system. (a) Response time for bare bulb systems. (b) Comparison of rates of response of various temperature measuring devices. Courtesy Taylor Instrument, Consumer Products Division, Sybron Corp.*

The materials most used in bimetal thermometers are Invar, which is an alloy of nickel and iron, as the low expansion metal, and brass or a nickel–chrome alloy as the high expansion metal.

Invar has the same modulus of elasticity over a wide range of temperatures and very low thermal expansion. Brass is useful up to approximately 300°F, and nickel–chrome alloys have to be used for higher temperatures.

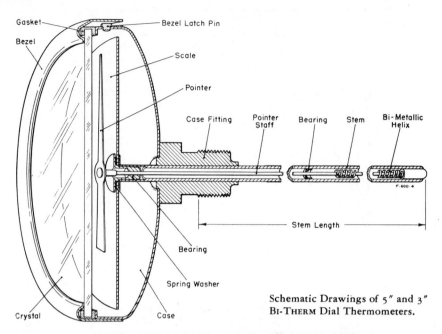

Schematic Drawings of 5″ and 3″
Bi-Therm Dial Thermometers.

Figure 4.12 *Typical bimetallic thermometer construction. Courtesy Taylor Instrument,
Consumer Products Division, Sybron Corp.*

Temperature wells can be used with bimetallic thermometers as protective devices against wear and corrosion. They cause the same order of
time lag as for other types of thermometers.

Bimetallic thermometers may be used in refineries, oil burners, tire
vulcanizers, hot solder tanks, coffee urns, hot water heaters, tempering
tanks, electric dipping tanks, diesel exhausts, and impregnating tanks.
These thermometers are available in both straight and angle stem forms.

Calibration of bimetallic thermometers can be made by a comparison
method using heat sinks, water baths, or calibrating furnaces where adequate immersion space is available.

4.8 RESISTANCE THERMOMETERS

Industrial resistance thermometers are essentially coils of wire wound
into or around frames of insulating material capable of withstanding
the temperature for which the thermometer is designed. Usually, the
coils are made up of fine wire so wound on the frame that there is negli-

gible physical strain exerted as the wire expands and contracts with temperature changes. The wires are arranged on the frames so that there is good thermal conductivity and a high rate of heat transfer.

A resistance thermometer is basically an instrument for measuring electrical resistance, which has been calibrated to read in degrees of temperature instead of units of resistance. Industrial resistance thermometers have historically been made of platinum, copper, or nickel but, with the advances made in the semiconductor field, some semiconductor materials have been found suitable for use in the measurement of temperature. The thermistor is one type now in use in resistance thermometry.

When a material changes resistance with a change in temperature, the change is called the *temperature coefficient of resistance* for the material. This coefficient is expressed as ohms per degree of temperature at a specified temperature, and is positive for most metals.

It is highly desirable to have the greatest resistance change per degree for a given value of resistance to obtain the highest sensitivity of measurement, but the material must also have good stabiltiy over a long period of time (years) and over a wide range of temperature without changing its electrical characteristics.

Metals chosen for resistance thermometers have a high degree of linearity over the resistance temperature range for which the thermometer is designed. Most pure metals have a practically linear change in resistance with temperature for at least a portion of their resistance–temperature curve. The relationship between resistance and temperature can be expressed mathematically as:

$$R_T = R_0(1 + \alpha T) \tag{4.6}$$

where

$$\alpha = \frac{R_T - R_0}{R_0 T} \tag{4.7}$$

or differentially,

$$\alpha = \frac{dR}{R\,dT} \tag{4.8}$$

where R_0 = resistance at a reference temperature in ohms
R_T = resistance at some temperature T in ohms
α = coefficient of resistance of the material used

In cases in which the resistance–temperature coefficient is not linear, a general equation is often used:

$$R_T = R_0(1 + aT + bT^2 + cT^3 + dT^4 + \cdots) \tag{4.9}$$

and, the further the equation is carried out, the more accurate the particular resistance–temperature curve becomes. The coefficients *a*, *b*, *c*, *d*, and so on, can be calculated on the basis of three or more resistance–temperature values uniformly spaced over the desired working temperature range.

Some of the most used types of resistance windings for industrial applications include: (1) spaced windings of bare wire over a cylindrical insulating arbor, as shown in Figure 4.13a; (2) insulated wire smoothly wound on a metal, ceramic, or plastic arbor, as shown in Figure 4.13b; (3) bare wire wound on a narrow, evenly notched mica strip; (4) bare wire wound in a small helix around crossed, evenly notched mica strips; (5) bare wire wound into a small helix or coil which is inserted into double-spiral grooves of a thin walled insulated metal or ceramic tube; (6) a strip of very thin metal foil; and (7) woven wire mesh cloth. These windings are carefully placed so that neither the wire nor the mounting support breaks under the temperature cycling in the range for which they

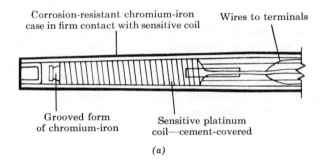

(a)

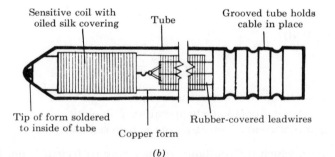

(b)

Figure 4.13 *Resistance thermometer windings. (a) Spaced windings of bare wire over a cylindrical arbor. (b) Insulated wire smoothly wound on metal, ceramic, or plastic arbor. Courtesy Leeds and Northrup Co.*

are designed. There is always a danger of breaking or stretching the wire or strip if the temperature range is exceeded.

The output of resistance thermometers is usually measured by some type of resistance bridge. While the Wheatstone bridge is the most common type used, other bridges, such as the Muller bridge, double-slidewire bridge, Collendar-Griffiths bridge, and capacitance bridge can be used with either dc or ac power. These bridges can be of either the null balance or deflection type. Bridges are discussed in detail in Chapter 5.

Resistance thermometers are available in three types, and each type requires a different method of connection from the resistor bulb to the measuring instrument. Three methods are commonly used for making the electrical connnection between a resistance winding and the measuring instrument. These three methods use two wires, three wires, or four wires.

Two-Lead Method. As shown in Figure 4.14*a*, two relatively low resistance leads *a* and *b* are used to connect the bulb resistance winding to the bridge measuring instrument. Usually, the leads are copper, and a Wheatstone bridge arrangement is used.

In this arrangement the resistance R_x comprises the resistance of the bulb plus the resistance of the leads *a* and *b*. This means that the leads *a* and *b*, unless of very low resistance, can add appreciably to the resistance of the resistance thermometer bulb. Such an arrangement will result in an error in the temperature reading, unless there is some type of compensation or an adjustment of the bulb or bridge apparatus to balance this difference in resistance. It should be noted that, although the resistance of the leads may be well known, and that an allowance has been made for this value in the measurement, the leads, as well as the resistance bulb, are subject to ambient temperature changes. Since they are copper, they thus introduce another possible source of error in the measurement.

The two-lead method should be used only when the lead wire resistance can be kept to a minimum and when a moderate degree of accuracy is adequate for the measurement being performed. Since the extent of this error is related to the resistance of the bulb winding, it is obvious that it will be proportionately higher with windings of a low resistance than with windings of a high resistance.

Three-Lead Method. The most practical and widely used method in industrial resistance thermometry is the three-lead method, shown in Figure 4.14*b*. In this circuit the two leads *a* and *c* are connected directly to one end of the resistor bulb winding at a common point. The third lead

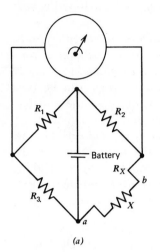

(a)

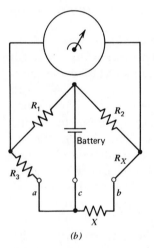

(b)

Figure 4.14 *Multiple-lead methods for bridge connec-tions of resistance thermometers. (a) **Two-lead method.** (b) Three-lead method.*

b is connected to the other end of the winding. By keeping the lengths of leads a and b equal, the resistance of lead a is added to bridge arm R_3, while the resistance is divided into both arms and retains a balance in the bridge circuit. This method compensates for the effect of lead re-sistance, thus permitting the use of relatively long resistance leads, but the ultimate accuracy of the circuit depends upon leads a and b being of equal resistance. Since lead resistance does not enter into the balance of the bridge in this circuit, leads of relatively long length and resistance can be used.

Four-Lead Method. In areas where the highest degree of accuracy is required, the four-lead method is used. Such a system is used with a platinum resistance thermometer employed as a laboratory standard for calibration purposes. Two circuit arrangements are used, and these are shown in Figures 4.15a and b. Actually, both circuit arrangements are required to make a measurement. A single measurement involves first taking a reading when the circuit of Figure 4.15a is in use, and then taking a reading when the circuit of Figure 4.15b is used. These readings are then added and divided by two to give the average reading. For the circuit of Figure 4.15a,

$$R_a + C = X + T \tag{4.10}$$

For the circuit of Figure 4.15b,

$$R_b + T = X + C \tag{4.11}$$

Addition of these two equations results in

$$R_a + R_b + C + T = X + X + T + C \tag{4.12}$$

and

$$X = \frac{R_a + R_b}{2} \tag{4.13}$$

The time required to carry out the procedure, and the inconvenience of taking the two readings, make it obvious that the method is used only when the ultimate in accuracy is required.

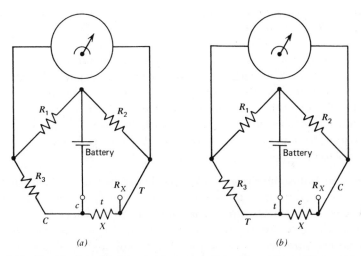

Figure 4.15 *Four-lead method for bridge connections of resistance thermometers. (a) First arrangement needed. (b) Second arrangement needed.*

In Figure 4.15*a* note that two separate leads are connected directly to each end of the resistance winding. These are leads *C* and *c* to one side, and *T* and *t* on the other side. Leads *C* and *T* are in opposite legs of the bridge, with the bridge supply connections made to leads *c,* as in the case of the three-lead method described in the last section. In Figure 4.15*b* leads *T* and *C* are in the reverse leg of the bridge, and the potential connection is made to lead *t.*

In a laboratory setup, normally a triple-pole, double-throw mercury contact switch is used to transfer the resistor bulb leads from one form of connection to the other. For the most accurate results, the leads should be well balanced, although this circuit reduces the effect of unbalanced leads and theoretically cancels out the effect.

In general, lead resistances should not exceed 2.5 Ω per lead, and the distance or wire size should be adjusted so that this resistance is not exceeded. Individual manufacturers' instruments may require less resistance than this, but in general 2.5 Ω is representative. Should a manufacturer's instrument require less resistance, either the wire size will have to be increased or the distance decreased to meet the individual requirements.

Resistance thermometers have a high degree of accuracy. Precision laboratory units can be calibrated for measuring temperatures to within $\pm 0.01\,°C$, and industrial units can be calibrated to detect actual temperatures to within $\pm 0.5\,°F$ up to $250\,°F$, and $\pm 1.0\,°F$ from 250 to $1000\,°F$ in such applications as cooling processes, heating ovens, drying ovens, kilns, process vessels, baths, quenches, refining, precisely controlled cold storage, steam and power generation condensates, steam exhausts, pickling and plating, bakery doughs, injection molding, compression molding, transfer molding, and air inversion measurements. See Figure 4.16 for some typical units available.

Resistance thermometers are normally designed for fast response, as well as accuracy, to provide close control of processes in which narrow ranges or small temperature difference spans must be maintained. Each type of resistance thermometer is interchangeable in a process without compensation or recalibration, because each type is calibrated to a standard resistance–temperature curve so that if a unit is damaged it can be easily replaced.

Three-Wire Industrial System. A typical three-wire instrument measuring system for resistance thermometers is shown in Figure 4.17, which uses a Leeds and Northrup Thermohm in a Leeds and Northrup recorder or controller Wheatstone bridge circuit. In this system, two slide-wire resistors *S* and S_1 are so adjusted and mounted on a common disc that the ratio of *a* and *b* is always unity for all practical purposes. The

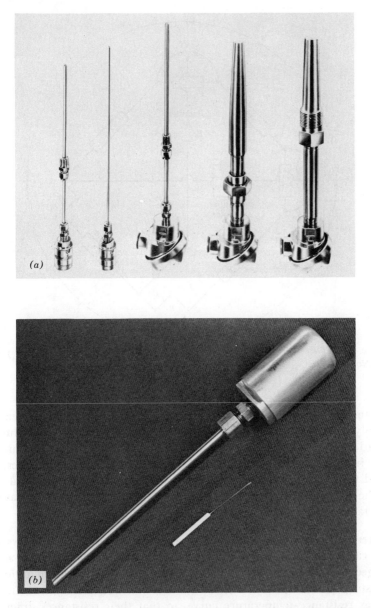

Figure 4.16 *High accuracy resistance thermometers. (a) 100 Ω Thermohms. These units extend the reliable resistance sensor range to 1500°F. They use standard three-lead construction and are available for both direct immersion and well applications. Courtesy Leeds and Northrup Co. (b) Platinum resistance detector available in 50, 100, 200, and 400 Ω models for temperature ranges from —200 to 1000°C. Courtesy CGS/Thermodynamics, Division of CGS Scientific Corp.*

135

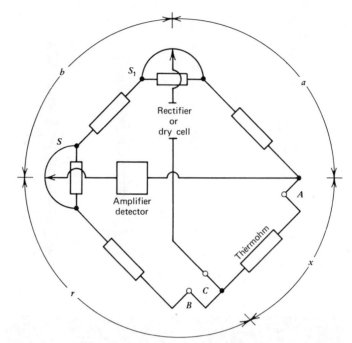

Figure 4.17 *Three-wire industrial system for resistance thermometers. Courtesy Leeds and Northrup Co.*

resistance of the resistance thermometer [Thermohm] arm is measured simply by the length of slidewire S needed in the r arm to produce an identical resistance. There is actually one wire in the r arm and another wire in the x arm whose resistances are equal and cancel each other. The third wire is used only for the current supply and has no practical effect in the resistance measurement. In this system the temperature measuring resistance can be 25, 50, or 100 Ω.

Differential Temperature Four-Wire Systems.   Figure 4.18 shows schematically how two four-wire Leeds and Northrup Thermohms are arranged in a Leeds and Northrup recorder for temperature difference measurements. The two resistance thermometers are calibrated to the same linear resistance–temperature curve, so that their resistances differ in a constant ratio with the temperature difference. For accurate measurement of even small differences, lead wires having four wires of equal resistance connect these resistances into the circuits, so that the arms a and b are always at a ratio of unity. The other circuit resistances are de-

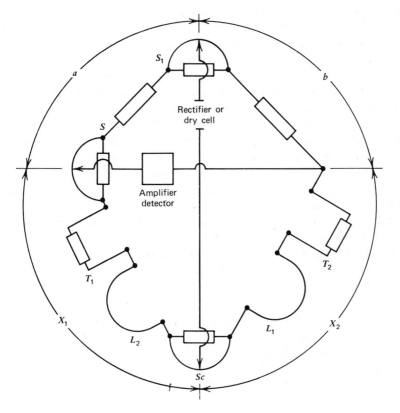

Figure 4.18 *Differential temperature system using a four-wire connection method. Courtesy Leeds and Northrup Co.*

signed to bring contact V to L on slidewire S at zero temperature difference; the arms X_1 and X_2 are balanced by the setting of slidewire S_1.
With two wires connecting each Thermohm to the lower arm, and the
other two connected to the opposite arm, the lead wire lengths and resistances cancel each other. This means that the contact V moves only the
amount from the low end of slidewire S necessary to balance the difference between the low (Thermohm$_1$) and the high (Thermohm$_2$) temperatures being measured.

Thermistors. Thermistors are thermally sensitive resistors (*therm*, from
thermally, and *istor*, from resistor). They are electronic semiconductors
whose electrical resistance varies with temperature and are useful industrially for the automatic detection, measurement, and control of physical

energy. Thermistors are extremely sensitive to relatively small temperature changes. Actually, thermistors permit the measurement of 1°C spans, which is not feasible with a normal resistance thermometer or thermocouple elements. Thermocouples are discussed in Chapter 5.

Thermistors have large negative temperature coefficients, in contrast to metal resistance thermometers which have small positive temperature coefficients of resistance. The large negative temperature coefficients and the nonlinear resistance characteristics of thermistors enable these devices to perform many unique regulatory functions. Figure 4.19 illustrates the changes in resistance in two types of thermistor materials, as compared to platinum metal used in resistance thermometers. With some types of thermistors it is possible to double the resistance with as small a temperature change as 17°C.

Thermistors are available as beads as small as 0.015 in. in diameter, as discs ranging from 0.2 to 1.0 in. in diameter, and as rods ranging from 0.03 to 0.25 in. in diameter in lengths up to 2 in. These units are made up of metal oxides and their mixtures, which are pressed or extruded into

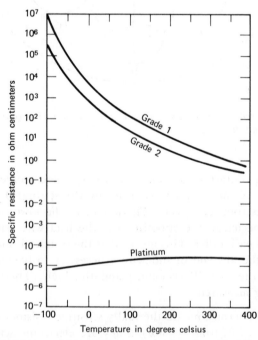

Figure 4.19 *Thermistor resistance changes as compared to platinum. Courtesy General Electric Co.*

the desired shape and sintered to produce a dense ceramic like body. The electrical contact may be made by wires embedded in the material during the pressing or extruding operation, by plating, or by metal-ceramic coatings. Some typical shapes are shown in Figure 4.20.

The resistance–temperature relationship can be expressed mathematically as

$$R = R_0 e^{\beta(1/T - 1/T_0)} \tag{4.14}$$

where R = resistance in ohms at temperature T in kelvins
R_0 = resistance in ohms at temperature T_0 in kelvins
β = a constant, in kelvins
e = Naperian base = 2.718

β may be considered constant over a limited temperature range, and its value depends on the composition and construction of the semiconductor, and the manufacturing process. β can be approximately related to the temperature coefficient α, normally expressed as a percent change in resistance per degree temperature change, by the equation

$$\alpha = \frac{\beta}{-T_0^2} \tag{4.15}$$

where T_0 is in kelvins, and at room temperature α is approximately 10 times greater than the value for metals used in resistance thermometers.

Thermistors, when properly aged, have good stability, as shown in Figure 4.21. As the curves show, the most aging effects occur during the first week and, after preaging for a month, the age rate per year thereafter is approximately 0.2%. In terms of temperature, this resistance change rate represents a change in temperature of 0.05°C. When thermistors are operated in a vacuum or shielded with a coating of thin glass, they age at a slower rate. This means that the stability with time of the resistance–temperature relation depends on both thermistor construction and use conditions.

Response time can vary from a fraction of a second to minutes, depending on the size of the detecting mass and the thermal capacity of the thermistor; it varies inversely with the dissipation factor. The power dissipation factor varies inversely with the degree of thermal isolation of the thermistor element, and may have a range of values from 10^{-5} to W per degree Celsius rise in temperature.

The upper operating limit of temperature is dependent on physical changes in the material or solder used in attaching the electrical connections, and is usually 400°C or less. The lower operating temperature limit is normally determined by the resistance reaching so great a value that it cannot be measured by standard methods.

Consideration must be given to maintaining as low a measuring current as possible to avoid heating in the detecting unit, so that any resistance change depends only on the temperature change of the surrounding area. Where this feature is a problem, thermistor stability can be

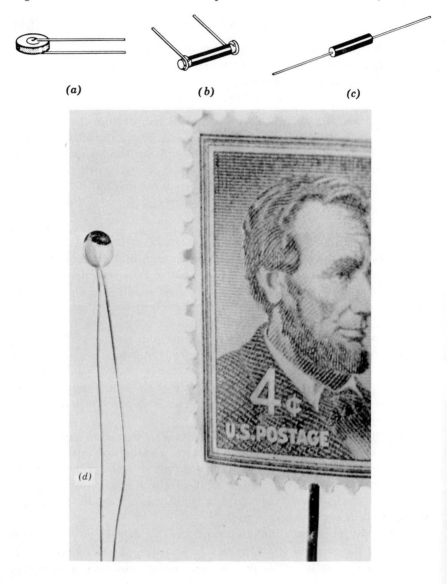

(a)   (b)   (c)

(d)

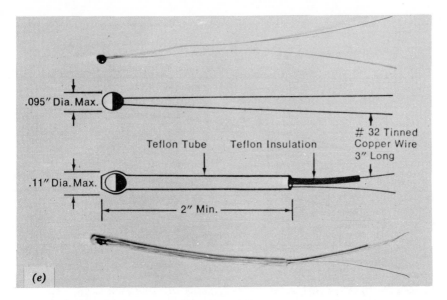

Figure 4.20 *Typical shapes of thermistors. (a) Disc with butt soldered joint. (b) Rod with wrapped soldered lead. (c) Rod with platinum fired in lead. Courtesy General Electric Co. (d) Relative size comparison of a thermistor. (e) Two actual thermistors and their cross sectional configuration. Courtesy Yellow Springs Instrument Co., Inc.*

maintained at a given temperature with an auxiliary heating element. One way of effectively reducing the average power dissipation and retaining the highest sensitivity of the thermistor is to energize the element with pulses of measuring power.

Thermistors can be located remotely from their associated measuring circuits if units of sufficiently high resistance values are chosen so that lead resistance errors become negligible. This feature, their availability in small sizes, and their mechanical simplicity give thermistors advantages over conventional thermocouples, resistance thermometers, and filled system thermometer sensing elements. Two typical installation assemblies are shown in Figure 4.22.

Thermistors may be used for compensation of changes in resistance in electrical circuits (a major application), as a safety and warning circuit switch, to stabilize output voltage in circuits with a wide variation in input voltage, as the sensitive element of a vacuum gage, in flowmeters, as time delay devices, and as sequence switching devices. A graph and schematic circuits showing these applications are included in Figure 4.23. In Figure 4.23a, a copper coil is temperature compensated by a thermistor. The thermistor is connected in parallel with a shunt resistor to obtain the desired negative temperature characteristics. In Figure 4.23b, two thermistors are used. One delays the start or operation of machinery after power is applied, and the other operates visual or audible warning signals. In Figure 4.23c, the thermistor is placed in series with a selected

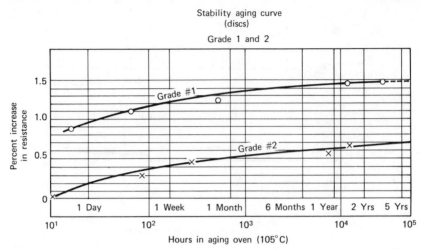

Figure 4.21 *Stability curves for thermistors. Courtesy General Electric Co.*

value of resistance R to maintain the output voltage over a wide range of current. In Figure 4.23*d*, thermistor T_1 is placed in the vacuum system, and thermistor T_2 is outside the system. As the vacuum chamber is evacuated, T_1 becomes hotter than T_2, for the same current flow, since its heat is not dissipated as rapidly in the thinner air. The resultant difference in current then flows through the galvanometer which is calibrated to read in units of pressure. Figure 4.23*e* shows thermistor T_1 in the flow path and thermistor T_2 in the gas or liquid, but outside the line of flow. In this case T_2 dissipates heat less rapidly than T_1, and current flows through the galvanometer in the bridge circuit which is calibrated in units of flow. Figure 4.23*f* shows the thermistor in series with a resistor. Because of their voltage-current characteristics, which depend on inherent thermal inertia, thermistors function as time delay devices with cycles ranging from a fraction of a second to minutes. As current flows through the thermistor, it self-heats until enough current flows, as a result of reduced resistance, to actuate the relay. Sufficient time must be allowed for cooling to obtain reliable time delay periods. A set of relays to bridge the thermistor is recommended for those cases in which cycle times are shorter than the cooling time required for the thermistor to reach ambient temperatures. Figure 4.23*g* shows several loads connected to the same voltage source, each thermistor having a different time delay before permitting the loads to be energized on a sequential basis.

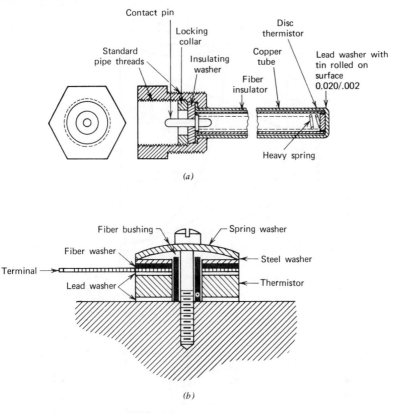

Figure 4.22 *Typical thermistor installation assemblies. (a) Disc thermistor assembly. (b) Washer thermistor assembly. Courtesy General Electric Co.*

Thermistors are most useful in compensating for temperature changes of other components, in giving a large negative resistance change for a small temperature change, and for applications in which their small size and mechanical simplicity are essential. They are relatively expensive and do not possess the long range accuracy of metal resistance thermal detectors.

Electronic Thermometers. The latest breakthrough in the measurement of temperature in the 0 to 350°F range with very high accuracies and above average linearity is the use of electronic components and integrated circuits. A typical circuit is shown schematically in Figure 4.24 for a linear thermometer covering the 0 to 300°F temperature range. A sili-

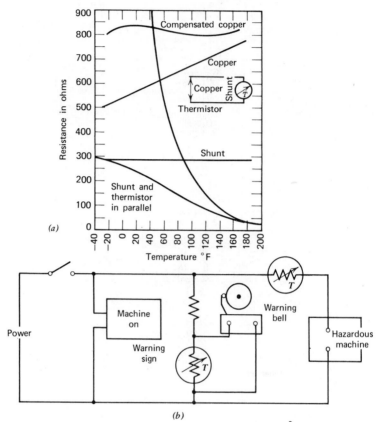

(a)

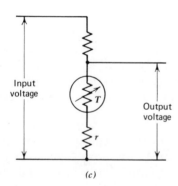

(b)

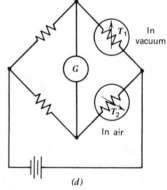

(c)

(d)

144

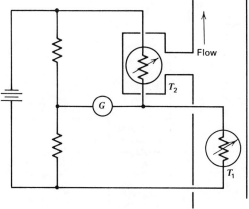

(e)

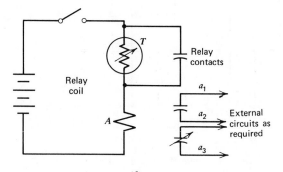

(f)

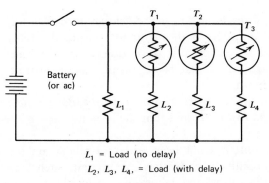

L_1 = Load (no delay)

L_2, L_3, L_4, = Load (with delay)

(g)

Figure 4.23 *Typical thermistor applications. (a) Temperature compensation. (b) Safety and warning circuit. (c) Voltage regulation. (d) Vacuum gage. (e) Flowmeter. (f) Time delay. (g) Sequence switching. Courtesy General Electric Co.*

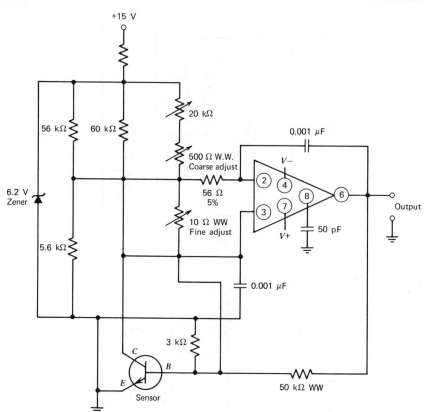

Figure 4.24 *Electronic thermometer schematic diagram.*

con planar transistor is shown as the temperature detector, and a type 741 operational amplifier is shown as the output device of the circuit. This output can be read on any conventional digital voltmeter with adequate display capabilities. The power supply can be separate, or designed into the final circuit.

We have used the schematic shown to verify that temperatures can be detected and read to 0.01°F, and that the responses are in the millisecond time span.

This type of unit has not been widely available on the commercial market and is relatively expensive. At present, such units find their widest use in the calibration laboratory as highly accurate and linear temperature standards. In addition to the normal linear range, the electronic thermometer can be designed for differential and offset temperature measurements.

4.9 SUMMARY

In summary, the different types of thermometers and their useful measuring ranges are shown in Table 4.1.

The sensing devices for temperatures above 1000°F are discussed in Chapter 5.

Review Questions

4.1   Name the different kinds of thermometers used in industry.

4.2   Why are the scales on liquid-in-glass thermometers not perfectly linear?

4.3   Give the reason for providing an expansion chamber at the top of the glass tube in a liquid-in-glass thermometer.

4.4   What determines the speed of response of a liquid-in-glass thermometer?

4.5   Name the advantage of using a filler wire in a hermetically sealed liquid-in-metal thermometer system.

4.6   Where do organic liquid actuated recording thermometers fill the greatest need?

4.7   What is the significance of the classifications class I, class II, class III, and class V as applied to thermometers?

4.8   Under what conditions should a vapor actuated thermometer not be used?

4.9   What is peculiar about saturated vapor? What effect does it have on a temperature measurement in a vapor actuated thermometer system?

4.10 What is Charles's law? How does it apply to temperature measurements?

4.11 What is one of the proofs of the accuracy credited to gas filled thermometer systems?

4.12 What type of transmitting device is used with gas filled thermometers? Why?

4.13 When do gas filled thermometer systems require compensation? How is compensation accomplished?

4.14 How much overrange can a gas filled thermometer withstand? How does this compare to mercury actuated and vapor actuated systems?

4.15 What is the principle of operation of the bimetallic thermometer?

4.16 What establishes the upper limit of temperature at which bimetallic thermometers can be used?

4.17 What materials are used in bimetallic thermometers for the 300°F range? The 800°F range?

Table 4-1 Thermometers and Useful Measuring Ranges

Type of Thermometer	Temperature Range (°F)	Accuracy	Speed of Response	Pressure Range
Liquid-in-glass	−80 to 950	Medium to high	Medium	25 lb/in^2
Liquid-in-metal	−38 to 1200	Medium	Slow	Vacuum to 5000 lb/in^2
Vapor actuated	−40 to 600	Medium	Medium (bare bulb)	Atmospheric
Gas actuated	−125 to 1000	Medium to high	Fast (bare bulb)	Atmospheric
Bi-metal	−40 to 1000 (800 constant)	Low to medium	Medium to slow	100 lb/in^2
RTD	−100 to 1000	High	Medium to fast	Vacuum to 350 lb/in^2
Thermistor	−180 to 750	Medium to high	Fast (bare)	Vacuum to 350-lb/in^2
Electronic	0 to 350	High	Fast	Vacuum to 350 lb/in^2

4.18 How are bimetallic thermometers calibrated? What is their accuracy range?

4.19 How are resistance thermometers constructed?

4.20 What is the principle of operation of the resistance thermometer?

4.21 What types of metals are used in resistance thermometers? Give the characteristics most desirable for the metals used.

4.22 Name the three types of resistance thermometers and describe how they are used in measuring circuits.

4.23 How are resistance thermometers calibrated? What is their accuracy range?

4.24 What is a thermistor? What is its principle of operation?

4.25 What advantages does the thermistor have compared to the metal resistance thermometer? What disadvantages?

Problems

4.1 If a thermometer registers a temperature of 25°C, what is the temperature in degrees Fahrenheit?

4.2 If a thermometer registers a temperature of 125°F, what is the temperature in degrees Celsius?

4.3 How is a temperature bulb installed to obtain the maximum accuracy and speed of response?

4.4 Assuming the same volume at two temperatures in a liquid-in-metal thermometer, what can be predicted about the pressure in the system? Show the equations used.

4.5 Why does a vapor actuated thermometer have two possible modes of operation? Explain the action of each.

4.6 Prepare sketches of the setup and explain how a gas actuated thermometer is calibrated.

4.7 What precautions should be considered when the rate of response of a thermometer actuating device is being chosen?

4.8 Show how bimetallic thermometers are constructed and explain how they work.

4.9 If the resistance of a resistance thermometer is 2.50 Ω at 32°F and 7182 Ω at 527°F, what is the coefficient of resistance of the material used in the thermometer?

4.10 When the resistance–temperature coefficient is not linear, how is the resistance for a temperature t determined?

4.11 Show how thermistors can be used in a sequence switching system.

Bibliography

Considine, Douglas M., *Process Instruments and Controls Handbook*, McGraw-Hill, New York, 1957.

General Electric Company, Metallurgical Products Dept., *Thermistor Manual*, No. TH-13A, Detroit, August 1956.

Honeywell, Inc., *Thermometers, Indicating, Recording, Controlling*, Catalog C 60-2a 10M, Ft. Washington, Pa., February 1962.

Leeds and Northrup Company, *Thermohms, Assemblies, Parts, and Accessories*, Catalog EN-S4, Warminister, Pa., 1952.

Minneapolis-Honeywell Regulator Company, Industrial Division, *Fundamentals of Instrumentation for the Industries*, 1955–1957, G00003-00 10M, Ft. Washington, Pa., January 1958.

Taylor Instrument Companies, *Taylor Bi-Therm Bimetallic Thermometers*, Bulletin 98267, Rochester, N.Y., January 1963.

Taylor Instrument Companies, *Taylor Industrial Thermometers*, Catalog E, File 6-1, Rochester, N.Y., August 1961.

Taylor Instrument Companies, *Taylor Recording Thermometers*, Catalog 76J, Rochester, N.Y., September 1955.

Upton, E. F., Jr., *Measuring Industry's Hot Spots, Plant Engineering*, October 1961, pp. 117–120; November 1961, pp. 109–112.

CHAPTER 5

Pyrometry

Industrial measurements of temperature from -300 to over $1000°F$ are normally accomplished by means of thermocouples, radiation pyrometers, optical pyrometers, and two-color pyrometers. Thermocouples are the only detectors in the group that can be used at very low temperatures, especially in applications in which their accuracy is adequate. The general spectrum of temperature measuring instruments, including those discussed in Chapter 4, are shown in Figure 5.1.

5.1 THERMOCOUPLES

A thermocouple is made up of two dissimilar metal conductors joined at one end, usually called the *hot* or *detecting* junction, and connected to some emf measuring instrument such as a millivoltmeter or potentiometer at the cold end of the conductors. The measured emf is normally compared to some reference such as the *ice point*. A typical system is shown schematically in Figure 5.2. This basic circuit contains all the essential elements for making a temperature measurement. The thermocouple T shows the joined dissimilar conductors A and B at the hot or detecting junction, the connecting wires C from the hot junction connecting head to the measuring junction, including the meter M and the reference junction T_R. The meter M measures the emf difference between the hot junction and the reference junction, $T - T_R$. This shows that thermocouples are actually differential temperature measuring detectors, which simply means that they measure the difference in temperature that exists between the hot junction end and the reference junction. For all practical purposes this means that the emf developed is dependent on two temperatures. The emf is actually proportional to the difference between the two temperatures.

151

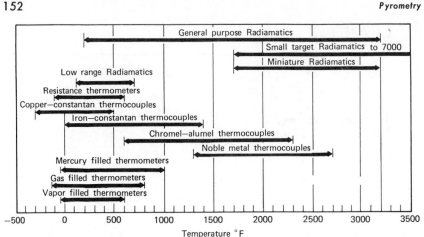

Figure 5.1 *General spectrum of temperature measuring instruments. Courtesy Honeywell, Inc.*

While Figure 5.2 shows all the essentials necessary for making a temperature measurement, several things are required to obtain accuracy for the best measurement and control. The conductors should generate as high an emf as possible for a unit change in temperature, and should possess certain characteristics to make this possible.

In 1821, Seebeck observed that, when he fused a copper wire to an iron wire and heated the fused end, he obtained an emf. He also found that current flowed from the copper to the iron at the heated end. This was the first known observation of the phenomenon of thermoelectric current. Later discoveries revealed that the flow of current observed by Seebeck was apparently the result of two separate causes. Each was named after the scientist who discovered it. Today we designate these two sources of emf as *Peltier emf* and *Thomson emf*.

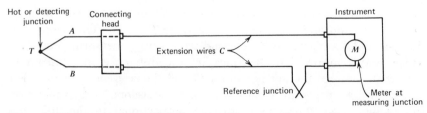

Figure 5.2 *Schematic of a typical thermocouple system. Courtesy Honeywell, Inc.*

Peltier EMF. The portion of the total emf of a thermocouple caused by a potential difference at the junction of two dissimilar conductors or wires is the Peltier emf. This potential difference varies with the junction temperature, but there is no guarantee that it varies uniformly.

Thomson EMF. The portion of the total emf of a thermocouple that exists because of a difference in potential in a single section of wire or conductor having a temperature gradient is the Thomson emf. This means that there is a potential in a wire of homogeneous material if one end is at a higher temperature than the other. The complex theory used to explain this phenomenon does not exactly agree with all the observed experimental effects. For example, it is possible to fabricate a thermocouple that produces emf curves which indicate that the Thomson emf actually reverses itself at certain temperatures. A bismuth thermocouple indicates this type of reversal in the 3000°F range.

Thermocouple Materials. Thermocouple wires are chosen so that they will produce a large emf that varies linearly with temperature. Ideally chosen thermocouple material should have:

1. The Thomson emfs of the two wires additive in the circuit
2. Thomson emfs that vary directly with temperature
3. Peltier emfs that develop potentials at the hot junction that are in the same direction as the Thomson emfs
4. Peltier emfs that vary directly with temperature
5. Thermoelectric power as high as is possible

No known metals or alloys have all these desirable characteristics, although some approach them quite closely. Since there are no perfectly behaving thermocouples, all emf curves deviate from a straight line or linear response to some degree. In an effort to satisfy the experimentally observed results, empirical equations have been developed. One such empirical equation was developed for the platinum–platinum–rhodium noble metal thermocouple. It can be expressed as:

$$\sum_{0}^{T} e = a + bT + cT_2 \tag{5.1}$$

where e = emf of the thermocouple in millivolts
$\quad T$ = temperature in kelvins
$\quad a, b, c$ = constants depending on the type of wire metal or alloy

The equation for a platinum–platinum–rhodium thermocouple is

$$e = 0.323 + 0.000827T + 0.000001638T^2 \quad \text{(millivolts)} \tag{5.2}$$

In industrial applications the choice of materials used to make up a thermocouple depends on the temperature range to be measured, the kind of atmosphere to which the material will be exposed, and the accuracy required in the measurement. Thermocouple material should be selected to have good resistance to oxidation and/or corrosion in the atmosphere and temperature range where it will be used, resistance to change in characteristics that will affect its calibration, freedom from parasitic currents, and reproducibility of readings within the accuracy limits required.

Several combinations of dissimilar metals make good thermocouples for industrial use. These combinations of wire must have reasonably linear relationships between temperature and emf, they must be able to develop an emf per degree of temperature change that can be detected with standard measuring instruments, and in many applications they have to be physically strong enough to withstand high temperatures, rapid temperature changes, and the effects of corrosive and reducing atmospheres. With such severe requirements to be met, there is no one combination of wires that can satisfy all the conditions. Based on experience gained from years of application, industry has standardized on a few wire combinations to meet the majority of its needs. Not only are different wire combinations used, but different wire sizes in the same wire combinations may be necessary to obtain the physical strength needed for a given application. The chart in Figure 5.3 shows the temperature limitations of four combinations with appropriate wire sizes.

Copper–constantan (type T) thermocouples are commonly used in the −300 to 600°F temperature range and appear superior for measurement of relatively low temperatures, especially subzero temperatures. They stand up well against corrosion and are reproducible to a high degree of precision.

Iron–constantan (type J) thermocouples are used in reducing atmospheres where there is a lack of free oxygen. They are useful in the 0 to 1600°F temperature range. When used at temperature above 1000°F the rate of oxidation increases rapidly. Heavier wire is recommended for 1000 to 1600°F applications, and protection wells are used to cover the thermocouple. Unprotected iron–constantan thermocouples are used extensively up to 550°F in reducing atmospheres.

Chromel–alumel (type K) thermocouples are used extensively in oxidizing atmospheres where there is an excess of free oxygen. These thermocouples can be used to measure temperatures up to 2400°F, but are most satisfactory at temperatures up to 2100°F for constant service. Reducing atmospheres have a tendency to change the thermoelectric characteristics of these wires and reduce their accuracy.

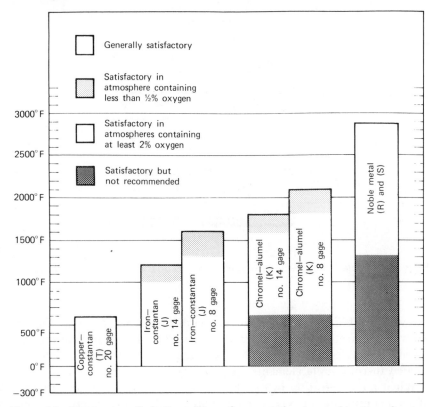

Figure 5.3 *Temperature limitations of four thermocouple wire combinations. Courtesy Honeywell, Inc.*

Platinum–platinum–rhodium (types R and S–10 and 13% rhodium) thermocouples are normally designated noble metal thermocouples, and are used for higher temperature ranges. These thermocouples, like chromel–alumel couples, are adversely affected by atmospheres containing reducing gases and should be protected by an impervious tube when used at temperatures above 1000°F when such gases are present.

Thermocouples are seldom used as bare wire, except for the detecting junction. The wire covering may consist of a heat resistant enamel or varnish, heat resistant rubber, waxed cotton braid, asbestos braid, silicone impregnated glass braid, silicone impregnated asbestos, glass fiber braid, Teflon-glass braid, extruded nylon, high temperature silica braid, ceramic tubing, ceramic beads, alumina or molybdenum oxide. Combinations of the coverings are also used by having a covering on each wire and the

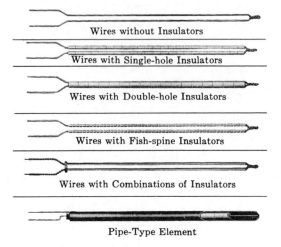

Figure 5.4 *Typical insulated thermocouple elements. Courtesy Leeds and Northrup Co.*

two-wire combination encased in an overall protective covering. The industrial application with respect to atmosphere and temperature range normally dictates the requirements for the protective covering. Some insulated thermocouple elements are shown in Figure 5.4.

When the application of the thermocouple requires the measurement of temperature in corrosive atmospheres or other atmospheres that are highly detrimental to the metals used in the thermocouple to form the exposed junction, a protective thermocouple well or tube is used in addition to the protective covering on the wires above the fused junction. These protective tubes or wells are constructed of a variety of materials, and the material employed depends on the usage. Wells may be constructed of wrought iron, alloy coated wrought iron, cast iron, seamless steel, stainless steel, nickel, Inconel, Fyrestan, ceramic bonded silicon carbide, or some other material that prolongs the life and accuracy of the thermocouple for the particular application. In applications in which high pressures are used, the protecting wells are usually drilled from solid bar stock or built up welding a tube, a plug, and a hex head together. These two types of construction are shown in Figure 5.5. These protective wells may be either straight or angled to meet the installation requirements. Some typical protective tubes are shown in Figure 5.6.

Thermocouples requiring the use of protective tubes or wells are customarily made up as an assembly. Some typical assemblies are shown in

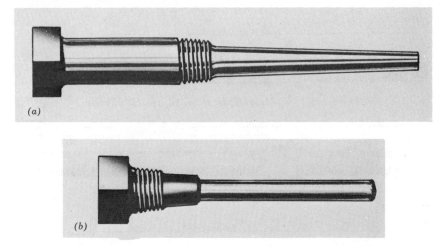

Figure 5.5 *Two types of protective well construction. (a) From solid stock. (b) Built up. Courtesy Leeds and Northrup Co.*

Figure 5.7. A cross sectional view of a high speed response thermocouple assembly is shown in Figure 5.8.

The sensitivity of a thermocouple can be increased by reducing the mass of the measuring junction. One method of accomplishing this mass reduction is to butt weld the two thermocouple wires as shown in Figure 5.9*a*. Where the physical strength of a butt weld is inadequate, the two wires are twisted together as shown in Figure 5.9*b* and the ends are welded. Three turns of wire are widely used, and as many as five turns are required for some applications. A good clean weld or braze is necessary to obtain a solid junction for accurate and repeatable measurements. A thermocouple also responds to a change in temperature more rapidly when its protecting tube or well has the smallest possible diameter and the thinnest walls. Larger diameters and thicker walls cause slower response.

Thermocouple Systems. When a thermocouple is combined with a millivoltmeter or potentiometer, which measures the emf generated and indicates or records this emf in terms of temperature, we have a thermocouple pyrometer. Other components may be added to this simple thermocouple to provide various modes of automatic control, as discussed later.

Millivoltmeter Pyrometer. The basic components of a millivoltmeter pyrometer are shown in Figure 5.10. The sensitivity, accuracy, and auto-

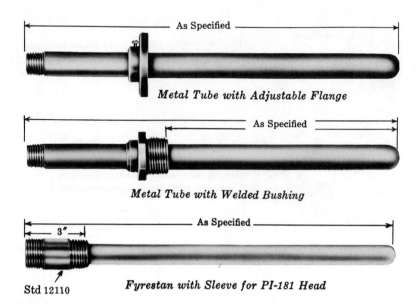

Metal Tube with Adjustable Flange

Metal Tube with Welded Bushing

Std 12110   *Fyrestan with Sleeve for PI-181 Head*

Figure 5.6 *Typical protective tubes. Courtesy Leeds and Northrup Co.*

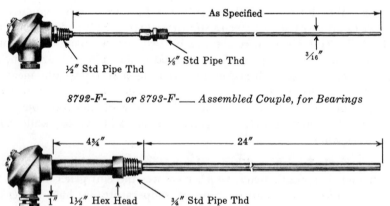

8792-F-___ or 8793-F-___ Assembled Couple, for Bearings

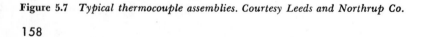

8783 Assembled Couple for Ventilating Air

Figure 5.7 *Typical thermocouple assemblies. Courtesy Leeds and Northrup Co.*

158

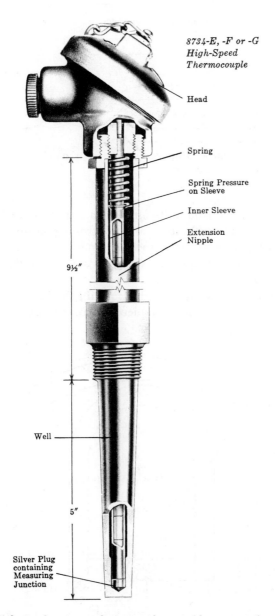

8734-E, -F or -G
High-Speed
Thermocouple

Head

Spring

Spring Pressure
on Sleeve

Inner Sleeve

Extension
Nipple

9½″

Well

5″

Silver Plug
containing
Measuring
Junction

Figure 5.8 *High speed response thermocouple assembly cross section. Courtesy Leeds and Northrup Co.*

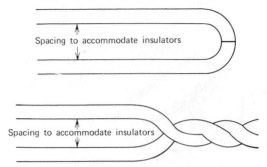

Figure 5.9 *Methods of joining dissimilar metals for a couple. (a) Butt weld. (b) Twisted pair. Courtesy Honeywell, Inc.*

matic control features of the millivoltmeter are quite suitable for many industrial applications. Indicating and controlling millivoltmeters are used widely on furnaces, ovens, autoclaves, and kilns. They are also frequently used as excess temperatures alarms and cutoff devices on the majority of heating equipment.

The millivoltmeter is essentially a d'Arsonval galvanometer. A magnetic field, set up by the magnet and pole pieces, surrounds a coil which is sus-

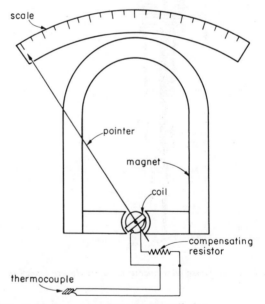

Figure 5.10 *Millivolt pyrometer. Courtesy Honeywell, Inc.*

pended by pivots and jeweled bearings. The indicating pointer is attached to the coil. The electrical current generated by the thermocouple passes through the coil and sets up an opposing magnetic field. The opposing magnetic field is proportional to the current passing through the coil and causes the coil to turn. The turning of the coil moves the pointer along the scale. The coil and pointer are deflected against hairsprings. These hairsprings retard the movement of the coil and pointer and return them to the zero setting when there is no current flow. They also carry the current to the coil. A bimetal spiral is often attached to the hairspring to provide reference junction compensation. As stated earlier, a thermocouple has two junctions, a measuring or hot junction and a cold junction. Since the instrument measures the millivoltage generated by the temperature difference between the measuring junction and the reference junction, it is important to maintain the reference junction at a constant temperature or that the measuring instrument be automatically and accurately compensated for temperature changes at this junction.

The meter movement of the millivoltmeter pyrometer has a fixed resistance and, to obtain accurate readings, it is essential that the thermocouple and the connecting wires have the same resistance as the meter movement. It is extremely important that all connections be clean and solid, because a poor or dirty connection or poorly soldered, brazed, or welded joint can create a false cold junction and cause a gross error in the indication. It is equally important that the extension wires used be of the proper material to match the materials used in the junction, because the thermocouple actually is made up of the couple plus the extension wires. Thermocouples cannot be connected to ordinary lead wire, because each connection would act as an additional thermocouple with dissimilar characteristics. Extension wires are chosen so that they have very nearly the same thermoelectric characteristics as the couple material. Quite often the regular thermocouple wire is used as extension wire.

The size or gage of extension wire may be varied to help match the resistance of the meter, especially when multiple measurements are made from thermocouples located at widely varying distances from the measuring instruments.

In the discussion thus far we have considered only one thermocouple in the measuring system. There are applications in which it is an advantage to use more than one thermocouple. Thermocouples can be used in series or parallel to meet the requirements of the particular application.

When it is necessary to obtain high sensitivity, the thermocouples can be placed in series. In such an arrangement the total emf developed is the sum of the individual thermocouples used. The total resistance is the sum of the individual resistances. This is shown algebraically by equations 5.3 and 5.4:

$$E_m = E_1 + E_2 + E_3 + \cdots + E_n \tag{5.3}$$

$$R_T = R_1 + R_2 + R_3 + \cdots + R_n \tag{5.4}$$

where E_m = difference in potential across the millivoltmeter
$E_{1,2,3n}$ = potential developed for each thermocouple used
R_T = total resistance
$R_{1,2,3n}$ = resistance for the individual thermocouples

The average temperature of the system or process is found by dividing the total emf by the number of thermocouples used. This series arrangement provides more sensitivity, but may be less accurate because uncertainties can be introduced by inhomogeneity of the couples.

Multiple thermocouples in a parallel arrangement must be connected directly to the two common terminals across which the net emf is measured. This arrangement measures only average emfs, so care must be exercised to keep the thermocouples as nearly alike as measurements can make them. In a parallel arrangement the resistance of the indicating meter and the load have an effect on the potential indication as shown by equation 5.5:

$$E_m = \left(\frac{e_1}{r_1} + \frac{e_2}{r_2} + \frac{e_3}{r_3} + \frac{e_n}{r_n} \right) \left[\frac{R_m}{1 + (R_M + R_L)(1/r_1 + 1/r_2 + 1/r_3 + 1/r_n)} \right] \tag{5.5}$$

where E_m = difference in potential across the indicating meter
R_M = resistance of the indicating meter
R_L = resistance of the line in the circuit
$r_{1,2,3n}$ = resistance of the separate units
$e_{1,2,3n}$ = potential generated by each unit

A reference junction must be established in both the series and parallel arrangements to obtain accurate temperature measurements.

To illustrate the use of multiple thermocouples in series, an arbitrary value of 10.2 Ω will be chosen as the millivoltmeter resistance to be matched for the most accurate measurements. The ice point will be used for the reference junction, since most thermocouple emf tables are based on the ice point temperature reference. Using a single iron–constantan thermocouple to measure an arbitrary temperature of 700°F, the thermocouple, if properly matched in resistance, should develop an emf of 20.26 mV using a reference junction of 32°F. Therefore, if we used three thermocouples in series, a generated emf of 60.78 mV would be expected using the same reference. This emf represents a temperature of 1905.4°F, from the emf tables for iron–constantan. Under these circumstances the emf from the series is divided by the number of couples to obtain the average temperature, because the millivolts per degree temperature change at the higher temperatures is greater than at the lower temperatures. This creates an error of approximately 200°F in this case, if the readings are taken in terms of temperature. Also, the 10.2 Ω resistance of the meter would have to be matched by the resistance of the three thermocouples. The use of multiple thermocouples in applying the series arrangement in temperature measuring instruments is discussed later in this chapter under radiation instruments.

In general a millivoltmeter thermocouple measuring and control system is more economical on first costs (instruments, thermocouples, and installation) if their accuracy, response, and control features are adequate to satisfy the application. These units are predominantly on-off galvanometer actuated mechanisms with high and low set points. The pointer arm carries some type of vane or other type of actuating device which controls the on-off action by changing an oscillator frequency, breaking a light beam, or some similar control device. Two typical systems are shown in Figures 5.11 and 5.12.

Several variations of the vane millivoltmeter pyrometer are available for industrial applications. A simple on-off type has one set point and one vane. In this case, when the vane changes the frequency of the oscillator, the amplifier unit turns off the power system generating the heat

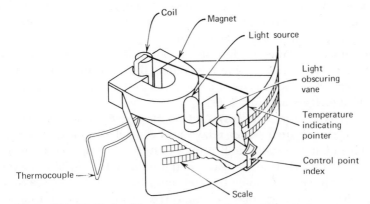

Figure 5.11 *Typical millivoltmeter pyrometer control system.*

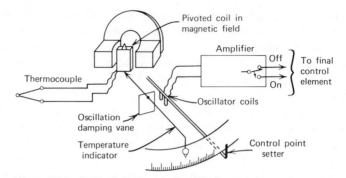

Figure 5.12 *Typical dual-control millivoltmeter pyrometer system.*

until the temperature drops below the set point. This moves the vane out of the off position and restores the control device to its original position. The vane is usually made wide enough to permit enough pointer movement above the cutoff point to indicate the overshoot in temperature above the set point. A mechanical stop is used to prevent the vane from going above the oscillator coils, because if the temperature overshoot carries the vane through the oscillator coils on the upscale side, it will restore the power and cause overheating or burnout. In other words, unless the pointer that carries the vane is restrained mechanically, the power to the system can be turned on by the vane moving either upscale or downscale through the same coils. The width of the vane determines the temperature band where there is no control. The point at which the vane turns on the power establishing the lower point of the band, and the mechanical stop establishes the upper readable point. If the temperature drops off rapidly enough, the pointer and vane may not reach the mechanical stop. This narrows the band of no control and gives a true readable upper value of the temperature.

The vane can be used to shut off the power if the temperature exceeds a certain predetermined value and retain the off condition until manually reset. This type of control is used as a safety control on such equipment as autoclaves and furnaces.

Another type of pyrometer has set points where contacts are closed to actuate or deactivate a power relay. This control pyrometer has the disadvantage of stopping the indicating pointer when the contact is made. This means that there is no way for the pointer to indicate how much higher the temperature rose after the contacts closed. There is also some danger that the contacts may become corroded during use and not make adequate contact to give proper control. Such meters are useful where short runs are made and economy is of high importance. There is also some danger that the contacts may stick and shut down a process. In fact, in any applications in which contacts are used, they must be cleaned and adjusted periodically to ensure positive action and control.

When contact meters are employed, a maximum of two sets of contacts can be used. One set can be used as a low limit, and the other as a high limit. The pointer movement is then limited to the span between the two contact settings. These meters also have the disadvantage of not being able to read the undershoot or overshoot of the temperature.

Vane type millivoltmeter pyrometers are often equipped with burnout features so that, if a thermocouple burns out or breaks, it will sound an alarm, shut down the system or sound an alarm and shut down the system. One undesirable feature usually associated with these meters is that if there is a power source failure they will not shut off, and must be reset manually when the power is restored. A second vane setting or other control device is required to obtain the burnout features so that when there is no generated emf the meter pointer will be prevented from moving downscale and activating the power source. A common method is to drive the pointer upscale to keep the control vane in the off position. A typical dual-control unit is shown in Figure 5.12.

Some industrial millivolt pyrometers are shown in Figure 5.13. Figure 5.13*a* is a Honeywell Pyr-O-Vane controller, Figure 5.13*b* is a Honeywell Protect-O-Vane controller, Figure 5.13*c* is a General Electric Model H indicator-controller. Each of these instruments has plug-in indicator-controller elements which can be removed for servicing and repair. Where large numbers of one type of instrument are used in an industrial installation, it is desirable to have an auxiliary test box into which the plug-in- units can be placed for repair and calibration. One feature that saves time in repair and checkout is to have the plug-in panel section rotate 180° so that the plug-in indicating control unit can be worked on from either the top or bottom by merely turning the unit through the 180°.

5.2 POTENTIOMETER PYROMETERS

A potentiometer pyrometer essentially operates on an error signal principle in which the emf generated by the thermocouple can be considered an error signal. This generated emf is then matched by the potentiometer system to obtain a null condition, and the error signal needed to obtain the null condition is indicated or recorded by the potentiometer system as the generated emf. A simple potentiometer system is shown in Figure 5.14. These systems are generally more expensive than millivoltmeter pyrometers, but they offer advantages that justify their cost in many applications. In a potentiometer pyrometer system, it is not necessary to match the resistances of the thermocouples to the potentiometer, because a counter emf is used to produce a null current. In a manually operated potentiometer system, an emf from the potentiometer is bucked against

(a)

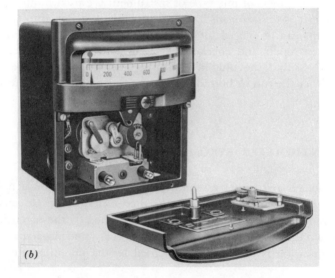

(b)

(c)

Figure 5.13 *Industrial millivolt pyrometer controllers. (a) Pyr-O-Vane. (b) Protect-O-Vane. Courtesy Honeywell, Inc. (c) General Electric Model H. Courtesy General Electric Co.*

the emf generated by the thermocouple to produce a zero galvanometer reading. A precision thermometer is usually mounted on the potentiometer to indicate the ambient temperature, so that a correction can be made in the emf reading if an ice point reference is not used. If a universal type of indication potentiometer is used, any type of emf generating thermocouple for which temperature versus emf has been developed can be used as the temperature difference detector. However, the counter emf used and read from the potentiometer slidewire dial has to be compared to a conversion table for the particular combination of wires used in the thermocouple and corrected for the ambient temperature if the ice point reference is not used. The manual potentiometer is a laboratory instrument and is used for taking precise temperature measurements and checking the performance of other types of instruments.

Small, compact potentiometer pyrometers are made for measuring the output of specific types of thermocouples such as chromel–alumel, iron–constantan, copper–constantan, and platinum–platinum–rhodium. These pyrometers are also made in combinations for two or more types of thermocouples. These units cover specific ranges or combinations of ranges within the millivolt span of the specific instrument. They are also in the same price range as millivoltmeter pyrometers.

A continuously balanced type of potentiometer pyrometer system is widely used for industrial applications. This continuous balance is per-

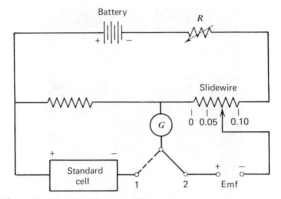

Figure 5.14 *Schematic of a simple potentiometer system.*

formed by a system such as the one shown in Figure 5.15. The potentiometer slidewire in this system has a definite span and covers a specific range. These systems are usually designed to provide for reference junction compensation, and are designed not only for a specific type of thermocouple but for a definite temperature range. The designs also provide for circuit and wiring modifications to accommodate both changes in thermocouple types and in temperature range.

In the null balance system shown in Figure 5.15, the emf being measured is detected by an electronic amplifier which drives the balancing motor. This is a servo type motor that can run in either direction, depending on the polarity of the error current detected by the electronic amplifier. The balancing motor turns in the direction needed to move the arm of the potentiometer, so that the adjustable calibrated emf exactly matches the emf generated by the thermocouple to give a zero difference or null balance. The emf needed to balance the system is then shown on the indicator and/or recorder as the emf produced by the thermocouple. These units can be calibrated to read directly in degrees Fahrenheit or Celsius.

In older model potentiometer pyrometers, the emf used to provide the bucking voltage is supplied by a dry cell battery, and the calibrated emf is periodically compared to a standard cell. A standard cell is either the Weston normal with an emf of 1.0183 V at 20°C, or the Clark standard with an emf of 1.4328 V at 15°C. The Weston normal standard cell and a 1.5 V dry cell are the combination used by major instrument manufacturers. This is termed *standardizing* the instrument. Periodic standardizing is necessary, because battery emf decreases with use. In some models provisions are made for automatic standardization. This provision was

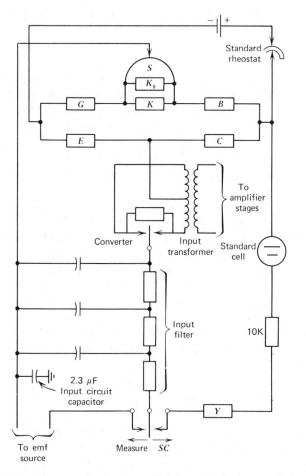

Figure 5.15 *Potentiometer circuit of a null balance method as applied to emf measurements. Courtesy Leeds and Northrup Co.*

made because people would forget to standardize the instrument manually and the readings would be in error. Provisions were also made to indicate the condition of the battery, so that it could be changed before its value dropped below the value of the standard cell. One disadvantage of automatic standardization is that on long runs, which extend through several standardization periods, the readings or recording is interrupted while the standardization takes place. In the circuit shown in Figure 5.15, the value of resistor C is chosen to be 407.8 to produce a voltage of 1.0195 V as generated by a current flow of 2.5 mA. This is exactly the standard cell voltage. Standardization occurs when the standardizing switch is in the SC position.

Normal operation occurs with the switch in the measure position M. A motor maintains the balance condition by positioning the slidewire S contact so that the emf being measured is balanced by the potential difference existing between the slidewire contact and the junction of resistors C and E. For a zero range start, resistors G and E are made equal. For zero suppression, where the range starts at some value above zero, the value of resistor G is made larger than the value of resistor E. The resistors K and K_s are slidewire shunts which determine the span of the range. Resistor B is chosen to provide a balance with the selected values for S, E, and G with C.

Figure 5.16 shows a bridge balance circuit which provides the same measurement function as the potentiometer, but does not depend on standardization for accurate measurements. The two slidewires S_1 and S_2 are mounted on a common molding, and their contacts move together. The resistor relationships are so chosen that, for any contact setting, the resistance between points 1 and 2 is always equal to that between points 2 and 3. This means that the ratio arms are always equal, and that when the bridge is balanced the unknown resistance R_x is equal to the resistance of the arm containing R. In this circuit the lead C is in the battery circuit and does not affect the accuracy of the reading.

Newer models of potentiometer pyrometers are being designed to use a zener reference voltage. A zener voltage is obtained from a reverse biased silicon or other suitable diode which produces a very stable reference voltage. One type of zener reference system is shown in Figure 5.17, which operates on the avalanche effect in the zener region as shown in Figure 5.18. The diode voltage is slightly dependent on the diode current, which in turn makes it also dependent on line voltage. One method of minimizing resistance effects is to place the zener diode in one leg of a Wheatstone bridge arrangement so that the resistive component cancels out. Equation 5.6 shows this relationship, where the dynamic resistance is expressed as R_D:

$$\frac{R_4}{R_D} = \frac{R_2}{R_3} \tag{5.6}$$

This leaves only the pure avalanche voltage of 8 to 9 V, which depends on the particular diode used in the circuit. Unless the pure avalanche voltage value is acceptable for the measurement system, it is customary to add a trimmer potentiometer or other type of dropping resistor in the circuit to establish a constant voltage to be used as a reference for the measuring system. This voltage may range from 1.5 to 7.5 V dc in value.

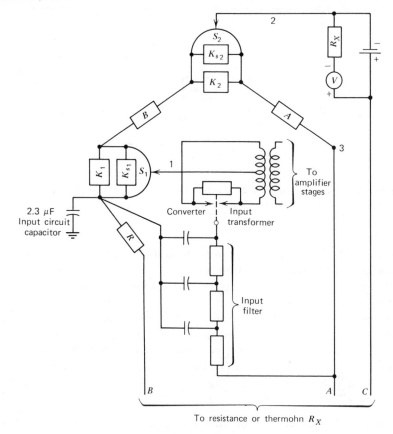

Figure 5.16 *Schematic diagram of a bridge null balance measuring system. Courtesy Leeds and Northrup Co.*

In the circuit shown, temperature compensation is obtained by the use of the two additional diodes in the zener leg, used in the forward direction. This design provides a maximum overall temperature coefficient of 0.001% in degrees Celsius. Many newer zener reference voltage systems use 7.5 V, in contrast to the 1.5 V formerly used, which apparently was chosen simply because batteries of a convenient size with adequate operating capacity were available with that potential.

5.3   RADIATION PYROMETERS

When temperatures must be measured and physical contact with the medium to be measured is impossible or impractical, use is made of

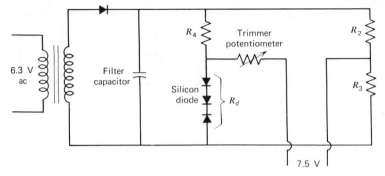

Figure 5.17 *Schematic diagram of a zener reference system to replace a battery and standard cell. Courtesy General Electric Co.*

thermal radiation or optical pyrometry methods and equipment. Optical pyrometry is discussed in Section 5.4.

Radiation pyrometry measures the radiant heat emitted or reflected by a hot object. Practical radiation pyrometers are sensitive to a limited wavelength band of radiant energy, although theory indicates that they should be sensitive to the entire spectrum of energy radiated by the object. The operation of thermal radiation pyrometers is based on blackbody concepts and has made possible the measurement and automatic control of temperature under conditions not feasible with other temperature sensing elements.

Theory of Radiation Measurements. A perfect radiating body, traditionally called a blackbody, is used as the comparative standard to measure quantitatively the energy radiated by a hot object. If blackbody con-

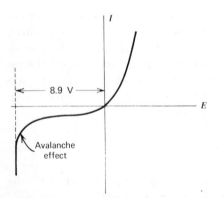

Figure 5.18 *Silicon diode curve showing voltage-current characteristics. Courtesy General Electric Co.*

ditions are approximated, meaning that either the body absorbs all the thermal radiation it intercepts or it radiates more thermal energy over a wider wavelength band that any other body with the same physical parameters and at the same temperature, the radiation detected by the thermal radiation cell will vary as the fourth power of the absolute temperature of the source. The thermal radiation energy and temperature relationship for a blackbody condition can be expressed as

$$W = kT_0{}^4 \tag{5.7}$$

where W = Radiant energy emitted per unit area from the blackbody
k = Stefan-Boltzmann constant
T_0 = absolute temperature in kelvins

This is the Stefan-Boltzmann law and assumes the blackbody to be radiating to a receiver which is at absolute zero.

In practical applications for thermal radiation pyrometers, the transfer of radiant thermal energy takes place at temperatures above absolute zero. Thus equation 5.7 has to be modified to express the radiant energy and temperature relationship under these conditions. The new equation can be expressed as

$$W = \sigma(T^4 - T_0{}^4) \tag{5.8}$$

where σ = a constant
T = absolute temperature of the blackbody
T_0 = absolute temperature of the surroundings

Emissivity. In general a rough black surface radiates more heat than a smooth bright surface at the same temperature. This effect is called *emissivity* and is expressed in numbers from 1 to 0. A blackbody or perfect radiator of thermal energy has an emissivity of 1. Less perfect radiating bodies have an emissivity of less than 1.

Some very thin transparent surfaces have emissivities very close to zero. Other surfaces such as plastics, rubber, and textiles have emissivities close to 1. Metals have varying emissivities depending on their surface and composition.

Emissivity is important, because many types of thermal radiation pyrometers require a correction to compensate for emissivity effects.

Industrial Thermal Radiation Pyrometers. Industrial applications requiring thermal radiation pyrometers for measurement and control may employ infrared techniques, so-called total radiation methods, or the two-color method. Each method or technique has advantages and disadvantages with respect to specific applications.

Infrared Pyrometry.   Infrared energy is invisible to the human eye, but can be felt. It is the radiant heat energy you feel when you hold your hand near a hot surface. As the surface temperature increases, there is a proportional increase in radiated infrared energy. Above temperatures of approximately 1000°F, a surface starts to radiate visible light energy, and simultaneously there is a proportional increase in the infrared energy. This proportional increase in infrared energy with surface temperature makes infrared pyrometry possible by combining a suitable detector, electronic circuitry, and a means of indication and/or control. The infrared spectrum ranges from 0.22 μm (1 μm $= 10^{-4}$ cm) to 17 μm, and the most used portion of the range is from 2 to 7 μm. Some manufacturers use only the 5 to 7 μm portion of the range for pyrometry measurements.

Infrared principles using bolometers, vacuum thermocouples, and thermopiles have been successfully employed in infrared spectrometers and as total radiation spectral pyrometers.

A bolometer is a thermal device that changes its electrical resistance with temperature (in the same manner as a resistance thermometer). The resistance of the bolometer changes in response to the intensity of the thermal radiation focused on it. The bolometer is usually made of a thin ribbon of platinum or nickel wire, depending on its application and the speed of response required. A platinum ribbon gives a more rapid response.

A vacuum thermocouple is a thermocouple with a low thermal mass, which can respond rapidly to changes in thermal radiant energy. The thermocouple is enclosed in an evacuated housing with a suitable window to admit the radiant energy.

In industrial applications in which a vacuum thermocouple is needed for speed of response, it is current practice to assemble several thermocouples in series to form a thermopile. A thermopile has a longer speed of response, but normally speeds of response of less than several seconds are not important in many industrial applications.

The use of solid state detectors and a selected narrow bandwidth spectral response minimizes undesirable total radiation effects in at least one type of infrared pyrometer. Combined with rugged construction, solid state circuitry, an integral calibration source, and control features, an accurate and practical infrared pyrometer can be made. These instruments can be used to measure infrared radiant energy in the temperature range from 150 to 4650°F with a minimum target size of 0.060 in.

Industrial infrared thermometers have been developed to make use of photoconductive lead sulfide as a detector with response times of 0.25 s as a standard or with a response time of 0.04 s for special applications.

Such an instrument system of detector and control is shown in Figure 5.19*a*. This system is available in six models to cover a temperature range from 65 to 4650°F with fields of view of 0.6, 1.0, or 1.4°. A typical application is shown in Figure 5.19*b* for detecting vacuum tube element temperatures, and in Figure 5.19*c* for vacuum processing.

Another type of infrared thermometer detector used in the system shown in Figure 5.20*a* is the thermistor, to be discussed in detail later in this chapter. This infrared industrial thermometer system has a response time of 50 ms as a standard, and is also available with a 500 ms response time. This model covers the temperature range from −50 to 2000°F and has a 2 or 20° field of view. This particular system is sensitive to long wavelengths of infrared radiation, which makes it useful for accurate low temperature measurements. Long time repeatability is assured by precisely controlling the temperate of an internal reference cavity. An industrial application for measuring the temperature of a moving sheet of material is shown in Figure 5.20*b*.

A portable infrared thermometer using a thermopile detector is shown in Figure 5.21*a*. This instrument is battery powered and has no moving parts. Typical applications include spot checking process temperature, material testing, equipment maintenance, trouble shooting and quality assurance. An industrial application is shown in Figure 5.21*b*.

An infrared system now used in the steel industry is shown in Figure 5.22. These systems can be specified with analog or digital output displays, time responses in four choices ranging from 10 ms to 10 s, and outputs of voltage, millivoltage, or current. A binary coded decimal output for digital data acquisition systems is also available. The system shown has an analog display, a peak picker for rapid response to temperature rises where smoke, steam, water spray, or other types of interferences may cause an intermittent signal to the detector. This feature is desirable when proportional control is used under intermittent process conditions.

On-off and time proportioning controllers for one and two points of control are available with one- and three-mode proportional controllers having rate-before-reset features. These modes of control are discussed in detail in Chapter 8.

Each of the systems discussed depends on the transmission of the infrared energy being emitted by the heated object to the detector in the measuring system through the surrounding atmosphere. There is no direct contact with the surface whose temperature is being measured. In the systems shown the sensor head is focused on a spot on the object whose temperature is being measured and/or controlled. The infrared energy falling on the detector either changes the detector resistance in proportion to the temperature, as in the case of a thermistor, or generates

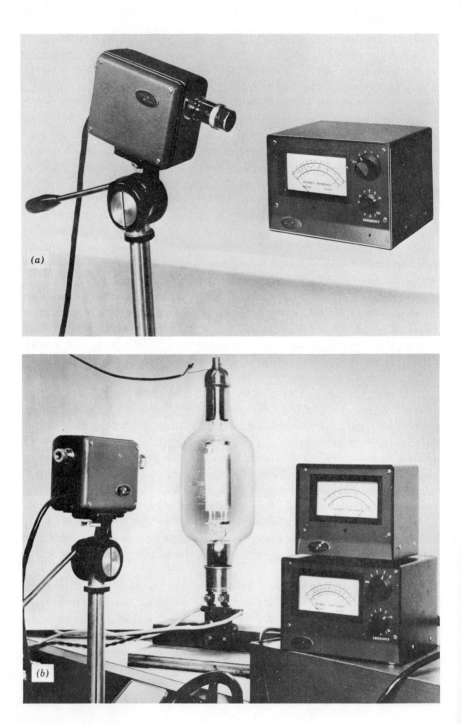

Figure 5.19 *Infrared thermometer. (a) Infrascope Mark I. (b) Infrascope measuring vacuum tube element temperature. (c) Infrascope measuring vacuum process. Courtesy Barnes Engineering Co.*

an emf in the detector, such as a thermopile. The change in resistance or generated emf is then indicated on a meter or digital display, or is used to operate a controller for the process.

Thermal Radiation Pyrometry. Thermal radiation pyrometers essentially operate according to the Stefan, Boltzmann law as expressed in equation 5.8. Since a blackbody at 1500°F radiates a wavelength spectrum of thermal energy from 1 to 17 μm, and approximately 90% of the spectrum is longer than 2 μm, a pyrometer must be sensitive to wavelengths longer than 2 μm to measure temperatures of 1500°F or less. This means that the pyrometer must be designed to have a wavelength response for the desired temperature range. The essential parts of a basic radiation pyrometer are shown in Figure 5.23. The lens or mirrors used must be capable of passing or reflecting the wavelengths of the radiant energy emitted by the hot object, and focusing them on the receiving detector which gives either an emf output or undergoes a resistance change. This output can then be measured in terms of temperature by the measuring device.

The most widely used detector in industrial pyrometers is the thermopile which is discussed in detail later in this chapter. Other detectors include thermistors, platinum and copper element bolometers, barrier photovoltaic cells, phototubes, and cesium vacuum photocells. The thermopile is more rugged than the other types, although the response time is usually 2 s or longer. When response time must be less than 2 s, the bolometer can be used, but it is expensive to construct and is less rugged than other detectors.

The two most widely used commercially available industrial radiation detectors with their associated automatic control devices are the Leeds and Northrup Rayotube and the Honeywell Radiamatic.

Leeds and Northrup. A cross sectional view of the Leeds and Northrup Rayotube detector is shown in Figure 5.24. Radiant energy from the heated object entering the window A is focused by the mirror B to form

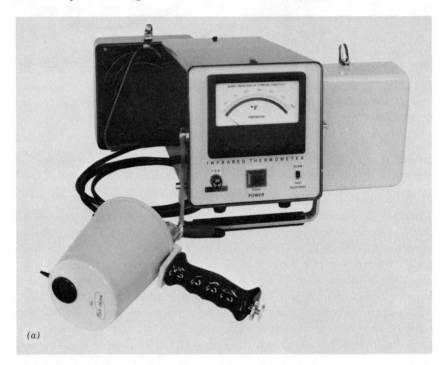

(a)

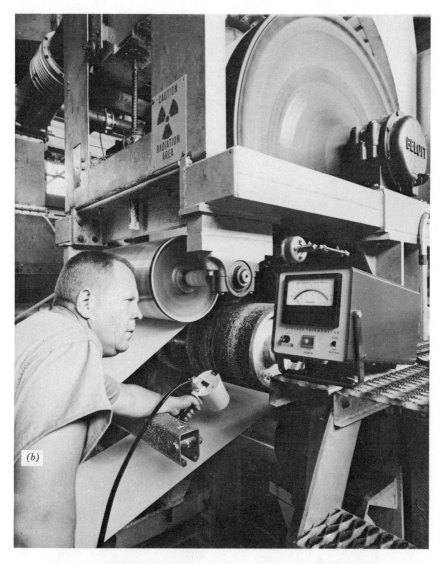

Figure 5.20 *Industrial infrared thermometer. (a) Barnes Model IT-3. (b) Model IT-3 measuring the temperature of a moving sheet of material. Courtesy Barnes Engineering Co.*

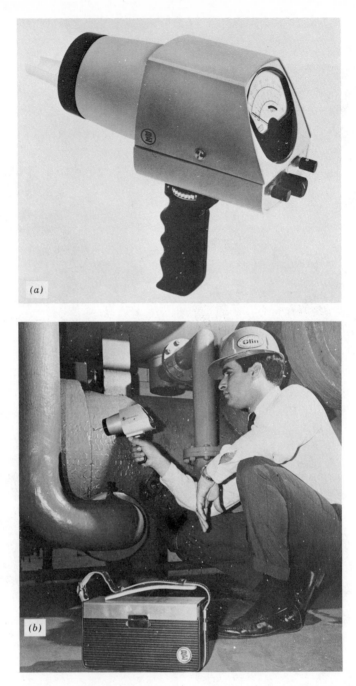

(a)

(b)

Figure 5.21 *Portable infrared thermometer. (a) Models PRT-10-L/PRT-11. (b) Process application of a portable infrared thermometer. Courtesy Barnes Engineering Co.*

180

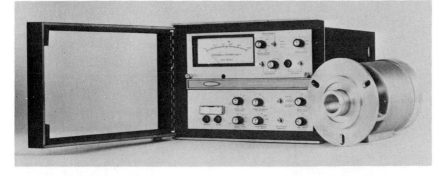

Figure 5.22 *Modline infrared temperature and control system. Courtesy Ircon, Inc.*

a sharply defined image on the surface *G*. A small opening *C* in this disc permits a portion of the radiated heat energy to pass through to mirror *D* which focuses it on the measuring junction of thermopile *E*. The image of the radiating source can be seen through lens *F* on surface *G*. To properly position the Rayotube detector to measure a specific area, the image of the area is selected to cover the opening *C*.

The Rayotube detector shown uses a double-mirror optical system to focus the heat, radiated from the source whose temperature is being measured, upon a blackened disc. The measuring junctions of the thermopile are attached to this disc. The temperature of the disc rises until the rate of heat loss from the disc equals the rate of energy absorption from the source. The millivoltage developed by the thermocouples in the thermopile represents a temperature difference between the hot and cold junctions. The reference junctions of the thermopile are at the housing temperature, and the black disc is at the temperature generated by the radiating source. The measuring instrument indicates the generated millivoltage in terms of the temperature of the source.

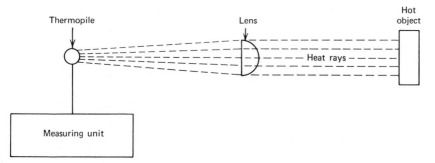

Figure 5.23 *Schematic of a basic radiation pyrometer. Courtesy Honeywell, Inc.*

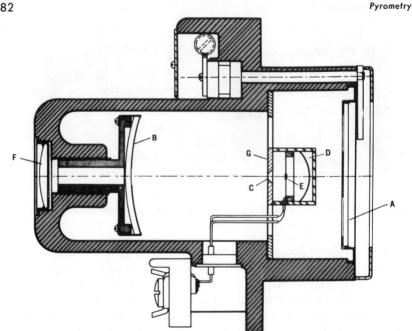

Figure 5.24 *Cross-sectional view of Leeds and Northrup Rayotube®. This cutaway view of the 8890 Series Rayotube detector shows the basic details of construction. Rayotube is shown full size. Courtesy Leeds and Northrup Co.*

The speed of response or *time constant* required for a Rayotube to reach 63.2% of the difference between initial and final millivoltage output, when suddenlly exposed to a source of higher temperature, ranges from 0.15 to 6.6 s. The 0.15 s time constant is for the *high* speed response unit, and the 6.6 s time constant is for the *low* speed response unit. The *normal* speed of response is 1.8 s. The *fast* speed of response is needed to measure sources with temperatures that vary very rapidly, or when temperatures are measured along a rapidly moving body. The *slow* speed of response is used when it is undesirable to follow rapid, erratic changes in temperature such as those generated by a swirling flame, which would create excessive measuring instrument and control action.

A response curve for an instantaneous step change in temperature from t_1 to t_2 for the theoretical and actual time is shown in Figure 5.25. This curve shows the change in millivolt output E for the actual response represented by the dashed line. As shown, E has actually increased by 63.2% of the full response value at time τ.

Rayotube detectors are available to cover temperature ranges from 450°C (800°F) to 2200°C (4000°F), a target emittance range from 0.28 to 1.0, and target sizes from 0.32 in. at 6 in. from the source to 12.7 in. at 12 ft from the source. The small target Rayotube has a target size of 0.10 in. at a distance of 4 in. from the source to 0.60 in. at 36 in. from the source. Different Rayotubes are required to cover the indicated range of temperature, each covering a span of approximately 550°C (800°F) to 1200°C (2150°F) with the smaller spans in the 450°C (800°F) to 1000°C (1800°F) temperature region.

The lens of the Rayotube may be open mounted or closed mounted, depending on the application. When the ambient temperature is high, either air or water cooling of the housing is available. Rayotubes are also used with target tubes of 12 to 48 in. lengths constructed of a material, such as silicon carbide, Fyrestan or an alloy of nickel and chromium, to meet the requirements of the atmosphere in the application. When there is danger of damage to the Rayotube, a shutter assembly is available for temporary protection. Opening the shutter actuates the control circuit, and tripping the shutter opens the control circuit.

Honeywell Radiamatic. The standard Brown radiation pyrometer is a lens type unit and uses a thermopile detector. A cross sectional view is shown in Figure 5.26. Lens *A* is used to focus the radiant energy, emitted by the source, on the thermopile *D*, through the aperature *C*. Aperture

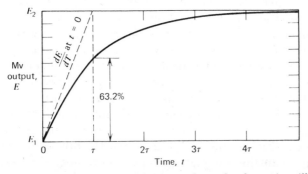

Figure 5.25 *Rayotube response curve. This curve shows the change in millivolt output (E) of a Rayotube detector when the source temperature is instantaneously increased from t_1 to t_2. The time constant is the time τ at which E would be attained if the initial rate of change (dE/dt a t = 0) had been maintained; because of Newton's law of cooling, E has actually increased by 63.2% of $(E_2 - E_1)$. Courtesy Leeds and Northrup Co.*

B, which may be moved parallel to the pyrometer axis, functions as a sensitivity adjustment and also intercepts the cone of rays produced by lens *A.* Adjustment of aperture *B,* by means of pinion *F,* is used to match the pyrometer output to a selected emf value for a particular target temperature. The thermopile leads are connected to terminals *H.* These terminals are accessible by removing the end cap *J* containing the sighting lens *G.* The pyrometer is sighted on the target through the sealed sighting lenses *G* and *I.*

The thermopile used in the Radiamatic is shown in Figure 5.27. This is a group of very small thermocouples connected in series, as described earlier, whose outputs are additive. The tiny thermocouple junctions, about pinpoint size, are flattened and blackened so that they absorb all the radiant energy reaching them. Varying ambient temperatures, which may affect the thermopile, are compensated for by a nickel resistance spool. This spool provides a variable shunt across the emf produced. As the ambient temperature increases or decreases, the corresponding resistance changes of the nickel coil varies the emf output of the head, resulting in accurate compensation over the entire range of the instrument.

A phantom view of the Radiamatic head is shown in Figure 5.28. This shows the construction described for the cross sectional view.

Low Range Radiamatic Unit. The low range Radiamatic unit shown in Figure 5.29 can measure temperature from 125°F to approximately 700°F and responds to 98% of the temperature change in 4 to 6 s. This model differs from standard models. It has a separate control box and heater coil to maintain a constant reference junction temperature at the thermopile as a means of temperature compensation. This temperature control system can maintain the housing temperature t_0 within ±1°F of a preselected value. The heater coils are recessed in the pyrometer shell as shown. The resistance thermometer windings are connected to an ac Wheatstone bridge whose unbalance signal is amplified and applied to the control grid of a thyratron tube whose plate circuit is in series with the heater winding. As the housing temperature varies, the resistance of the resistance thermometer varies. This resistance change is used to raise or lower the thyratron grid voltage to permit the tube to conduct or cutoff. During conduction the plate current flows through the heater coil to restore the housing temperature. This pyrometer also contains a concave mirror which reflects radiation, otherwise lost, back to the thermopile, thereby increasing the sensitivity of the detector. A calcium flouride lens is used because at low temperature the energy radiated by the source is composed mostly of wavelengths longer than 6 μm, although the unit has a sensitivity range from 0.3 to 10.0 μm.

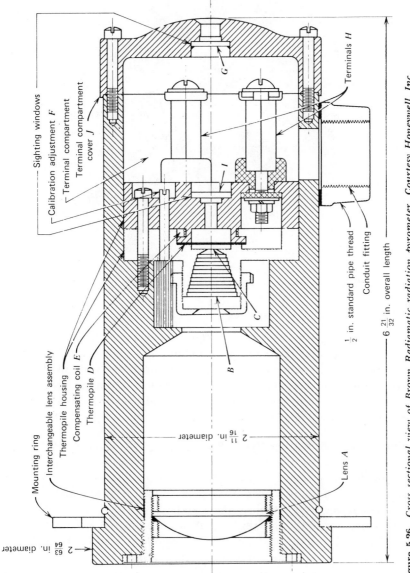

Figure 5.26 Cross sectional view of Brown Radiamatic radiation pyrometer. *Courtesy Honeywell, Inc.*

Sighting windows

Calibration adjustment F

Terminal compartment

Terminal compartment cover J

Terminals H

$\frac{1}{2}$ in. standard pipe thread

Conduit fitting

$6\frac{21}{32}$ in. overall length

Mounting ring

Interchangeable lens assembly

Thermopile housing

Compensating coil E

Thermopile D

$2\frac{11}{16}$ in. diameter

$2\frac{63}{64}$ in. diameter

Lens A

185

Figure 5.27 *Radiamatic thermopile. Courtesy Honeywell, Inc.*

The low range Radiamatic unit requires a relatively large target close to the lens, because relatively little radiation is emitted at lower temperatures. This instrument has a chart and arbitrary scale of 0 to 100 even graduations, because blackbody calibration is not practical under the application conditions. A sighting path and field-of-view diagram for the low range Radiamatic is shown in Figure 5.30.

Calibration of Radiation Pyrometers. For industrial applications calibration is normally obtained by comparison with a radiation pyrometer that has had a primary calibration. Primary calibrations are rarely of any value to a radiation pyrometer user. A manufacturer makes use of blackbody targets to establish the standards for use in an industrial application. Normally, the application in an industrial process does not offer blackbody conditions, and emissivity and emittance values do not permit high accuracy corrections to an actual pyrometer reading.

The use of secondary standards with a primary calibration permits the calibration of other radiation pyrometers without a blackbody radiation source. By exposure of the secondary standard and the pyrometer to the

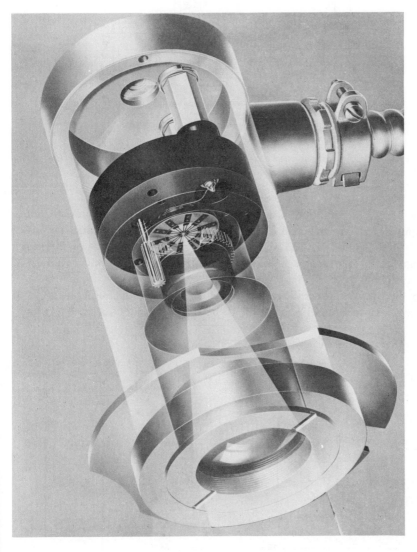

Figure 5.28 *Phantom view of Radiamatic head. Courtesy Honeywell, Inc.*

187

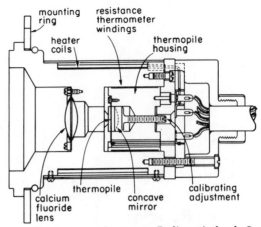

Figure 5.29   *Cross sectional view of low range Radiamatic head. Courtesy Honeywell, Inc.*

same source of radiation, which has at least graybody characteristics, the output signals are adjusted to read the same value. These values will also hold for blackbody conditions, if the reference pyrometer is of the same design as the pyrometer being calibrated and has the same calibration curve shape.

In many cases the calibration is made at only one temperature, and quite often an optical pyrometer, to be discussed later, is used to estab-

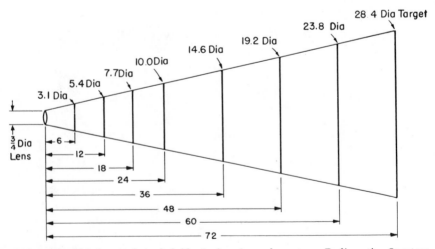

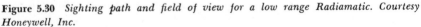

Figure 5.30   *Sighting path and field of view for a low range Radiamatic. Courtesy Honeywell, Inc.*

lish the temperature when radiation pyrometer reference standards are not available. However, optical pyrometers are subject to the same errors, but of different magnitude, as radiation pyrometers.

Another common method used in lower temperature regions is to establish the blackbody or graybody source temperature with a certified thermocouple or resistance thermometer. The important factor is to be certain that all the errors remain constant during the calibration. Otherwise no method will give a reliable value of comparison. It is also important to remember that under nonblackbody conditions the temperature measured is not the true temperature. In industrial processes this is of secondary importance to the reproducibility of the reading of the pyrometer.

Radiation Pyrometer Applications. Radiation pyrometers are used industrially where temperatures are above the practical operating range of thermocouples, where thermocouple life is short because of corrosive atmospheres, where the object whose temperature is to be measured is moving, inside vacuum or pressure furnaces, where temperature sensors would damage the product (such as growing crystals), and for obtaining the temperature of a large surface when it is impractical to attach primary temperature sensors.

When an application is being considered, the following data should be available: the target temperature, the normal operating temperature, and the high and low limits; the minimum target size and its distance from the pyrometer lens; the target material and its emittance; the angle of observation of the emitted radiation; the target condition, moving or stationary and, if moving, the speed; the atmospheric condition between the target and detector; the ambient temperature or range of ambient temperature if it changes; whether the unit can be sighted directly on the target or must be sighted through a port or window; the mechanical details involved in the application; the type of chart and scale needed; the type of instrument desired.

5.4 TWO-COLOR PYROMETRY

The two-color or ratio pyrometer is a temperature measuring device which uses the radiation emitted by an incandescent object to produce a meter indication of the object's temperature. Radiant energy in watts per square centimeter available for emission by an incandescent body is expressed by the Wein-Planck law:

$$J\lambda = \frac{C_1\lambda^{-5}}{e^{-C_2/\lambda T - 1}} \qquad (5.9)$$

Equation 5.9 relates the amount of energy $J\lambda$ radiated by a perfect black-body radiator per square centimeter per second at wavelength λ to the absolute temperature T of the radiator.

In the visible portion of the spectrum and at temperatures below $7000°F$, Planck's law reduces to Wein's equation to within 1%:

$$J\lambda \cong c_1\lambda^{-5} e^{-C_2/\lambda T} \qquad (5.10)$$

This is the region of the spectrum and the temperature range in which the two-color pyrometer operates, and equation 5.10 is adequate for this discussion.

Experimental results show that surfaces measured in industrial prac-tices are not perfect radiators. Internal reflection at the surface reduces the intensity of the radiation passing through the surface, so that the emitted radiation is less than that given by the theoretical expression. The diverse and poorly defined surface conditions encountered make it impossible to account for the surface property in a theoretical discussion of emitted radiation. The formal expression of Wein's equation can be retained, and use made of emissivity to account empirically for the sur-face properties expressed by

$$U_\lambda = \epsilon_\lambda C_1\lambda^{-5} e^{-C_2/\lambda T} \qquad (5.11)$$

which can be expressed in a modified form as shown in equation 5.12:

$$U_\lambda = C_1\lambda^{-5} e^{-C_2/\lambda T_B} \qquad (5.12)$$

Equation 5.12 contains a variable temperature T_B called the brightness temperature, which is the apparent temperature of the observed surface. Wein's equation is a single-valued function of temperature, so there is a temperature T_B at which a perfect radiator emits the same amount of energy as a surface with emissivity ϵ_λ and temperature T. ϵ_λ is determined in practice by taking, at a given wavelength, the ratio of the energy emitted by the surface being measured to that emitted by a blackbody at the same temperature. ϵ_λ is dimensionless and has a value of 1 only for blackbody conditions. For nonblackbody conditions ϵ_λ is always less than 1. Thus T_B is always less than the true temperature. Equation 5.12 em-bodies the principle of operation of the optical pyrometer to be discussed later in this chapter.

The ratio of the energy emitted by a surface at two different wave-lengths λ_1 and λ_2 can be expressed as:

$$R = \frac{\epsilon_{\lambda_1}C_1\lambda_1^{-5}e^{-C_2/\lambda_1 T}}{\epsilon_{\lambda_2}C_1\lambda_2^{-5}e^{-C_2/\lambda_2 T}} = \frac{C_1\lambda_1^{-5}e^{-C_2/\lambda_1 T_R}}{C_1\lambda_2^{-5}e^{-C_2/\lambda_2 T_R}}$$

$$= \frac{(\lambda_2/\lambda_1)^5 \, e^{-C_2}}{[(\lambda_2 - \lambda_1)/(\lambda_1\lambda_2)] \, (1/T_R)} \qquad (5.13)$$

The ratio R expressed by equation 5.13 is also a single-valued function of temperature, so there is a temperature T_R at which a blackbody has the same ratio R_1 as the surface with the emissivity ratio $\epsilon_{\lambda 1}/\epsilon_{\lambda 2}$, and temperature T. Equation 5.13 embodies the principle of operation of the ratio or two-color pyrometer. The apparent temperature T_R is thus defined as the ratio temperature. C_1 and C_2 are the first and second radiation constants. The surface appears only through the ratios of the emissivities at the two different wavelengths. Thus the ratio temperature T_R which is indicated for a surface at temperature T, if the ratio of the emissivity values at the two wavelengths is the same, has the same value whether both emissivities have a value of 0.9 or 0.09.

This means that the two-color pyrometer depends on the emissivity-versus-wavelength characteristics of the surface. The slope of the ϵ-versus-λ curve may be positive or negative. When it is positive, the ratio temperature is lower than the real temperature. When it is negative, the ratio temperature is higher than the actual temperature. If the emissivity is independent of the wavelength, the possibility exists that the ratio temperature will be the true temperature even though the emissivity may be far from unity. Also, the surface dependence cancels out of the expression if the emissivities remain constant. The existence of this possibility makes the ratio pyrometer a device to measure a surface's temperature from its emitted radiation. The possibility of emissivity being independent of wavelength also makes the ratio pyrometer theoretically free from emissivity errors. Surfaces for which this is true are as rare as those with true blackbody characteristics.

Fortunately, the emissivity of most surfaces varies slowly with wavelength, so that the ratio of $\epsilon_{\lambda 1}/\epsilon_{\lambda 2}$ is close to unity. The practical ratio (two-color) pyrometer makes use of this slow variance with wavelength. Under good industrial measuring conditions, the ratio temperature T_R can be much closer to the true temperature T than the brightness temperature T_B or the radiation temperature derived from a total radiation device.

Some probable advantages gained through the use of the ratio measurement are: (1) relatively dense smoke can be tolerated; (2) dirty lenses, dusty atmospheres, and the presence of steam or vapors in moderate amounts have little or no effect on accurate temperature measurements; (3) sighting distances can range from inches to hundreds of feet.

With recent developments and refinements in infrared and thermal radiation detectors and controls, and the relatively high cost of ratio pyrometers, industrial application of the latter has not been too widespread.

5.5 OPTICAL PYROMETRY

The optical pyrometer is the official device recognized internationally for measuring temperatures above 1063°C. It has been used to establish the *International Scale of Temperature above 1063°C.*

The optical pyrometer measures the intensity of radiant energy emitted in a narrow wavelength band of the visible spectrum. Both manual and automatic units are available, and both have approximately the same accuracy.

Spectral radiant intensity is related to temperature and wavelength as discussed in Section 5.4 and expressed in equation 5.9.

An optical pyrometer is a device for measuring the temperature of a hot object by determining the brightness of the surface of the object. The unaided human eye was the first optical pyrometer used to determine the temperature of glowing objects. This method was crude and at best only an estimate, but it was the only method available for high temperatures. The human eye still plays an important role in optical pyrometry. It is used to match the brightness of one object against that of another with a high degree of accuracy. The optical pyrometer is actually an instrument with which the unknown brightness of an object is measured by comparing it with the known brightness of a fixed source. The instrument can also be calibrated accurately against the known source.

The intensity of light in the visible spectrum emitted by a hot object varies rapidly with its temperature. This can be best shown by the fact that the visual effect of red radiation at 2500°F varies 12 times as fast as the temperature. So with a small change in temperature there is a much larger change in brightness, which provides a natural means for determining temperatures with good precision and accuracy.

A red filter is used in all optical pyrometers between the object and the observer's eye. The monochromatic light chosen has a red wavelength of maximum sensitivity to the eye to minimize or eliminate the factor of individual differences in color judgment or color sensation. It should be noted that persons suffering from color blindness can obtain the same results with an optical pyrometer as those with perfect color vision.

Optical pyrometers can be roughly divided into two general types or groups. The group type shown in Figure 5.31 optically matches the light from the hot object with the light from a constant comparison lamp in the instrument. The light output of the comparison lamp is kept constant by maintaining a constant current through the filament. The comparison to the hot body is made by rotating a graduated optical absorption wedge to change the apparent brightness of the hot body until a small luminous test mark, which appears in the field of vision, disappears.

The second group type shown in Figure 5.32 varies the intensity of the light from a calibrated comparison lamp to match the intensity of the light emitted by the hot object. In both types the light emitted by the hot object and the calibrated comparison lamp must be of the same wavelength to obtain accurate measurements.

Industrial optical pyrometers are available in both types and are reasonably rugged, dependable instruments for measuring temperatures in the 1400 to 5200°F temperature range, and at least one unit can be used with external filters to measure temperatures up to 18,000°F. These optical pyrometers are available as single-scale or multiple-scale units with scale spans ranging from 850 to 3800°F.

Pyro Optical Pyrometer. The Pyro optical pyrometer is shown in cross section in Figure 5.33, and as a self-contained operating instrument in Figure 5.34. As shown in Figure 5.33, the Pyro optical pyrometer uses an optical wedge. The optical system of lenses and prisms acts as a telescope and permits a clear and enlarged view of the hot object whose temperature is to be mesaured. The light emitted by the hot object passes through the optical system and the optical or photoscreen wedge which varies from light to dark intensity. As the wedge is rotated, by means of a scale drum ring, the hot body emitted light is filtered to an intensity equal to the light intensity produced by a standard incandescent lamp contained within the instrument. When the two light sources are equal, the test mark disappears, and a photometric match has been made. The optical wedge is calibrated in terms of temperature, so that when the match is made the temperature can be read on the direct reading scale contained within the housing of the instrument.

In this instrument a constant current is maintained through the service bulb to produce the calibrated intensity. A master lamp is furnished

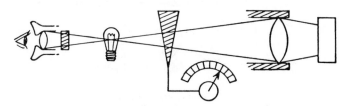

Figure 5.31 *Schematic of an optical pyrometer in which the brightness of the source image is varied by an optical wedge. Courtesy Pyrometer Instrument Co., Inc.*

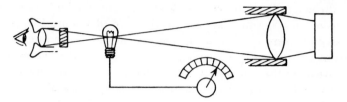

Figure 5.32 *Schematic of an optical pyrometer employing a variable filament temperature. Courtesy Pyrometer Instrument Co., Inc.*

for calibration of the service lamp, and a calibration can be made on the job in a few minutes by checking the service lamp against the master lamp. Temperature values are based on blackbody conditions and are reproducible to within a few degrees and with an accuracy of approximately 0.5% of the temperature being measured.

The normal maintenance on Pyro units consists of changing the battery and calibration of the service bulb with use. It is essential that the current through the service bulb be adjusted periodically, as the battery is used to maintain the calibrated intensity. The filament in the service bulb is also consumed during use, so that current adjustments are required during calibration to maintain the calibrated intensity.

Pyro Microoptical Pyrometer. The microoptical pyrometer is primarily a laboratory instrument with a high degree of precision, and can be used

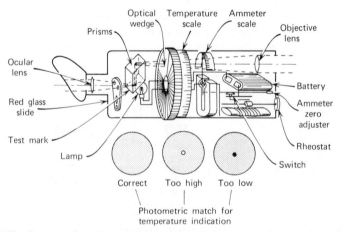

Figure 5.33 *Cross section view of a Pyro optical pyrometer. Courtesy Pyrometer Instrument Co., Inc.*

Figure 5.34 *Pyro optical pyrometer. Courtesy Pyrometer Instrument Co., Inc.*

for measuring targets of less than 1 mil in size at a distance of 5.5 in. The complete instrument setup is shown in Figure 5.35. The telescope T is mounted on the tripod and is connected to the meter M and battery B with a wiring harness. The auxiliary objective lenses are in lens case L to meet the focal distance and target size requirements. Sighting is accomplished by means of vernier horizontal and vertical worm gears on the mounting.

The microoptical pyrometer is available in several models covering the temperature range from 700 to 5000°C in three spans. By means of externally mounted filters, the range can be extended to 10,000°C (18,000°F). These higher readings are not direct, but must be determined from comparative charts. Calibration must be made individually with higher temperature filters, so that these filters are not interchangeable with other than the instrument used during the calibration.

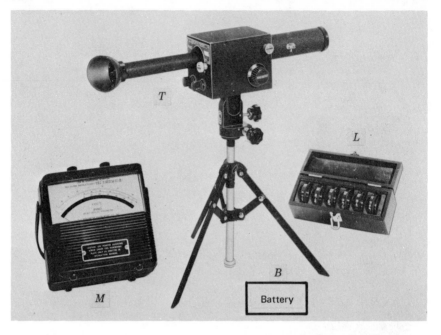

Figure 5.35 *Pyro microoptical pyrometer. Courtesy Pyrometer Instrument Co., Inc.*

This type of optical pyrometer is especially adapted for measuring small grids and filaments, and uses an apex disappearing filament as shown in Figure 5.36. It has an accuracy of 2 to 3°C.

Leeds and Northrup Optical Pyrometer. The Leeds and Northrup potentiometer optical pyrometer is shown schematically in Figure 5.37. The circuit consists of the following components. Four flashlight batteries supply the filament current of the lamp. This current flows through the rheostat R which is manually adjusted to rotate R_1 until the brightness of the lamp matches the brightness of the object being measured. When R_1 is moved, the potentiometer contact P is moved through a frictional clutch arrangement, so than an approximate balance is maintained. With this arrangement, when the standard cell circuit is closed by pressing the potentiometer knob P_1, only a minor adjustment is needed to zero the galvanometer. Knob P_1 moves only the potentiometer contact P. The spring switch S, located in the hand grip of the telescope, automatically opens both the battery and the standard cell when the telescope is released from the hand. A second switch S_1 associated with the knob operating the potentiometer contact keeps the standard cell circuit open, ex-

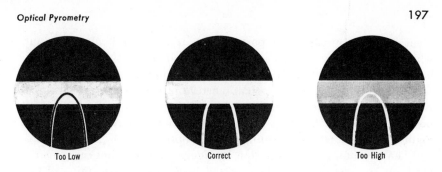

| Too Low | Correct | Too High |

Figure 5.36 *Disappearing filament used in microoptical pyrometer. Courtesy Pyrometer Instrument Co., Inc.*

cept when the position of the contact is being moved to obtain a balance for a galvanometer (*GA*) zero reading. This arrangement conserves the batteries and provides a double safeguard against accidental drain of the standard cell.

A drum scale is attached to the potentiometer contact and reads directly in temperature to three, or by estimation, to four significant figures on the scale provided. The resistor *G* provides the required suppression to establish the temperature range.

A unique feature of the circuit just described is that the current measurement is made by a potentiometer method without the usual standardization of current in the potentiometer circuit. Although the measure-

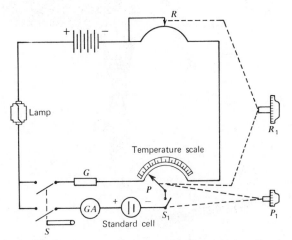

Figure 5.37 *Basic circuit of a potentiometer optical pyrometer. Courtesy Leeds and Northrup Co.*

ment is made by varying a portion of a resistor in the lamp circuit, which is bridged by a standard cell, until a balance is obtained, the calibration is in terms of equivalent blackbody temperature.

The Leeds and Northrup potentiometer optical pyrometer consists of two units, a telescope portion used for sighting the target and a control box containing the batteries and the potentiometer circuit as shown in Figure 5.38. The telescope, containing the optical system, filters, screens, and lamp is shown in cross section in Figure 5.39. The optical system A produces an erect image of the object sighted. A perfect merging of filament with image is made possible through the use of a long, thin ribbon filament in a lamp with flat windows. This is shown in B. The screen shifting device permits easy range changing. The red filter for intensifying the red spectrum and the switch discussed in the basic circuit are also shown.

Leeds and Northrup Automatic Optical Pyrometer. The optical pyrometers discussed so far are manually operated instruments. A block diagram of the Leeds and Northrup automatic pyrometer is shown in Figure 5.40, and a photograph of the actual instrument is shown in Figure 5.41.

Figure 5.38 *The Leeds and Northrup potentiometer optical pyrometer. Courtesy Leeds and Northrup Co.*

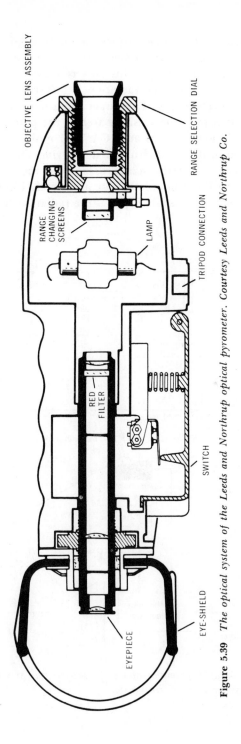

Figure 5.39 *The optical system of the Leeds and Northrup optical pyrometer. Courtesy Leeds and Northrup Co.*

199

The instrument must be manually and visually aligned on the target, and the area or object whose temperature is being measured must be large enough to fill a field stop. This area is viewed on a ground glass screen or with an eyepiece. The radiation is focused onto the field stop by means of an objective lens. The target radiation traverses an optical path which directs it onto the sensitive area of the photodetector. In this instrument the photodetector is an RCA 7265 multiplier phototube. A similar optical path, displaced from that carrying the target radiation, directs the radiation from the standard lamp to the same spot on the detector. A rotating light modulator disc alternately blocks radiation, at 90 cycles/s, from one path, while it passes the radiation from the other.

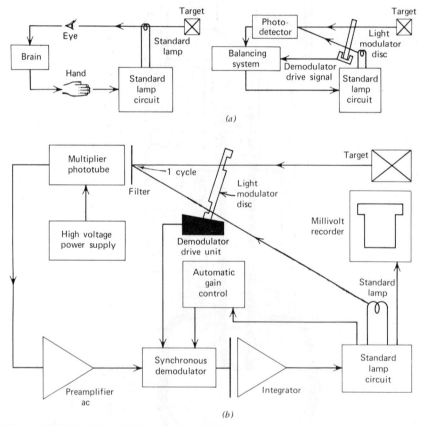

Figure 5.40 *Operational diagrams of optical pyrometers. (a) Manually adjusted. (b) Automatic. Courtesy Leeds and Northrup Co.*

Figure 5.41 *Leeds and Northrup Mark I automatic optical pyrometer. Courtesy Leeds and Northrup Co.*

The rotating disc modulator thus permits the photodetector to alternately see the radiation from the target and the standard lamp. The standard lamp operates at a wavelength of 6530 Å, which is the monochromatic principle of the Leeds and Northrup optical pyrometer discussed in the preceding section. We assume that the target brightness is greater than that of the standard lamp. Under these conditions, the photodetector output current takes the form of square dc pulses of alternating amplitude. This signal is fed into an ac preamplifier and then to the synchronous demodulator. A unit triggered by the light modulator disc furnishes the synchronous feature of the demodulator. The demodulator signal is fed into the integrator. This integrator drives the standard lamp circuit so that the lamp current increases to a point where there is a match in brightness temperature between the target and the lamp. When the system reaches this balance point, an indication of the temperature may be obtained by measuring the standard lamp current. Actual temperature of the target is obtained by reference to a calibration curve.

Because radiant energy emitted by a blackbody target increases very rapidly with temperature, as discussed in the section on optical pyrometers, an automatic gain control is incorporated into the balancing system. This control reduces the circuit gain to prevent the measuring system from becoming oversensitive at high target temperatures. To maintain a fairly constant sensitivity, the gain must be changed by a factor of 20 over the normal operating range of the instrument. The method employed is to vary the width of the demodulator gates. A change in gate width changes the amount of signal passed through the system, which results in a change in the gain of the system. Changes in the standard lamp current also vary the width of the gate. A function generator is used to correct for the nonlinear relation of radiant power to temperature.

Actual temperature readout can be provided by passing the lamp current through a calibrated resistor and reading the voltage drop across the resistor with a potentiometer or a millivolt recorder calibrated in degrees Celsius or Fahrenheit. The sensitivity of the instrument gives detection of $\pm 0.2\,°C$ at the gold point ($1063\,°C$) with a 1 s response time. Through the use of appropriate filters, temperatures from $1400°$ to $10,470\,°F$ (775 to $5800\,°C$) can be measured. As the other pyrometers, the total span is broken up into smaller spans for better accuracy and to keep within the range of the filters.

Calibration is normally carried out by the manufacturer, and calibration checks or recalibrations can be made by comparison to an optical pyrometer standard. Overall accuracy is dependent on the calibration and stability of the standard lamp.

5.6 SUMMARY

Temperature measurements by means of thermocouples, infrared pyrometers, thermal radiation pyrometers, two-color pyrometers, manual and automatic optical pyrometers, and their associated measuring and control equipment have been discussed. These instruments normally cover the 900 to $18,000\,°F$ temperature range, and each type has a place in industry in the production of a quality product. Together with the various types of thermometers discussed in Chapter 4, all industrial temperature measurement and control areas as we know them today, have been discussed. Our space age requirements make further demands on this type of measuring and control equipment, which will undoubtedly be reflected in the industrial requirements of the future. These demands will be for more exotic alloys to better withstand the reentry temperatures and control units to maintain livable temperatures in manned space stations and space vehicles.

Table 5-1 shows the sensing devices discussed and their useful temperature range for the − 300 to + 18,000°F temperature span.

Table 5.1   Pyrometer Sensors and Useful Range

Type of Sensor	Temperature Range (°F)
Thermocouples	−300 to 2700
Copper–constantan	−300 to 600
Iron–constantan	0 to 1400
Chromel–alumel	600 to 2300
Noble metals	1300 to 2700
Radiation	65 to 4650
Infrared	65 to 4650
Thermal Ray-O-Tube	800 to 4000
Thermal Radiomatic	125 to 700
Two-color	125 to 3200
Optical	1063 to 18,000
Optical wedge	1063 to 10,000
Comparison lamp filament	1292 to 18,000
Microoptical	1292 to 18,000
Pyro optical	1063 to 10,000
Leeds and Northrup	1400 to 10,470

Review Questions

5.1   Why is a reference junction required to make a temperature reading with a thermocouple?

5.2   Why does a thermocouple have to have two dissimilar wires to make up the couple?

5.3   Define hot junction and cold junction.

5.4   In what types of applications would you use the following thermocouples: (1) iron–constantan, (2) chromel–alumel, (3) platinum–platinum–rhodium, (4) chromel–constantan? Give reasons for your choices.

5.5   When is the resistance value of a thermocouple very important as part of the measuring system? Give examples. What type of equipment can be used to eliminate the problems related to resistance? What is the principle involved in the system you chose?

5.6   Explain why reference junction compensation is desirable in a thermocouple system.

5.7   What function does wire size determine when a thermocouple is being chosen for a particular application?

5.8   Explain how thermocouples can be used in series and how one interprets the measured value.

5.9   Explain how several thermocouples in parallel can be measured by one readout device individually and collectively.

5.10 How can the millivolt output of a thermocouple be used as a temperature control medium?

5.11 What is the limitation on the indicated temperature of a contact type millivolt meter controller? How can it be overcome? What is the danger that must always be recognized in using free motion indicating controllers?

5.12 What is the operating principle of a continuous balance potentiometer system? Why is it not used exclusively for temperature measurements?

5.13 State the difference between a potentiometer system using a reference junction and a universal type using an ice point reference.

5.14 How is a continuous balance system standardized? What measurement problems does this cause and how are they minimized or eliminated? How is this problem eliminated in solid state circuitry?

5.15 Explain the meaning of thermal radiation pyrometry. How is thermal radiation measured?

5.16 What factors cause errors in infrared temperature readings and what technique has been developed to overcome them?

5.17 Where is fast speed of response desirable? Normal speed of response? When is slow speed of response acceptable for temperature measurements?

5.18 What is the total temperature range over which thermal radiation instruments are capable of measurement? How many different kinds of thermal radiation instruments are needed to cover the total possible range?

5.19 What is two-color pyrometry? Why do we need it? Where is it used?

5.20 What significance does the ϵ-versus λ curve slope have in two-color pyrometry?

5.21 What characteristic related to temperature change and brightness makes optical pyrometers useful temperature measuring instruments?

5.22 Explain the two basic methods used in industrial optical pyrometers to obtain temperature measurements.

5.23   What is the highest temperature and the smallest object size that can be measured for temperatures values with optical pyrometers?

5.24   Explain how to calibrate an optical pyrometer to guarantee its accuracy.

Problems

5.1   Determine the millivolt output at a temperature of 350°F for thercouples constructed of :(1) iron–constantan, (2) chromel–alumel, (3) platinum–platinum–13% rhodium, (4) chromel–constantan. Which thermocouple material would you choose for use in a dry kiln application?

5.2   What wire sizes would you select to obtain the fastest and most accurate response in a reducing atmosphere at a temperature of 2300°F?

5.3   Sketch the setup you would use for a two-wire resistance thermometer system.

5.4   If the ambient temperature is 70°F and the radiant energy source is 1020°F, what is the radiant energy emitted per square centimeter when $\sigma = 2.5$?

5.5   Sketch the setup and explain how to calibrate thermocouples.

5.6   Sketch the calibration setup and explain how to calibrate a two-color radiation pyrometer.

5.7   Prepare a sketch and explain how an automatic pyrometer functions.

Bibliography

Considine, Douglas M. (ed.), *Process Instruments and Controls Handbook, Radiation and Optical Pyrometry,* Forsythe, W. E., and Horner, J. F., McGraw-Hill, New York, 1957.

Gray, W. T., *Precision and Accuracy in Radiation Pyrometry,* paper presented at the Symposium on Precision Electrical Measurements, June 3–5, 1963.

Harrison, Thomas R., *Principles for Applying the Brown Radiamatic Pyrometer under Non-Blackbody Conditions,* Technical Data Bulletin No. TB-930-1, Honeywell, Inc., Ft. Washington, Pa.

Honeywell, Inc., *Fundamentals of Instrumentation for the Industries,* G00003-0010M, Ft. Washington, Pa., January 1958.

Latronics Corporation, *Design, Operation and Application of Latronics Color ratio Pyrometer,* Bulletin 5-63; *Latronics Automatic Optical Pyrometer,* Bulletin CP-1-1064, Latrobe, Pa.

Leeds and Northrup Company, *L & N Automatic Optical Pyrometer*, A1.4111-1965; *L & N 8630 Series Optical Pyrometers Single Adjustment, Potentiometer Type*, Data Sheet B1.4121, 1964; *Optical Pyrometer, Potentiometer Type*, Catalog N-33D, 1957; *Directions for Nos. 8621, 8622, and 8623 Optical Pyrometers*, Direction Book 77-1-0-3, 8-251; *Manual for Speedomax H*, 077990, issue 6, North Wales, Pa.

Magison, E. C., The "WHYS" of Radiation Pyrometry, Honeywell, Inc., Industrial Products Group, *Instrumentation*, Vol. 17, No. 1.

Milletron, Inc., *Therm-O-Scope Two Color Pyrometer. Theory and Its Advantage*, Pittsburgh, Pa.

Pyrometer Instrument Co., *New Micro-Optical Pyrometer*, Catalog 96; *Optical Pyrometer*, Catalog 85, North Vale, N.J.

Robertson, D. R., Frock, H. N. and Shreve, W. T., *A Demonstration of an Automatic Optical Pyrometer*, paper presented at the Symposium on Precision Electrical Measurements, Leeds and Northrup Company, June 1963.

Soisson, Harold E., *Electronic Measuring Instruments*, McGraw-Hill, New York, 1961.

CHAPTER 6

Liquid and Dry Level Instrumentation

Measurement of the level of a liquid, slurry, or dry material in a container may look simple, but can become relatively difficult when the material is corrosive, abrasive, maintained under high pressures, radioactive, or in a sealed vessel in which moving parts are undesirable or practically impossible to maintain. Difficulties are encountered when there is a need for high accuracy in measurements for both very small and very large containers, open vessels, and bins.

Control of the level between two preset levels, a high point and a low point, is one of the most common applications for level measuring and controlling instruments. Measurement of the interface level between two liquids or a vapor also presents some interesting applications of instruments and controls. Levels can be measured and maintained by mechanical, pressure drop, electrical, or electrical-electronic devices. The type of device depends on the type of material in the container, the type of container, and the accuracy needed in the measurement or control. The use of radioactive sources and detectors for level measurements is discussed in Chapter 10.

6.1 MECHANICAL LEVEL MEASURING INSTRUMENTS

Mechanical measuring and control instruments for level or pressure head measurements and control include visual and indicating devices.

The simplest level measuring device is a stick or rod, graduated in inches or other appropriate units, which can be inserted into the vessel; the actual depth of material is measured by the wetting action on the stick or rod. This method is used for measuring the depth of gasoline in tanks at a gasoline station. A special stick or rod may be used which will show if condensate has collected in the bottom of the tank below the level of the gasoline. Such measurements are simple but effective for the application. The gasoline and water example also illustrates an interface level measurement. Such a method is not too practical when the material is toxic or corrosive and an individual has to stand over the opening and handle the measuring stick or rod as it is withdrawn from the container. This method offers no means of controlling the level other than manually filling the tank or vessel to provide the volume needed for the application.

The sight glass is another simple visual method for measuring the level of noncorrosive and nonstaining or nonsticky liquid materials. The sight glass method is shown schematically in Figure 6.1 for both open and closed vessels. In the closed vessel application, relatively high pressures can be maintained by using armored sight glasses for safety in case of breakage. Normally, sight glasses are used to measure visually liquid levels with a variance of 3 ft or less. Where they must be used to measure greater variations, the technique of employing several short sections can be applied; the sections are so arranged that the liquid level is always visible in one of them.

Sight glasses can be found on steam irons, coffee urns, steam boilers, tanks, and other open or closed vessels for which visual indications are adequate. This generally means that there is someone in attendance who can manually control the level of the liquid within the limits chosen for the application, and remote indications are not necessary. In some cases special lighting is placed behind the sight glasses and is so arranged that

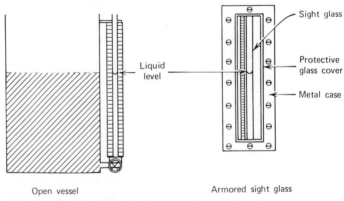

Figure 6.1 *Schematic arrangements for sight glasses.*

different colors are seen for different level limits. Such an arrangement is made to alert an operator to take the appropriate action, such as stopping an incoming flow or starting a flow. Photoelectric sensors can be added to sight glasses to obtain automatic level control. In this type of installation, the light sensitive detector and relay provide the signal to initiate action to pumps or valves. This could result in adding liquid or stopping the incoming flow.

Care must be exercised to keep the bore of the sight glass as large as possible because, if the bore is too small, capillary action occurs and the indicated level will be higher than the actual level. Liquids also have a meniscus where the liquid adheres to the glass. Water has a concave or cup shape, and the level should be read from the base. Mercury has a convex or dome shape and the level is read from the top. The same reading problems are encountered in the sight glass as in manometer tubes used for low pressure measurements.

A sight glass indicates changes in level to an accuracy of $\frac{1}{64}$ in. under the best conditions, but does not indicate the actual volume of liquid in the vessel. For example, a 1 in. level change in the water volume of a steam iron is far far smaller than a 1 in. level change in a 15,000 lb/h steam boiler. Therefore, if the level measurement is to establish the volume of liquid in the vessel, the head measurement can be expressed as

$$V = \int A \, dh + C \qquad (6.1)$$

where A = area of the vessel
dh = change in the head or level measurement

The shape of the vessel quite often determines the type of instrument needed to make the level measurement. A tall narrow vessel gives a more accurate level measurement with respect to volume than a short broad vessel, but a small level change in a short broad vessel represents greater capacity in a controlled system.

Float Level Measuring Systems. When an indication or recording of the level measurement is needed, a float system using a float and tape or float and chain can be used for open vessels. On closed vessels, under vacuum or pressure where seals are mandatory, torque arm floats, cage floats, magnetically coupled floats, and float operated hydraulic devices are used.

A typical float and tape or chain level measuring mechanism is shown in Figure 6.2. Either a tape or chain can be used as the connecting link between the float and the indicating or recording mechanism. Floats can be any desired shape, round, cylindrical, or a combination of the two.

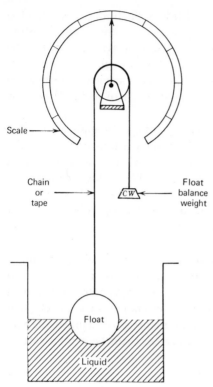

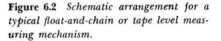

Figure 6.2 *Schematic arrangement for a typical float-and-chain or tape level measuring mechanism.*

They can also be of various sizes, depending on the size of the vessel in which they are used.

A float must be constructed so that it will float in the material. This means the float density must be lower than the material supporting it. The float material should also be chosen so that it will not be corroded or eroded by the material in which it floats. Otherwise it will change in density.

The only limitations on a float-and-tape or chain level measuring system are the height of level variation to be measured in the particular vessel or standpipe and a practical consideration of the float size for the application.

If remote indication and control are needed, pilot relays can be mounted on the rotating shaft carrying the chain or tape. The shaft rotation is translated into level indication pneumatically, electrically, or hydraulically for use at the remote location. In the latest equipment a transducer is used to convert the angular motion into an electrical signal for transmission.

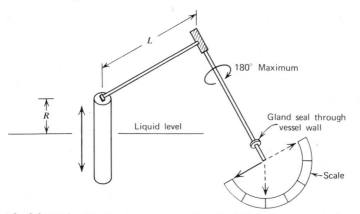

Figure 6.3 *Schematic arrangement demonstrating the ball float level measuring mechanism.*

Float-and-tape or chain mechanisms may be limited in some applications, because it is difficult to maintain clean and efficient operation. In most chain gages the chain is engaged in a sprocket wheel which turns the rotating shaft. In tape gages the tape is wrapped around a drum which turns the rotating shaft. A counterweight must be used to keep the chain or tape taut as the float rises and falls with the level of the medium being measured.

Ball Float Mechanisms. Ball float mechanisms can be used in either open or closed vessels. One such mechanism is shown in Figure 6.3. These level measuring mechanisms are normally used where there is a limited level change, and are definitely limited to measuring levels determined by the length of the arm holding the ball float. The maximum level range is twice the length of the arm for a 180° arc from empty or low level to full or maximum level. For practical measurements the arc should not exceed 60° for the most satisfactory response and accuracy of measurement. The level range represented by R is a function of the length of the arm L. If the angle of rotation is represented by θ, this relationship can be expressed as

$$R = 2L \sin \theta/2 \tag{6.2}$$

where θ represents the angle of shaft rotation in degrees, and $\theta/2$ is half the angle from the horizontal where $\sin \theta = 0$.

The ball float mechanism can be used on sealed vessels by bringing the rotary shaft out through a packing gland or so-called stuffing box. In this type of application the ball float must be able to withstand the vacuum or pressure in the vessel, and the mechanism cannot be repaired or maintained during operation of the system.

The rotary shaft can be used for operating a control valve to maintain the proper level within the vessel, but the level measurement and control valve action will be erratic if there are any ripples or rolls in the surface being measured. Boiling liquids give such erratic action, and are one type of application for which the ball float should not be chosen.

One of the most common uses of the ball float and control valve mechanism to control liquid level is in the water storage tank of toilets. The level of the ball can be adjusted to close the inlet water valve at any desired water level.

Remote control and indications can be arranged by the use of pilot relays on the rotating shaft to convert the level measurement to pneumatic, hydraulic, or electrical signals for transmission to the desired location.

Measurements of interface levels can be accomplished by means of the chain-and-float arrangement shown in Figure 6.4. Other methods are also used for interface level measurements, and are discussed later in this chapter.

The pancake float shown in Figure 6.4 is usually preferred because it offers the best sensitivity and accuracy in this type of measurement. The buoyancy of the float should be chosen so that it sinks in the lighter density liquid and floats on the heavier liquid, and ideally should be a mean

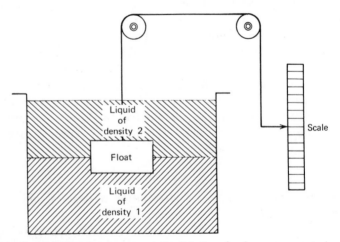

Figure 6.4 *Cross sectional arrangement for interface level measurements in an open vessel using a float.*

between the two liquids. This type of float is normally restricted to open vessels in which the total quantity of liquid has wide variations. A change in density in the heavier liquid causes a change in the reading of the level that will be in error by some proportion to the change in density. Such a system will have relatively low driving force on the readout device, because the float is balanced against the weight of the float assembly. This driving force is usually less than 1 lb.

The displacement float shown in Figure 6.5 has a height equivalent to the interface range to be covered. The float is normally small in diameter and long. It exerts a force on the transmitting element in proportion to the interface height. The float must always be submerged. These floats are calibrated so that a full tank of the lighter liquid gives a 100% reading. Any change in density of either liquid produces an error in the reading as compared to the density ratio for which the float was designed. This type of float can be used under high pressure for flammable or toxic liquids as well as other liquids.

The control action of the system shown in Figure 6.5 is as follows. When the float moves upward, it transmits motion through the bellows *A*

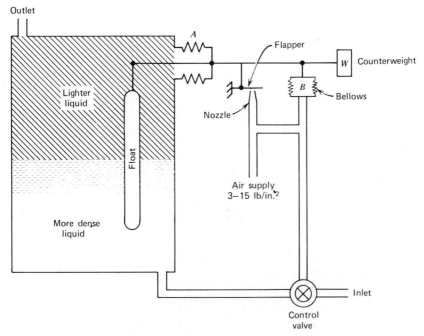

Figure 6.5 *Cross sectional arrangement for interface level measurements in open or closed vessels using a displacement float.*

and closes the flapper-to-nozzle distance, which causes the air supply to increase the pressure on the bellows *B*, which is balanced by the counter-weight *W*. When the pressure on bellows *B* increases, the bellows expands and transmits a downward motion to the control valve to close it. When the float moves downward as the level drops, it transmits motion through bellows *A* and opens the flapper-to-nozzle distance, which lowers the pressure on bellows *B* and allows the control valve to open and admit more liquid to fill the tank.

Magnetic ball float gages of the general type shown in Figure 6.6 can be used for level measurements where it is necessary to isolate the float from the material being measured. By using a nonmagnetic dip tube that can withstand the temperature, pressure, and other application conditions, the system can be completely sealed without packing glands or other seals through which rotating shafts must operate.

The two magnets are of opposite magnetic poles, so that they are attracted to each other. As the outer magnet moves up and down the outside of the dip tube, the inner magnet moves in the same direction because of the attraction to the outer magnet. The accuracy of the meas-

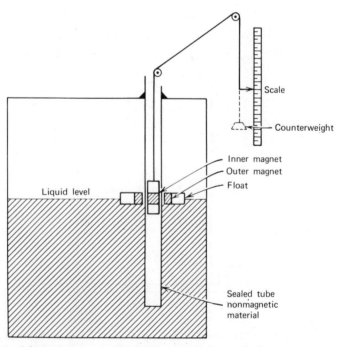

Figure 6.6 *Schematic cross sectional arrangement of a magnetic ball float liquid level gage.*

urement depends on the strength of the flux linkage and the friction in the system. With low friction and high flux linkage units, an accuracy of $\frac{1}{8}$ in. or less can be obtained over ranges from inches to feet.

Another variation for on-off control of relatively small level variations of a few inches is to attach a magnet to a float arm and have the magnet operate a switch. When the magnet is raised by the ball, it pulls the switch into the off position and when the float lowers the magnet below the switching position it releases the switch to the on position. This type of switch is mounted in a vertical position and can be used on any system for which the float and housing can be constructed. There is a minimum of moving parts and the system can be completely sealed.

In both types of magnetic level measuring systems the float is free floating, and anything that restricts this action or causes the float to change weight produces an error in the level measurement.

Cage float operated level measuring gages are a variation of the ball float gage in which the float is located on the inside of a closed unit attached to the vessel in which the level is to be measured or controlled. The float action is transmitted through a packed gland, and the attached unit operates on the same principle as the sight glass for equalization of the liquid level between the main vessel and the attached cage in which the ball floats. The unit is attached through valves and flanges, so that the valves can be closed and the unit removed for maintenance and repair without disturbing the integrity of the main vessel. This is shown schematically in Figure 6.7. This is one advantage over the conventional ball float mechanism which is placed inside the vessel.

Cage float gages can be used to operate pilots or relays for transmitting level information or the control of valves or pumps to control the level of the liquid automatically.

6.2 PRESSURE DROP SYSTEMS

In the pressure measurement of liquid level the height of the level is determined by two main factors. These are the pressure and the density of the liquid. This relationship can be expressed as

$$h = \frac{P}{\sigma} \tag{6.3}$$

where h = height
p = pressure
σ = density

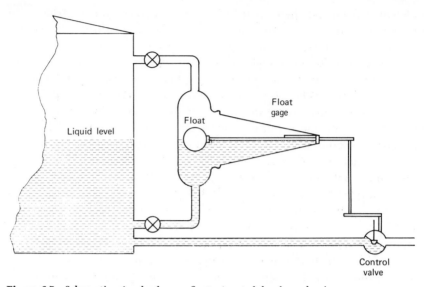

Figure 6.7 *Schematic of valved cage float operated level mechanism.*

The density can also be expressed as σw SG, where w is the density of water at 60°F, and SG is the specific gravity. When the pressure measurement is made in pounds per square inch and the density in pounds per cubic inch, the height of level is measured in inches.

A simple arrangement for the measurement of liquid level using a bubbler system is shown in Figure 6.8a. The pipe through which the air passes should be long enough to reach a predetermined system zero level. At the zero level there is no back pressure, so there is a free flow of air. For each inch of level added pressure is developed. The pressure developed is dependent on both the height of the liquid and its density. This means that there will be an indicated level change for either a change in level or density. An increase in density is indicated as an increase in level, and a decrease in density is indicated as a decrease in level. Any change in density of such a system, such as the settling out of particles in the base or a rise in the temperature of the volume, produces an error in the level reading. Figure 6.8b shows an all-pneumatic purge liquid level control system.

Manometer liquid level indicators can be used on either open or closed vessels. A well type manometer is used as shown in Figure 6.9a for open vessels and in Figure 6.9b for closed vessels. In *A* the tube leg of the

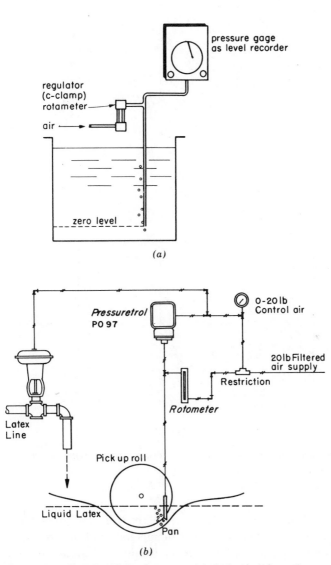

(a)

(b)

Figure 6.8 *Bubbler purge liquid level systems. (a) A simple schematic cross sectional arrangement for a bubbler liquid level measurement. (b) All pneumatic purge liquid level control system for maintaining a constant level in a latex pan maintains a constant flow. Courtesy Honeywell, Inc.*

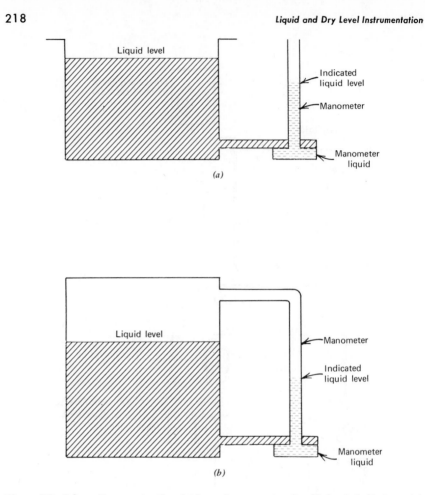

Figure 6.9 *Schematic cross sectional view of manometer liquid level indicators. (a) Open vessel. (b) Closed vessel.*

manometer is left open to the atmosphere. It is absolutely necessary that the liquid in the vessel and the manometer liquid do not mix. Mercury is commonly used as the manometer liquid. In practice the manometer inlet is mounted at least 2 in. above the bottom of the vessel, so that sediment does not collect in the connecting pipe or on the mercury well surface. The liquid in the vessel produces a pressure which in turn produces a certain difference in level between the two legs of the manometer, just as for the pressure measurements discussed in Chapter 3. The height of the liquid in the small arm of the well manometer varies inversely as the ratio of the specific gravities of the liquid in the vessels and the liquid in the manometer for a given height of liquid in the vessel. This can be expressed by the equation

$$h_m = H_v \frac{\sigma v}{\sigma m}$$

where h_m = manometer scale reading
   H_v = height of liquid in the vessel
   σv = density of vessel liquid
   σm = density of manometer liquid

The same equation is valid for both open and closed vessel applications. The method is limited to the length of manometer tube that can be tolerated, because the measurement of large variations in level necessitates the use of a tall manometer column. Also, the manometer must be mounted at the base of the vessel or below it. When it is mounted below the vessel, a correction must be added for the head or length of the connection between the manometer and the vessel.

The manometer can also be used as a pressure differential measuring unit by using a recording or indicating flowmeter manometer to measure the liquid level. This type of manometer is discussed in some detail in Chapter 7. A differential pressure is measured between the high and low pressure legs of the manometer. These manometers are usually constructed of a metal that can withstand high pressures, and use mercury as the liquid. Sealing chambers are used to isolate the manometer from the parent vessel. This chamber must be large enough to maintain a practically constant liquid level on the high pressure side to accommodate any movement of the mercury in the manometer body, as shown in Figure 6.10. In this way a constant pressure is maintained on the high pressure side, and the pressure differential is produced by any change in liquid level in the vessel by varying the pressure on the low pressure side of the manometer. The low pressure side of the manometer is connected to the vessel near the base where the zero liquid level has been established for the installation. The system is then calibrated using this level as the zero point for the differential pressure measurement, and the maximum differential as the high level point where the high level is determined by the vessel connection to the high pressure side of the manometer.

When a manometer is used for liquid level measurements, a manometer range tube must be designed for use with each particular application. Only a few applications are satisfied by the standard size range tubes used for the flow applications discussed in Chapter 7.

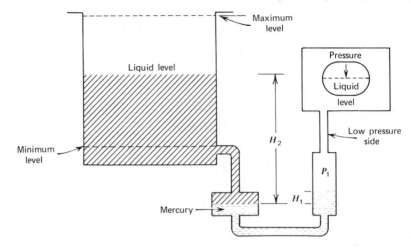

Figure 6.10 *Schematic representation of a recording or indicating differential pressure flowmeter manometer for measuring liquid level.*

The pressure differential manometer has one advantage in the measurement of liquid levels in closed vessels. The pressure above the liquid in the system has no effect on the instrument, because equal pressure is exerted on both the high and low legs of the manometer. When there are very high pressures and there is a vapor density above the liquid, the effect of the vapor density weight may require compensation.

Diaphragm liquid level indicators or recorders are used for measuring liquid levels of corrosive liquids or when it is undesirable to have gas bubbling through the liquid, which would cause contamination or crystallization. A type of diaphragm unit construction is shown in Figure 6.11. The diaphragm unit is placed at the minimum or so-called zero liquid level position. Air pressure from a compressor or other source is adjusted to balance the diaphragm position at the zero level. As the level increases, the diaphragm deflects and requires additional pressure to keep it from closing the pressure line to the system vent. The manometer system shown here is the same as those shown in Chapter 3 for pressure measurements. The fundamental principle is that a given head or height of liquid exerts a pressure in proportion to the height of the column and the density of the liquid. For example, each foot of height of water exerts about 0.5 lb/in². Since water is considered to have unit density, a fluid with a density of 2 exerts a pressure of 1.0 lb/(in²)(ft) of height as explained in Chapter 3. This simply means that the density and anything that changes the density of a fluid have to be considered when pressure

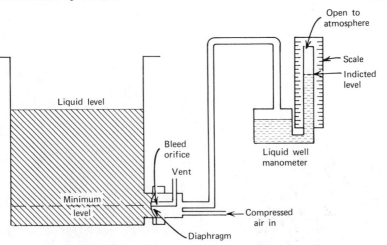

Figure 6.11 *Schematic representation of a diaphragm liquid level measuring system for open vessels.*

is used to measure the liquid level or the height of the liquid in a vessel.

Diaphragm gages can also be used for measuring the level of solids if the diaphragm is located on the side of the vessel or bin, as shown in Figure 6.12. The pressure of the solid causes the diaphragm to expand and operate a switch or other pressure sensing device. As the solid level is lowered, it permits the diaphragm to spring back and release the switch, or produce a signal to another type of sensing devices such as a strain gage.

6.3 ELECTRICAL LEVEL MEASURING INSTRUMENTS

In conducting liquids probes can be installed at the high and low level points. When the liquid level rises to the upper probe, an electrical circuit is completed and the signal can be used to actuate valves or pumps as well as lights or other types of signals. Such a system cannot be used where there is an explosive vapor over the liquid.

Where there is a need for a fairly continuous record of the level, more probes can be added, with separate signal ouputs for each probe. These systems use low voltage to eliminate danger to the operators and to prevent arcing at the probe contacts.

The advantages of these systems are that the signal can be transmitted to any desired location, and that they can be used on pressurized vessels without packing glands or shafts.

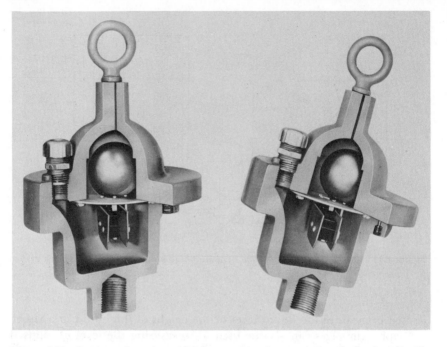

Figure 6.12 *Cutaway view of two diaphragm gages for measuring the level of liquids, slurries, or solids in vessels and bins. Courtesy Monitor Manufacturing, Inc.*

Some of the disadvantages, besides not being safe in explosive atmospheres, are that a large number of probes is needed for continuous indication, and corrosive liquids corrode the electrodes and the electrical charge promotes the corrosion. These systems can also present difficulties when there is saturated vapor above the liquid phase, or when a level measurement or control level can be upset by droplets of liquid forming a conducting path between the electrodes. The droplet situation can be avoided by offsetting the probes, but this makes the installation more difficult.

The probe units are simple to calibrate, since the spacing between the electrodes can be accurately measured and the indicated value checked for each measured value.

A single resistance or a series of resistance rods can be used in conducting liquids to give a continuous level measurement. As the liquid rises on the rod, a corresponding change in resistance occurs, which can be calibrated in terms of the liquid level. In such a system the rod must be in contact with the liquid at all times. Any changes in the conductivity of the liquid cause a corresponding error in the level measurement.

An electrical variation of the ball float system uses a suspended electrode to actuate a reversible motor. When the electrode is in contact with the liquid, the motor drives in one direction to pull the electrode out of the liquid, and when the contact with the liquid is broken, the motor reverses to lower the electrode to make contact. Essentially the motor hunts about the point of contact. A pen and recorder mechanism can be mounted on the drum used to raise and lower the electrode and produce a record of the liquid level. Such a system has the advantage of eliminating the weight of the cable, which has to be counterbalanced in ball float systems, and errors due to variations in the weight of the float caused by corrosion or sediment buildup.

Electrical level measuring units such as the unit and a typical installation shown in Figure 6.13 can be used for dry materials from powders to rather coarse material. The paddles on the indicating units rotate in the bin or vessel until the material reaches a high enough level to stall the paddle. When the paddles stall, the motor continues to rotate until the

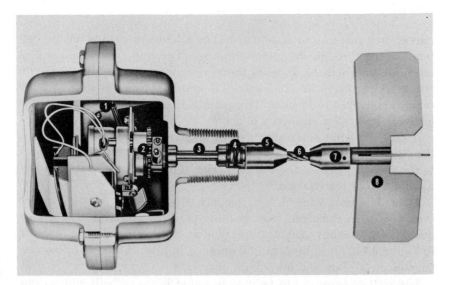

Figure 6.13 *Bin-O-Matic paddle solid level indicating and control mechanism. The numbers on the photograph show: (1) adjustable sensitivity; (2) gear drive with slip clutch; (3) stainless steel driveshaft; (4) stainless steel seal assembly with static O ring; (5 and 6) stainless steel flexible coupling; (7) Coupling threaded to $\frac{1}{4}$ in. paddle hub or shaft extension; and (8) stainless steel paddle. Courtesy Monitor Manufacturing, Inc.*

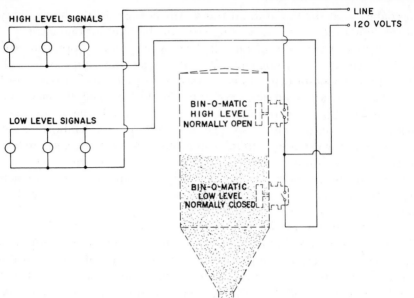

Figure 6.14 *Cross sectional view of a typical installation using paddle solid level controllers. Courtesy Monitor Manufacturing, Inc.*

microswitch in the unit is actuated. The microswitch shuts off the current flowing through the motor windings. The unit can remain stalled indefinitely. When the material drops away from the paddles, a special return spring reactivates the motor. The return spring actuating pressure is adjustable for 4 to 12 in.oz. This permits field adjustments to meet the material density for sensitivity of action.

A built-in signal delay allows small amounts of paddle motion when the unit is in the stalled position without initiating a control signal back to the control panel or the control device. This prevents false starts due to vibration or material shifts.

These bin level controls are compact, rugged, sensitive units which can be used in chemical, food, mining, plastics, feed, and ceramics industrial applications. A typical installation cross sectional view is shown in Figure 6.14. These units can be installed in bins, chutes, hoppers, concrete or metal silos, and other bulk material storage containers.

The control features can be used to sound horns or bells and operate indicating lights or other alarm devices. They also can start and stop conveyors, feeders, or other process functions through interlocks or relays. As the installation of Figure 6.14 indicates, these units are not continuous level indicating or control devices, but can be arranged to indicate high, intermediate, or low material levels.

6.4   ELECTRO-ELECTRONIC LEVEL MEASURING INSTRUMENTS

Electronic detectors and control devices include conductance, capacitance, eddy current, ultrasonic, and radiofrequency types with normally use an electrical control or indicating and/or recording mechanism.

Conductance Type. The conductance level measuring instruments shown in Figures 6.15 and 6.16 are a variation of the rod resistance system discussed in Section 6.3. Special units of this type can be used in services in which explosive vapors may be present. The conductance actuated liquid level controls shown are relatively inexpensive, versatile, reliable, and require minimal maintenance. These units have no moving parts in the liquid, are easy to install, and are adaptable to both large and small vessels. No adjustments are required, levels can be controlled

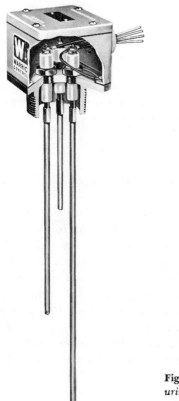

Figure 6.15 *Rod conductance actuated level measuring devices. Series 3E fitting with series 3R electrodes. Courtesy Charles F. Warrick Co.*

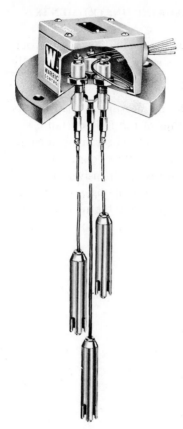

Figure 6.16 *Wire suspended electrode conductance actuated level measuring devices. Series 3F fitting with 3Z1B adapters, 3Z1A suspension wire, and series 3W electrodes. Courtesy Charles F. Warrick Co.*

to close limits, and they are unaffected by pressure, temperature, or corrosive liquids within certain limits.

Conductance actuated level controls can be used wherever it is desired to energize and/or deenergize automatically a signal or control device in response to an increase and/or decrease in the level of any electrically conductive solution, slurry, semisolid, or granular solid material. While this use sounds unlimited, it does not include organic liquids and animal, mineral, or vegetable oils. All aqueous solutions are not applicable, although the majority are. Conductance actuated controls can be used in dilute and concentrated acids and alkalies, alcohols, beer, blood, brine, fruit and vegetable juices, moist whole fruits and vegetables, milk concentrated sugar solutions, ordinary water, carbonated water, distilled water, and demineralized water.

Conductance actuated level controls must be chosen for the proper sensitivity to accommodate the specific liquids being measured. This value is measured in ohm centimeters. (derived from the resistance of the liquid in ohms and the distance over which the level is measured in centimeters). The resistance of the water is obtained by measuring its resistance with a conductivity meter whose output is in ohms. Values of 100,000 Ω·cm or less are considered medium sensitivity, and values ranging from 100,000 to 2,000,000 Ω·cm are considered high sensitivity. Each sensitivity range (medium and high) has a separate control requirement.

Other considerations in choosing conductance actuated level controls are (1) primary and secondary power requirements, (2) shunt capacitance tolerances, (3) mode of operation, (4) location of controls, and (5) temperature.

The primary power of ac supply line voltages can be 115, 230, 460, and 575 V nominal at 50 or 60 Hz. Power is 4 W with short circuited electrodes or 15 VA with a short circuited electrode circuit.

Secondary power in the electrode circuit can be 25, 75, 150, 300, or 500 V ac rms nominal with an open circuited electrode circuit, or 6 VA with a short circuited electrode circuit. Alternating current must be used so that the electrodes do not become polarized.

For installations using conductance actuation units, the maximum allowable distributed capacitance placed across the measuring circuit, consisting of the control-to-electrode conductor and ground, is related to the secondary voltage such that the values not to be exceeded are 4.0 μF at 25 V, 0.4 μF at 75 V, 0.1 μF at 150 V, 0.025 μF at 300 V and 0.0075 μF at 500 V.

When thermoplastic insulated wires in dry conduit are used, the limiting control-to-electrode distances for differential level service applications are 75,000 ft at 25 V, 7500 ft at 75 V, 1750 ft at 150 V, 500 ft at 300 V, and 150 ft at 500 V. Double the distance is possible for single-level applications.

Industrial applications can be carried out at temperatures ranging from -30 to 130°F ambient.

A typical conductance actuated liquid level control system is shown in Figure 6.17. This system consists of one or more electrodes, the electrode fitting, and the control. The control actuates and/or stops the pump to maintain the level of liquid in the tank between the two electrodes. In such an installation rod electrodes are used only when electrode lengths do not exceed 6 ft, as illustrated by Figure 6.15. Wire suspended elec-

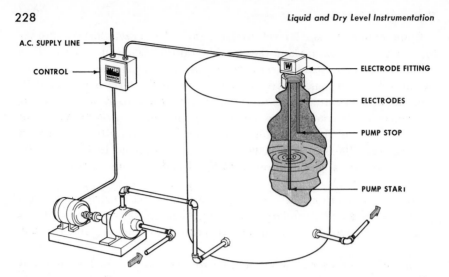

Figure 6.17 *Conductance actuated liquid level measuring and control system. Courtesy Charles F. Warrick Co.*

trodes as shown in Figure 6.16 are recommended for distances exceeding 6 ft. Wire suspended electrodes are adequate for cold and warm water and noncorrosive liquids.

Capacitance Type. Capacitance gages are available as either probe models or continuous level models. These gages can be used to indicate the level of alkaline materials, acids, chemicals, foods, fuels, grains, granulated solids, hydraulic fluids, oils, peroxides, powders, slurries, steam, and water. These gages can be used in environments having temperature ranges from −460 to 800°F and at pressures up to 6000 lb/in.² They are also built to withstand vibration and can be used on engine mounts.

The capacitance probe unit uses a glass or plastic covered probe and can be employed for high or high and low level detection. The unit detects level or interface level changes in both conducting and nonconducting liquids by capacitance changes detected by an oscillator and electronic circuitry. The capacitance change causes a corresponding change in the oscillator frequency to produce a signal through the electronic circuitry, which indicates the high or low level condition and either actuates or deactuates the appropriate control device. Such a unit is shown in Figure 6.18*a*.

One type of continuous level measuring capacitance gage is shown in Figure 6.18*b*. This system uses an insulated metal electrode inserted along the wall of the container as one side of a parallel plate capacitor.

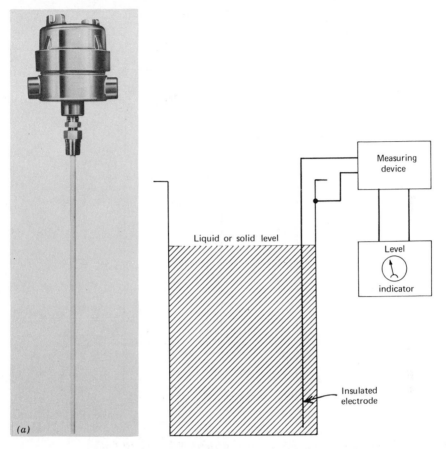

(a)

Measuring device

Liquid or solid level

Level indicator

Insulated electrode

Figure 6.18 *Capacitance level measuring instruments. (a) Level-Tek model 303 capacitance actuated level detection and control system. Courtesy Robertshaw Controls Company, Industrial Instrumentation Division. (b) Cross section of a continuous level capacitance measuring system.*

The container wall is the other plate of the capacitor. Variations in the dielectric characteristics of the material between the electrode and the wall, as the air material interface moves up and down, are measured on a capacitance bridge as the material level. This system can measure either liquids or dry materials, and can provide a signal output for remote indication and/or recording as well as a control device.

Ultrasonic Type. The ultrasonic level gage is illustrated schematically in Figure 6.19. The rods used are notched to produce a predetermined

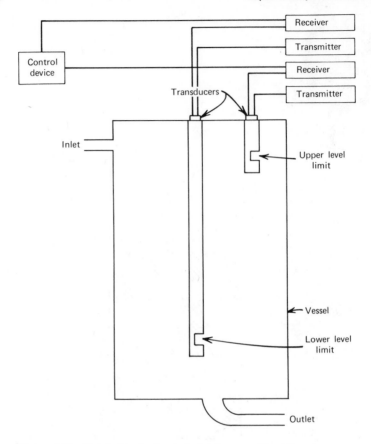

Figure 6.19 *Notched rod ultrasonic level measuring system.*

ultrasonic signal when clear of the material whose level is being measured and controlled. These rods can be positioned at various angles to produce the desired controlling action for the particular application.

Ultrasonic level gages require the use of an ultrasonic generator and receiver. The principle of operation is discussed in detail in Chapter 11 and is based on a change in the time required for a signal to reach the receiver and the attenuation of the transmitted signal with respect to the signal received. The signals are transmitted by an ultrasonic transducer into the notched rod and are detected by a receiving transducer attached to the same rod.

Ultrasonic level measuring systems can be used for both liquid and granular solids small enough to fill the notches and cause a change in the transmitted signal to the signal received.

6.5   NUCLEAR LEVEL MEASURING INSTRUMENTS

The theory of operation of a nuclear level measurement system is based on absorption of an emitted ray of energy by the material, slurry, liquid, or solid whose level is being measured. Here we discuss the use of gamma radiation sources and detectors. Other nuclear units are discussed in Chapter 10.

Here the level measurement is based on the principle of gamma radiation absorption whereby the material being measured absorbs or attenuates radiation from the source, which is sensed by the detector unit. As the material level in the vessel changes, a corresponding change occurs in the radiation detected. The output from the detector is conditioned so that it can be detected by means of a Geiger-Mueller tube output system. These detectors are discussed fully in Chapter 10. The number of pulses counted during a given time period is proportional to the level of radiation at the surface of the detector.

The detector unit shown in Figure 6.20 is electrically connected to a control unit, such as the one shown in Figure 6.21, by means of a single coaxial cable. The control unit converts the pulse signal from the detector to a high level dc current signal. The control shown generates either a 1 to 5 mA or a 4 to 20 mA signal for indication and/or control purposes. The control unit shown in Figure 6.21 is for panel mounting, and the control unit shown in Figure 6.22 is for surface mounting.

These control units are solid state devices which incorporate calibration adjustments for zero and span that are noninteracting for ease in initial calibration, an adjustment for response time which may be field adjusted for the specific application, and a calibrated compensation for source decay since the sources used have a finite half life. This means that the radiation emitted by the source diminishes with age, and the half life occurs when the source produces only one-half the original radiation. The half life time period is a function of the radioactive material used in the source.

It should also be noted that the radioactive source is supplied in a suitable holder and is adequately shielded to ensure that there are no personel hazards in handling, during the installation, or in taking measurements thereafter. The size and type of source depends on the particular application.

The unit shown is a continuous level measuring instrument available in 10 standard sizes for measurement spans of 9 to 90 in. The detector unit consists of one or more Geiger-Mueller tubes, depending on the measurement span, housed in a weathertight explosionproof enclosure. Nothing comes in contact with the process being measured, and applica-

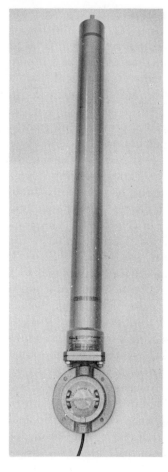

Figure 6.20 *Model 171 continuous level measuring detector. Courtesy Robertshaw Controls Company, Industrial Instrumentation Division.*

tions can involve high temperature, corrosive, sticky, or abrasive materials. The control unit can be remotely located up to 5000 ft from the detector and contains all the operating and calibration adjustments.

6.6 LEVEL MEASUREMENT BY WEIGHING

Both liquid and solid levels can be measured very accurately by weighing, providing the processing vessel does not place a practical limit on the application of the method. This method is commonly used where a continuous flow is carried by a conveyor, and controls the flow on to the conveyor by regulating a level in the feed hopper. Another applica-

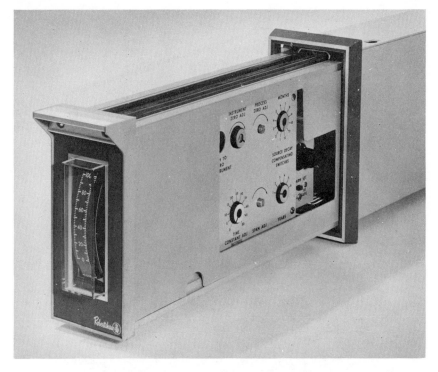

Figure 6.21 *Model 171 panel mounted control unit. Courtesy Robertshaw Controls Co., Industrial Instrumentation Division.*

tion is where material is fed into a dumping scoop or bucket and when the desired weight or level in the bucket is reached the flow is stopped and the bucket is dumped. It does not refill until the dump gage is closed to complete the cycle for maintaining the proper level in the receiving bin or hopper. This technique is used for maintaining the proper fuel level in pulverizers and stokers.

6.7 SUMMARY

Level measurements of liquids and dry materials can be made by mechanical methods, pressure drops, electrically, electroelectronically, nuclear methods, weighting techniques, and other special procedures for local or remote control. The desired level conditions can be maintained manually or automatically. The choice of methods and control depends upon

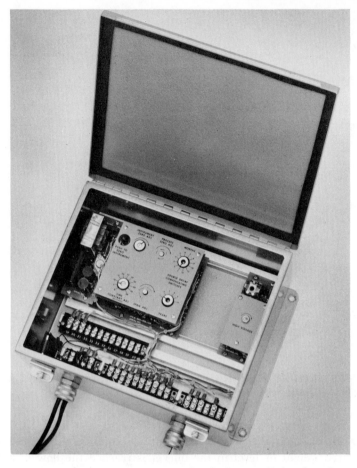

Figure 6.22 *Model 171 surface mount control unit. Courtesy Robertshaw Controls Co., Industrial Instrumentation Division.*

the application, accuracy of level to be maintained, and the economics of the situation.

The techniques used in the pressure drop method also has applications in the measurement of flow to be discussed in Chapter 7, ultrasonic techniques for nondestructive testing described in Chapter 11, and the use of radioactive sources and detectors for level measurements in Chapter 10.

A comparison of the various level measuring systems is shown in Table 6.1.

Table 6.1 Comparison of Level Measuring Systems

System	Components	Application	Operating Principles	Range	Advantages	Disadvantages
Stick or rod	Graduated wood or metal rod	Open vessels, closed tanks	Measure depth or level from wetted surface	Up to practical length of rod to be handled	Simple and inexpensive	Vessel must be opened for the measurement; operator exposed to fumes, and so on.
Sight glass	Glass tube, valves, sealants	Open or closed vessels	Level in sight glass matches level in vessel	Up to practical length of sight glass	Simple	Limited pressure to glass rupture values; noncorrosive and non-coloring liquids
Float	Float or ball, chain or tape, counterweight, indicator	Open vessels	Byoyant float or ball rises and falls with the level of liquid being measured	Minimum 0–6 in.; normal 0–3 ft; maximum 12 ft	Relatively simple	Limited height; difficult to clean and maintain
Float	Float, torque arm, indicator, seals, cage, air supply	Open or closed vessels	Byoyant float rises and falls with level of liquid; float rotates a torque arm	To apporoximately 60° of torque arm movement and twice the length of torque arm; minimum 0–6 in.; maximum 0–30 in.	Relatively simple; works under pressure and vacuum; can be cleaned without process shutdown	Limited to 60° of movement for good sensitivity and accuracy; measurement height limited to twice the length of torque arm
Magnetic float	Float, magnets, float tube	Open or closed vessels	Buoyant float carries a magnet which can operate switches or follow-on magnets	Unlimited	No seals on moving parts; continuous level measurement	Measurement lag; dependent on magnetic coupling strength
Pressure gage	Pressure gage, standard pressure components	Open vessels	Measures static head value	Minimum 0.1 in. water; maximum unlimited	Simple; inexpensive	Limited distance between instruments and vessels

(Continued)

Table 6.1 Comparison of Level Measuring Systems (continued)

System	Components	Application	Operating Principles	Range	Advantages	Disadvantages
Bubbler system	Pressure gage, air supply, regulator, rotameter	Open or closed vessels	Purge fluid or gas keeps measured fluid out of system; transmits liquid head pressure	Minimum 0.1 in. water; maximum unlimited	Ideal for corrosive fluids, viscous liquids, and slurries; pressure elements may be at any level	Limited by range of instrument and purge pressure; high maintenance
Mercury manometer	U-tube meter body or well meter body, indicator, recorder, mechanical or electrical transmission	Open or closed vessels	Pressure applied to U-tube or well to produce a level equivalent to level in vessel	From $2\frac{1}{2}$ to 700 in. water full scale	Interchangeable meters for flow or specific gravity service; will withstand 5000 lb/in^2 static pressure	Must have clean noncorrosive fluid in meter body
Diaphragm box	Chamber-diaphragm, diaphragm, air supply, well manometer	Open vessels	Static pressure transmitted through diaphragm.	Minimum range 1 in. water; maximum 50 ft water	Diaphragm seals measured fluid from manometer liquid	Needs air purge for corrosive or viscous fluids. Pneumatic runs should not exceed 250 ft and preferably be less than 30 ft
Electrical conduction	High level rod, low level rod, electrical circuit, switches and valves, pumps	Open or closed vessels 100,000 Ω·m 200,000 Ω·m, 0 to 100,000 Ω·m	Conducting liquid completes an electrical circuit and carries current	$\frac{1}{8}$ in. to limit of conducting liquid High level, medium level	Wide range; wide span; simple adjustment Minimal maintenance relatively inexpensive	Needs a conducting non-volatile liquid to conduct current Wire suspension needed for levels over 6 ft; separate controls per range
Capacitive gage	Probe, capacitance bridge, power, oscillator, electronic circuitry	Open or closed vessels; liquids or solids	Operates as a parallel plate capacitor	Limited only by practical heights dictated by bridge characteristics	Continuous or high and low level measurements	Variation in dielectric of materials is measured as a level change

Ultrasonic	Notched rods, ultrasonic generator, receiver, readout unit, transducer	Solids, granulars, liquids, slurries, grains, open and closed vessels	Ultrasonic signal attenuation in time response	Limited by practical length of rods	Adaptable to large vessels; continuous measurements; no seals to work through or maintain	Granular material must be small enough to fill notches; requires more equipment than other types
Resistance rod	Resistance rods, bridge circuit, readout meter	Open or closed vessels	Resistance is varied as the liquid rises on the rod	$\frac{1}{8}$ in. to limit of resistance of conducting liquid	Wide range; wide span, simple maintenance	Needs a conducting liquid to carry a current
Reversible motor	Reversible motor, electrodes, indicator	Open vessels	Motor drives electrode to liquid level; motor reverses when contact with surface is made	$\frac{1}{8}$ in. to 200 ft	Wide range; wide span, simple adjustment	When electrodes are too close the system will hunt between up and down; needs a conducting medium
Paddle wheel	Paddle wheel, power source, indicator	Large and small open or closed vessels, bins, and so on, for dry solid or granular materials	Material covers paddles and stalls them under power	Minimum limited to paddle width; maximum unlimited	Useful for powders to coarse materials in chemical, plastics, mining, and feeds; not too sensitive to vibration	Separate unit needed for each level of measurement
Conductance (rod and wire types)	Electrodes, primary power, secondary power	Large and small open or closed vessels	Measures the electrical conductance in ohm centimeter units in two ranges	Minimum 0.1 in.; maximum of 75,000 ft	Versatile; reliable; no moving parts; explosive vapor service	Not adaptable to organic liquids and animal, mineral, or vegetable oils
Nuclear	Source, detector, readout unit, power supply	Open and closed vessels; corrosive or volatile materials	Beta or gamma attenuation; gamma backscattering.	Beta $0-\frac{1}{4}$ in. aluminum, gamma 0–4 in. lead, gamma backscatter 0.1–3.0 in.	No physical contact with materials; no seals to work through	Source is radioactive. Areas must be posted. Personnel must be monitored
Weighing	Scales, balances, containers, diaphragms, air pressure, fixed volume	Conveyor feed systems; feed hoppers, pulverizers, and stokers	Material is weighed for volume and/or measured in a weight controlled bucket or dimensioned container	Limited by the processing vessel and supporting equipment	Accurate; good for moving systems; good for solids, granules, and powders	Weight of total process vessel must be supported and weighed

Review Questions

6.1   Give three good reasons why all level measurements cannot be made with a measuring stick.

6.2   Why is it important to keep the bore of a sight glass as large and as clean as possible?

6.3   What is the most easily controlled volume type of level measurement and control?

6.4   What has to be considered besides the change in level in a sight glass system for control of a level?

6.5   Why is the density of a float important in a float measuring system? What limits the size of the float?

6.6   Explain why there are certain applications for which a ball float system is a very poor choice.

6.7   In a magnetic level indicator why would you use opposing magnetic properties? Could you use attracting properties? If so, how? If not, why?

6.8   What type of level measuring system applications requires a bubbler? Could this type of system be used for interface measurements on immiscible liquids? Why or why not?

6.9   Upon what factors does a manometer reading depend when the manometer is used to measure liquid level? Where should the manometer be located for best results?

6.10 What purpose does a diaphragm serve in liquid level measurement systems? What are the two main causes of error when diaphragms are used to measure liquid level?

6.11 Why are resistive element probes sometimes considered dangerous for liquid level measurements? Where are they most frequently used? What advantages do they have in closed vessel applications? What are their disadvantages?

6.12 How does the paddle solid level indicator work to show a given level? How does one of the units become active when a bin empties?

6.13 Why should one consider capacitance level sensors? What do they have to offer in range and accuracy?

6.14 Why is mass weighing not more popular as a means of level measurement? Where does this method find desirable applications?

Problems

6.1   If a float arm is 2 ft long and is free to move 60°, what is the level range in terms of the arm length for a ball float?

6.2   What is the density of a liquid if the height of 30 in. represents a pressure of 7.3 lb/in^2?

6.3   What pressure would you read for a water level 48 in. high?

Bibliography

Barksdale Valves, P.S. Division, *Barksdale Level Control Switches,* Bulletin 621205, Los Angeles.

Considine, D. M., *Process Instruments and Controls Handbook,* McGraw-Hill, New York, 1957.

Controlotron Corporation, *Liquilite, Liquid Level, Sensing, Indication and Control,* 1001; 1002. Farmingdale, N.Y.

Hackman, J. R., How to Measure Liquid Interface Levels, *ISA Journal,* Vol. 4, No. 12, December 1957, pp. 554–557.

Honeywell, Inc., *Fundamentals of Instrumentation for the Industries,* G-00003-00 10M, Ft. Washington, Pa., January 1958.

Honeywell, Inc., *Liquid Level Controllers-Versa-Tran,* Technical Bulletin 95-2719, Fort Washington, Pa.

Honeywell, Inc., Industrial Products Group, *Measuring Liquid Level, Instrumentation,* Vol. 18, No. L, Ft. Washington, Pa.

Jo-Bell Products, Inc., *Liquid Level Safety Controls,* 3b, JO. Oak Lawn, Ill.

King Engineering Corporation. *Liquid Measurement,* Catalog 1010-A, Ann Arbor, Mich. March 1965.

Olive, Theodore R., and Danatos, Steven, Guide to Process Instrument Elements, *Chemical Engineering,* June 1957.

Soisson, H. E., *Electronic Measuring Instruments,* McGraw-Hill, New York, 1961.

CHAPTER 7

Flow Instrumentation and Measurement

Flow of material in a process or system can be measured by a variety of methods depending on the material, the volume, the accuracy needed, and the control required. The principal methods used in industrial applications include head flowmeters, area flowmeters, electromagnetic flowmeters, mass flowmeters, positive displacement flowmeters, and open channel flowmeters.

7.1 HEAD FLOWMETERS

Head flowmeters are the most common type used to measure the flow of gases, liquids, and slurries. These meters measure a pressure differential across a restriction to the flow. The pressure can be related to the force per unit area, and the head becomes a function of the flow velocity and the density of the flowing medium. All the applicable equations can be derived from Bernoulli's theorem which states that

$$\frac{\rho v^2}{2g} + p = \text{a constant} \tag{7.1}$$

where ρ = density of the liquid
 v = velocity of the moving liquid
 p = pressure
 g = gravitational constant

A pressure differential is created when a restriction is placed in a pipe, so that for the case of incompressible fluids the head h is defined as

$$h = \frac{p_1 - p_2}{\gamma} = \frac{v_2^2 - v_1^2}{2g} \tag{7.2}$$

where p_1 = pressure on the upstream side of the restriction

p_2 = pressure on the downstream side

γ = fluid specific weight

v_1 = average velocity on the upstream side

v_2 = velocity on the downstream side

This restriction to the flow, to create the pressure differential, is generated by a venturi tube, a flow nozzle, an orifice plate, or a pitot tube. The pitot tube is discussed, but it has limited industrial applications. Head flowmeters operate on the principle of conservation of energy by converting energy from one form to another. The venturi tube, flow nozzle, and orifice plate depend on the conversion of energy in the form of static pressure into kinetic energy of velocity. Pitot tubes depend on the conversion of kinetic energy to pressure at the impact tube.

Venturi Tube. A venturi tube is shown in Figure 7.1. This type of flow restriction is needed for slurries which are considered here as fluid flow containing solids in suspension. It may also be used for liquids when minimum restriction is necessary, and when the best accuracy for a head flowmeter is required.

The venturi tube is very expensive in comparison to the flow nozzle and the orifice plates to be discussed later.

The venturi tube is a flow measuring device which is normally inserted in a pipeline in much the same way as any other pipe section. The venturi consists of a main barrel section whose diameter is identical or very close to the internal diameter of the pipe connected to it. The main barrel leads into a cone of fixed angular convergence. The cone in turn connects to an accurately machined throat section with close design dimensions. The throat has a smaller internal diameter than the barrel. The throat then connects to an exit cone. The exit cone has a fixed angular divergence to match the exit portion to the internal diameter of the pipe used in the system. The design calculations for determining the

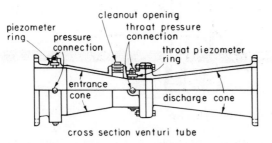

Figure 7.1 *A typical standard venturi tube in cross section. Courtesy Honeywell, Inc.*

fixed angles of convergence and divergence are beyond the scope of this book, but contribute significantly to the cost of a venturi tube.

The venturi barrel is equipped with piezometer connections in an annular pressure ring for averaging of the upstream pressure, and the high pressure meter connection is made at this point. A similar ring is designed for the throat where the low pressure connection is made. The differential pressure head is created by the difference in pressure between the barrel and the throat.

Insert Venturi Tube.   The insert venturi tube is a variation of the regular venturi tube. The insert venturi can be placed inside a high pressure pipe for high pressure applications at much less cost than that of designing a regular venturi for the high pressure application. The insert venturi tube is shown in Figure 7.2. It has all the desirable characteristics of the standard venturi tube for flow measurements.

The insert venturi tube is inserted into the pipeline and is held in place with a heavy ring which is an integral part of the tube, sometimes referred to as a dutchman. The low pressure connection is made to this portion which also contains the piezometer openings and the throat annular pressure ring. The high pressure connection is made through the high pressure pipe wall approximately one pipe diameter upstream from the entrance cone. As shown, the insert section contains the upstream entrance cone, the throat, and a recovery or exit cone. These parts re-

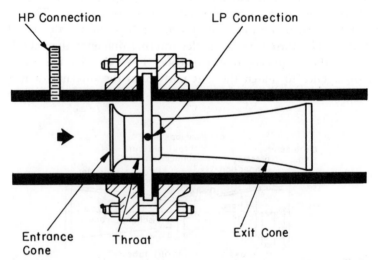

Figure 7.2   *An insert venturi tube in cross section. Courtesy Honeywell, Inc.*

quire very accurate and careful machining and are designed to give a minimum loss of pressure head.

Flow Nozzle. The flow nozzle is similar to the insert venturi tube, except that it does not have a recovery cone. One is shown in Figure 7.3. The flow nozzle is less expensive than the insert venturi tube and also has a lower head recovery. However, it has a higher head recovery than the orifice plate.

Orifice Plates. The orifice plate is the most widely used primary flow measuring device because of its simplicity, low manufacturing cost, and ease of installation. It also produces the highest head loss, but in the majority of flow measurements this loss is of less importance than the flow measurement. Well made and properly installed orifice plates give consistent readings and, if properly used with correct coefficients, can be as accurate as most other types of flowmeters.

Conventional orifice plates are similar to the one shown in Figure 7.4. This is a thin piece of metal with enough physical strength to prevent buckling under the pressure differentials it has to withstand in service. In conventional applications material thicknesses of $\frac{1}{16}$ in. are used for pipe sizes up to 4 in., $\frac{1}{8}$ in. thicknesses for pipe sizes up to 16 in., and $\frac{1}{4}$ in. thicknesses for pipe diameters over 16 in.

For consistent measuring results, the upstream face of the orifice plate must be flat and perpendicular to the axis of the pipe when installed. The upstream edge of the concentric hole must be square, sharp, and clean. The orifice edge thickness should not be greater than $\frac{1}{8}$ in. for openings greater than $\frac{1}{2}$ in. in diameter, and not greater than $\frac{1}{16}$ in. for openings smaller than $\frac{1}{2}$ in. in diameter. When heavier thicknesses are

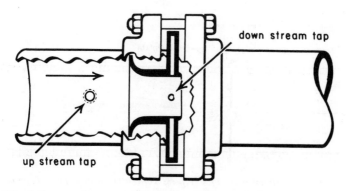

down stream tap

up stream tap

Figure 7.3 *A flow nozzle cross sectional view. Courtesy Honeywell, Inc.*

Figure 7.4 *A conventional orifice plate. Courtesy Penn Meter Co.*

required to withstand the pressures in the application, the wall thickness at the orifice opening can be reduced by beveling the downstream face at an angle of 45° or greater to obtain the proper thickness.

The concentric orifice plate has an especially high degree of predictable accuracy, primarily because the American Society of Mechanical Engineers (ASME) has accummulated extensive performance data covering a broad range of flow rates, pipe sizes, pressure differentials, and other factors related to its use.

A typical concentric orifice plate installation and the effluent flow contour are shown in Figure 7.5. Any fluid when passing through a square, sharp edge opening has the flow contour shown. The smallest diameter of the stream is known as the *vena contracta,* which is equivalent to the throat of the venturi tube or the flow nozzle.

The eccentric and segmental orifice plates shown in Figure 7.6 are two other types of orifice plates that are used when the measured fluid contains suspended material which may tend to build up back of a concentric plate. The tendency of such suspended matter to accumulate on the upstream side of concentric orifice plates leads to erratic and false read-

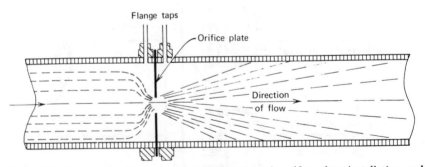

Figure 7.5 *Cross sectional view of a typical concentric orifice plate installation and the resultant flow diagram.*

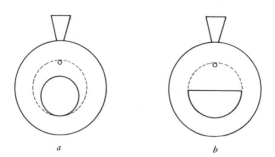

Figure 7.6 *Typical orifice plates. (a) Eccentric. (b) Segmented. Courtesy Penn Meter Co.*

ings. These orifice plates are usually installed so that the orifice is within 0.025 in. of the bottom of the pipe. This reduces clogging to a minimum. Calculations and flow factors for the concentric orifice do not apply to eccentric and segmental plates.

Orifice plates are usually mounted in flanges, as shown in Figure 7.5, and the flanges are bolted together. The flanges are either threaded to the pipe or welded to it, depending on the size of the pipe and the line pressure of the system. The upstream and downstream pressure taps are usually mounted in the flanges. The high pressure tap is on the upstream side, and the low pressure tap is on the downstream side. When extremely high line pressures are to be used, an orifice holding ring is used to carry the orifice plate. The holding ring is mounted between ring joint flanges. This ring gives additional physical strength to the orifice plate and to the flange seal.

When it is desirable to obtain the maximum differential pressure, use is made of vena contracta taps. The location of these taps depends on the pipe size and the ratio of the orifice to the pipe size. The distance is variable for different applications, so the downstream tap has to be drilled at the vena contracta and the upstream tap made at a point giving the true static pressure in the flow line. A typical location is shown in Figure 7.7. Tables are available from orifice plate manufacturers, which give the best location based on the orifice ratio and the pipe size. A few examples are shown in Table 7.1.

Pipe taps are also used in orifice plate installations, especially where it is desirable to use low differential manometers. One application is on large natural gas lines where it is more economical. The upstream pipe tap is usually installed $2\frac{1}{2}$ pipe diameters above the upper face of the orifice plate, and the downstream pipe tap is placed 8 pipe diameters below the downstream face of the orifice plate. Under these conditions

Table 7.1 Comparison of Flow Detectors

Flow Detector	Service	Relative Cost	Flow Volume	Head Loss	Comparative Accuracy
Venturi tube	Liquids and gases	High	Small to large	Very low	High
Insert venturi	Liquids and gases	Medium	Small to large	Low	Medium
Flow nossle	Liquids and gases	Medium	Small to large	Medium	Medium
Orifice plate	Liquids and gases	Low	Small to medium	Medium to high	Medium to low
Pitot tube	Liquids and gases	Low	Small to medium	Low to medium	Medium
Annubar element	Liquids and gases	Low	Small to large	Low to medium	Medium
Mechanical flowmeter	Liquids and gases	Medium	Medium	Medium	Medium
Electrical flowmeter	Liquids and gases	High	Medium	Medium	Medium
Transducer	Liquids and gases	High	Low to high	Low	High
Area flowmeter	Liquids and gases	Medium	Low to high	Medium	Medium
Electromagnetic	Conducting liquids	High	Medium	Low	Medium
Mass flowmeter	Liquids and gases	High	Medium	Low	Medium to high
Axial flow mass flowmeter	Liquids	Medium	Small to medium	Medium	Medium
Turbine flowmeter	Liquids and gases	Medium	Small to large	Medium to high	Medium to high
Piston pump	Liquids and gases	Low	Low	High	High
Nutating pump	Liquids and gases	Medium	Low	High	High
Rotary pump	Liquids	Medium	Medium	High	High
Diaphragm pump	Liquids and gases	Medium	Low	High	Medium
Peristaltic pump	Liquids	Medium	Low	High	Medium
Weir and Flume	Liquids	Low	Large	Medium to high	Low

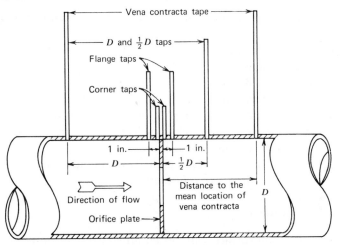

Figure 7.7 *Cross section view of vena contracta and other types of taps. Courtesy Fischer and Porter Co.*

only the friction loss is measured. This permits the measurement of a high flow with a low differential pressure meter. However, the majority of new installations make use of flange taps, since they meet the majority of applications without additional drilling or tapping. When threaded flanges are used, care must be taken to make sure the threads do not extent beyond the tap openings. In cases in which this occurs, it is necessary to drill the hole through the threaded pipe portion.

Orifice plates require careful installation to make certain that they are centered in the pipe. During installation the orifice plate is usually allowed to rest on the bolts, since this is a convenient way of holding it. However, the bolt holes may be $\frac{1}{8}$ in. oversize and meet standard specifications, so the orifice plate may be $\frac{1}{8}$ in. off center if not carefully checked during installation. This offset will cause an error in flow measurements. The gaskets used to make the seals on both sides of the orifice plate should have an oversize hole, so that they cannot slip and obstruct the orifice opening.

Orifice Sizing. The basic flowmeter equation shows that the orifice is a square root measuring device, and one equation used for relating flow rate and pressure drop across an orifice is:

$$W = KaYF_a \sqrt{2g_c (\Delta P)\rho} \qquad (7.3)$$

where W = gravimetric flow rate in pounds per second or grams per second

   a = cross sectional area of the orifice in square feet or square centimeters

   K = flow coefficient

   Y = gas expansion factor

   F_a = area factor for thermal expansion of an orifice

   g_c = a unit conversion factor in feet per square second or centimeters per square second

   ΔP = measured pressure differential in pounds per square foot or grams per square centimeter

   ρ = fluid density in pounds per cubic foot or grams per cubic centimeter

The area a of the orifice can be eliminated from equation 7.3 by equating it to the ratio β of the pipe diameter to the product of the orifice diameter and the cross sectional area A of the inside of the pipe, measured in square feet or square centimeters, to give equation 7.4:

$$W = K\beta^2 AYF_a \sqrt{2g_c \, \Delta P \, \rho} \qquad (7.4)$$

To obtain the flow rate in volumetric units of cubic feet per second the equation becomes

$$q = K\beta^2 AYF_a \sqrt{2g_c \frac{\Delta P}{\rho}} \qquad (7.5)$$

where q is the volumetric flow rate in cubic feet per second or cubic centimeters per second

In addition to the factors shown by the three basic equations, there are other factors such as the viscosity μ of the liquid, the molecular weight M of the material, the measured differential h_w in inches of water, the pipe Reynolds number R_D, the specific volume, v, the absolute pressure p, and the standard gas density ρ_{std} that are helpful in obtaining a complete analysis to measure an accurate flow rate using orifice plates.

One procedure for finding the orifice size is to solve for $K\beta^2$ in the basic equations 7.4 and 7.5:

$$K\beta^2 = \frac{W}{AYF_a \sqrt{2g_c \, (\Delta P)\rho}} \qquad (7.6)$$

$$K\beta^2 = \frac{q}{AYF_a \sqrt{2g_c \dfrac{\Delta P}{\rho}}} \qquad (7.7)$$

By using consistent units for the flow in pounds per hour, dimensions can be substituted in equation 7.6 so that it becomes

$$K\beta^2 = \frac{\text{lb/h}}{2837F_a\,(D)^2\,\sqrt{h(G)}} \qquad (7.8)$$

In a similar fashion, by the use of consistent units, equation 7.7 can be written for volumetric liquid flow in gallons per minute:

$$K\beta^2 = \frac{\text{gal/min}}{5.674F_a(D)^2}\sqrt{\frac{G}{h_w}} \qquad (7.9)$$

A new dimension D has been used in the last two equations, and it is the inside pipe diameter in inches. The dimension G is the specific gravity of gas compared to air or of liquid compared to water.

Evaluation of the $K\beta^2$ factor and establishment of the pipe Reynolds number R_D permits the use of curves such as the ones shown in Figure 7.8 for the evaluation of β for Reynolds numbers above 7000. The curves

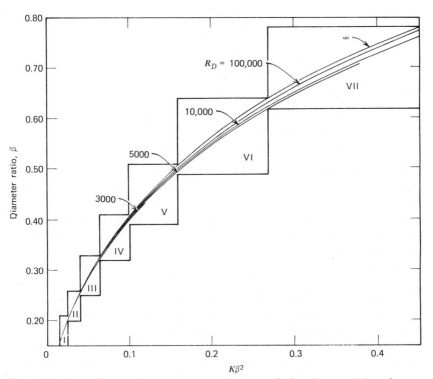

Figure 7.8 *Reynolds number curve composite from which values of $K\beta^2$ and β are found. This curve can be expanded into seven curves, as indicated on the graph. Courtesy Fischer and Porter Co.*

shown can be expanded into seven or more curves to provide higher accuracy in reading. For a given value of $K\beta^2$ and a given Reynolds number, the value of β can be read from the graph. The Reynolds number can be solved for in either gravimetric or volumetric units using the relationships

$$R_D = \frac{4W}{\pi D'\mu} = \frac{4q\rho}{\pi D'\mu} = \frac{6.316 \text{ (lb/h)}}{D' \text{ (cP)}} = \frac{3160 \text{ (gal/min) } (G)}{D' \text{ (cP)}}$$

where D' is in feet, fluid viscosity μ is in absolute units of centipoise, which is equivalent to mass in pounds per foot second or grams per centimeter second and $\pi = 3.1416$. After the value of β has been extracted from the curve, this value can be divided by the pipe diameter D to obtain the size of the orifice d.

Equations have also been developed for obtaining the value of $K\beta^2$ for perfect gases in standard volumetric units where either the molecular weight or the density of the gas is known.

$$K\beta^2 = \frac{\text{SCFM}}{707 Y F_a D^2} \sqrt{\frac{M \text{ (°F + 460)}}{h_w \rho}} \tag{7.10}$$

$$K\beta^2 = \frac{\text{SCFM}}{35.95 Y F_a D^2} \sqrt{\frac{\rho_{\text{std}} \text{ (°F) + 460}}{h_w \rho}} \tag{7.11}$$

where M = molecular weight in pounds per pound mole
ρ_{std} = standard density in pounds per cubic foot
SCFM = standard cubic feet per minute
(°F + 460) = absolute temperature in degrees Fahrenheit

While sizing of an orifice is straightforward, and care must be exercised in the choice of units, it is necessary to have the proper tables available, such as compressibility factors, Reynolds curves, and thermal expansion factors for the orifice plate, and to have an accurate sensing unit for industrial applications. It is not unusual for an industrial user to depend on the orifice manufacturer to furnish the proper orifice from the specifications the user supplies for his application.

Pitot Tube.   As stated earlier, the pitot tube has little importance as an industrial flow sensor. However, it is a very effective laboratory tool and can be used for spot checking flows. A typical basic type is shown in Figure 7.9a. The pitot tube has two pressure passages. One faces into the flowing fluid and is at the end of the tube in the center of the pipe. Here it intercepts a small portion of the flow and reacts to the total pressure of the liquid. The other passage faces perpendicular to the axis of flow and reacts to the static pressure of the liquid. The measure of the velocity is the difference between the two pressures.

From the basic physics of the system, the energy balance can be expressed as

$$\frac{V_1^2}{2g} + \frac{P_1}{\gamma} = \frac{P_2}{\gamma} \qquad (7.12)$$

where P_1 = static pressure in the flowing liquid in absolute units
P_2 = impact pressure in absolute units when $V_2 = 0$
γ = specific weight of the fluid
V_1 = velocity of the flowing fluid at a point in line with the impact opening on the pitot tube
g = gravitational force
V_2 = velocity at $P_2 = 0$

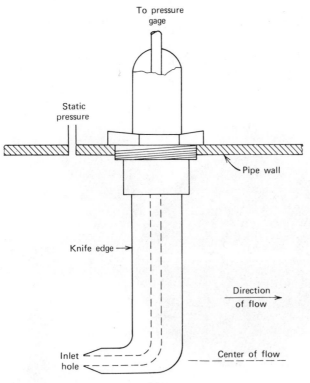

To pressure
gage

Static
pressure

Pipe wall

Knife edge

Direction
of flow

Inlet
hole

Center of flow

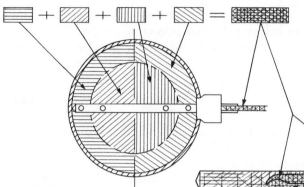

EQUAL ANNULI

Each annular segment represents an area of equal cross-sectional size. Accurate detection of flow within these annular segments is achieved by large sensing ports. These ports are precisely located at positions predetermined by computer. ANNUBAR'S® multiplicity of sensing points represents the various flows within the pipe regardless of flow profile or metered fluid.

DOWNSTREAM ELEMENT

This tube measures the downstream pressure which is the pipe's static pressure, less the suction pressure of the flow.

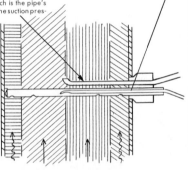

EQUALIZING ELEMENT

The sensing ports simultaneously detect the flow rates of their respective pipe segments. The various flow rates are averaged by the plenum of the upstream element. The internal interpolating tube* is incorporated on sizes over one inch so that the measured signal is received from the center of the plenum. This insures that both halves of the pipe are equally represented.

"ON-OFF" FEATURE*

Annubar's® upstream probe may be rotated by turning exterior portion of probe while system is under pressure. This allows sensing ports to be pointed fully downstream

This exclusive "on-off" feature* eliminates the need for purging when continuous readings are not required and flowing fluid is highly contaminated.

(b)

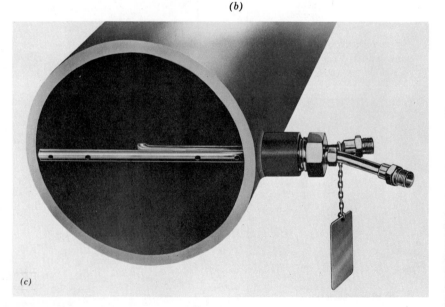

(c)

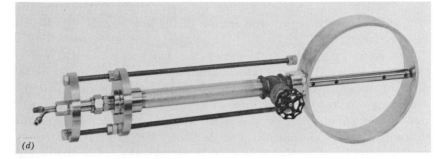

(d)

Figure 7.9 *Pitot tube and Annubar elements for flow detection. (a) Schematic cross section view of a basic pitot tube flow detector. (b) Schematic showing the Annubar operating principles. (c) Photograph showing the position of both the downstream orifice and the upstream orifices. (d) Photograph showing the relative diameter of the sensing orifice bars to the nominal pipe diameters. Courtesy Ellison Instrument Division, Dieterich Standard Corp.*

The limitations of the pitot tube's usefulness in commercial applications are its tendency to plug when the flowing fluid contains suspended solid particles, its velocity range for use with standard commercial flow indicating and recording flowmeters, and its sensitivity to velocity distribution effects in the fluid flow pattern.

Annubar Flow Element.   The *Annubar* flow element is a commercial version of the pitot tube principle and is shown schematically in Figure 7.9*b*. The Annubar is an innovation of the Ellison Instrument Division of Dietrich Standard Corporation and is applicable to liquid, steam, and gas flow measurements.

Although the design is new, it is based on the fundamental Bernoulli flow equation for energy balance. Very simply, this device consists of a small pair of pressure sensing probes mounted perpendicularly into the flow stream by means of a conventional threaded fitting or a pipe nipple and fitting.

As shown in Figure 7.9*c*, the downstream pressure sensing bar has one orifice which is positioned at the center of the flow stream to measure downstream pressure. The upstream pressure sensing bar has multiple orifices, versus one in the pitot tube. These orifices are critically located along the bar, so that each one detects the total pressure in an annular ring. Each of these rings has a cross sectional area exactly equal to the other annular areas detected by each of the other orifices. These areas are shown under the heading Equal Annuli in Figure 7.9*b*. On pipe sizes 1 in. in diameter and larger, another sensing tube is placed inside the upstream bar. It is precisely located between orifices to sense the true average head of the entire cross section of the flow stream regardless of flow regime.

The Annubar element is designed and has the tested and proven capacity to detect all normal rates of flow for gas, steam, or liquid for the pipe size in which it is properly installed. One element can accurately measure a wide range of flow rates without deviating from a true square root output signal which is compatible with all standard makes of flow instrumentation and controls. The Annubar flow element can handle pressures from −30 in.Hg up to 2500 lb/in². Another advantage is that the entire element can be reversed in direction, on pipe sizes over 2 in. in diameter, to detect reverse flows or act as an off position without a system shutdown.

As can be seen in Figure 7.9d, the flow detecting element offers a minimum of permanent pressure loss, since the diameter of the sensing bars is relatively small in comparison to the nominal pipe diameters. The flow restrictive effect of these sensing bars can be considered negligible, except for a more complex hydraulic or pneumatic installation demanding the highest possible accuracy. As a result, the Annubar element offers more freedom in design for many more balancing checkpoints and metering stations without adding appreciable restrictions or system costs.

This type of sensing system requires purging if used with slurries or liquids containing large amounts of suspended solids in the same manner as other types of flow detectors.

The Annubar element is relatively inexpensive as a flow detector, relatively simple to install, light in weight, accurate, and reliable for flow measurements through pipes or circular ducts from $\frac{1}{2}$ to 90 in. in diameter. These units are available as elements or as complete systems with element, indicating and/or recording instrumentation, and/or controls for the process flow.

Differential Pressure Flowmeters. The differential pressure or head meter is the most frequently used device for measuring the differential pressure created by a venturi tube, a flow nozzle, or an orifice plate. These head meters may range from a simple U-tube meter body to more complex devices using mercury, to complex differential converters and inductance bridge methods, depending on the application. All these meters measure the flow as a square root quantity from the basic relationship

$$Q = K \sqrt{\bar{H}} \qquad\qquad (7.13)$$

where Q = quantity flowing

K = a constant depending on the conditions specific for each installation

H = pressure drop or differential

When an evenly divided chart is preferable for recording purposes, a square root converter must be used and the differential pressure calibrated to read in units of flow. The same types of meters discussed in Chapter 3 for differential pressure measurements can thus be used to measure flow if they are calibrated to read in flow quantities such as gallons per minute or cubic inches per hour instead of inches of water or ounces per square inch or pounds per square foot.

Mechanical Flowmeters. A typical mechanical flowmeter is shown schematically in Figure 7.10. This meter uses a U-tube meter body which consists of a large chamber and a range tube connected at the base and filled with mercury. The large high pressure chamber contains a float which detects changes in the mercury level. With increases in differential pressure, the mercury is forced from the large tube into the smaller tube and causes the mercury level to rise. A full scale change in the differential pressure causes the float to travel from the top to the bottom of the

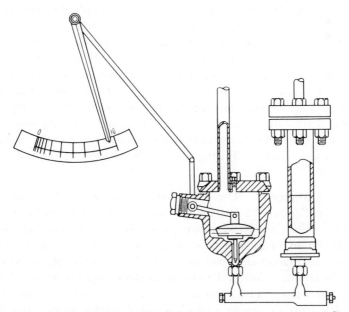

Figure 7.10 *Schematic of a mechanical flowmeter. Courtesy Honeywell, Inc.*

large chamber for a given range tube. Greater differentials cannot be measured when the float reaches the bottom, because of the physical limitations imposed.

One advantage of using range tubes is that, if a new range tube of smaller diameter and greater height is installed and the same amount of mercury used, a greater differential pressure can be measured. The smaller diameter range tube requires a greater displacement force by the float for the same mechanical displacement, because the mercury column must be forced higher in the smaller tube. This means that one basic meter body can be used for all flow ranges by changing range tubes. These meters can be used to cover the differential pressure range from 20 to 300 in. H_2O. As shown in Figure 7.10, there is a mechanical linkage from the bell to a pointer to indicate the flow. The manufacturer lists the scale as a square root graduation. The scale has to be changed to match each range tube, because the pointer has the same limited movement as the float to which it is attached.

An evenly graduated scale can be provided, as mentioned earlier in this discussion, by inserting a cam in the pointer movement, by proper shaping of one leg in U-tube meter body, or by use of a Ledoux bell construction. A Ledoux bell meter body is shown in a cross sectional view in Figure 7.11.

In the Ledoux bell meter the low pressure is applied to the inside of the bell and the high pressure to the outside. The mercury is the seal between the two pressures. The bell moves up and down as the differential pressure varies. The internal shape of the bell determines the position of the bell for a given differential, and different bell shapes are required for different ranges to obtain the same travel, just as range tubes have to be used for different differentials in the mechanical flowmeter. Here bell shapes are changed so that the same transmitting units can be used to send the signal to the indicator or recorder, while the range tube is changed to remain within the physical limitations of the indicator or recorder.

Electrical Flowmeters. The inductance bridge apparatus shown schematically in Figure 7.12 has a soft iron armature attached to the meter body float with a rigid rod which passes through a nonmagnetic alloy steel tube. Many of these meters are in use, but they are expensive and no longer manufactured. A pair of matched coils surrounds the armature, and as the armature moves it unbalances the bridge circuit to which the coils are connected. The unbalance in the bridge circuit causes the receiving armature to assume a position matching that of the transmitting armature. With the recorder arm attached to the receiving armature, the

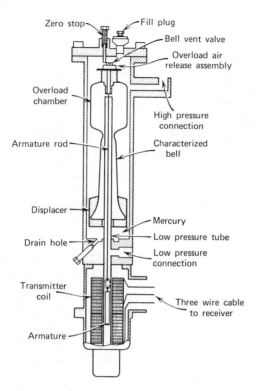

Figure 7.11 *Cross sectional view of Ledoux bell meter body. Courtesy Honeywell, Inc.*

flow is indicated and recorded as a function of the position of the armature. The armature, coils, and other electrical components are isolated
from any high pressures or corrosive action of the fluid whose flow is
being measured. This motion of the two armatures is similar to selsyn
action in rotary action for motor control used in processes for winding
paper or steel on spindles or spools. An electric flowmeter incorporating
an electronic integrating mechanism is shown in Figure 7.13. The electronic integrator shown in the lower left-hand corner is a direct reading
counter. This counter reads directly in the units of flow and automatically totals the flow. The electronic integrator shown consists of three
major assemblies: (1) a scanning unit which samples the flow rate indicated by the pen arm every 5 s.; (2) the electronic detector relay which
is actuated by the scanning unit and operates the counting mechanism;
(3) a motor driven counter assembly which totalizes the successive output
pulses generated by the detector relay.

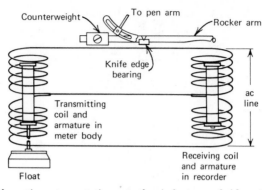

Figure 7.12 *Schematic representation of the inductance bridge principle used in electric flowmeters. Courtesy Honeywell, Inc.*

The operating principle of the integrator is a variation of that of an oscillating signal. Oscillator coils mounted on an arm driven by a precision cam are raised and lowered at 5 s intervals. These coils intercept a lightweight vane which is positioned by the flow indicating pen. As the leading edge of the vane is intercepted by the coils, the electronic detector relay is energized and current flows to the counter motor as long as the oscillator coils intercept the vane. When the vane and coils are separated, the motor stops instantly.

The integrator driving gears are designed to provide a decimal integration factor such as 0.1, 1, 10, or 100, and each digit records the flow times the factor. With a maximum count of 1500/h, the integrator can be made sensitive to small changes in flow rate. The electronic integrators just discussed can be furnished for use with either evenly graduated or square root graduated flowmeters.

Pressure Transducers. Pressure transducers used for measuring flow in head type measurements are of the differential pressure type discussed in Chapter 3. They are essentially the diaphragm type using strain gages, potentiometric, or variable inductance readout methods, depending on the application. Speed of response, temperature, differential pressure range, static pressure values, and signal conditioning accessory equipment determine whether transducers can be used as sensors and, if so, what types are applicable. For example, a bonded strain gage is not applicable if the temperature range is higher than the temperature the strain gage bonding material can withstand, or if the static pressure is higher than the design pressure of the housing in which the sensor and diaphragm are enclosed.

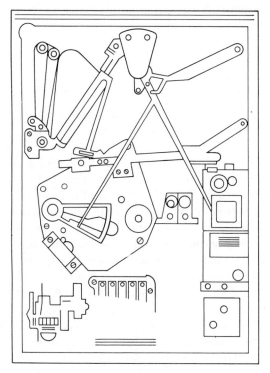

Figure 7.13 *Electric flowmeter with electronic integration. (Recording chart removed.) Courtesy Honeywell, Inc.*

A well made transducer unit is expensive, but offers the fast response times needed for control of variables to prevent damage to test specimens or products. Transducer units do not have the large inertial masses encountered in mechanical systems. Differential pressures can be measured and control action initiated for a change of flow in less than $\frac{1}{2}$ s. While this time is relatively slow in comparison to some microsecond measurement and control actions, it is relatively fast when we consider the time required to move a body of mercury and move a shaft and pointer to obtain a similar type of flow measurement.

In comparison to mechanical and electrical type flow measuring instruments, transducers must be provided with some kind of activating signal. The potentiometer type requires a power supply to maintain the voltage drop across the potentiometer. The strain gage type requires a bridge balancing circuit and a power supply, and the variable inductance type requires an oscillating signal so that a frequency change can be detected when the inductance changes, due to a change in flow and causes a change in the oscillator frequency.

7.2 AREA FLOWMETERS

The variable area flowmeter in one of its simplest and most elementary forms consists of two parts, as shown in Figure 7.14. The two basic parts are (1) a tapered glass tube set vertically in the fluid or gaseous piping system with its large end at the top, and (2) a metering float which is free to move vertically in the tapered glass tube. The fluid flows through the tube from the bottom to the top. When no fluid or gas is flowing, the float rests at the bottom of the tapered tube, and its maximum diameter is usually so selected that it blocks the small end of the tube almost completely. When a flow starts in the pipeline and the fluid or gas reaches the float, the buoyant effect of the fluid or gas lightens the float. Usually, the float has a density greater than that of the flowing material, so that the buoyant effect alone is not sufficient to lift the float. The float passage remains closed until the pressure of the flowing material, plus the material buoyancy effect, exceeds the downward pressure due to the weight of the float. The float then rises and floats within the flowing medium in proportion to the flow at the given pressure.

As the float moves upward toward the larger end of the tapered tube, an annular passage is opened between the inner wall of the glass tube and the periphery of the float. This forms a concentric opening through which the flowing material passes. The float continues to rise until the annular passage is large enough to pass all the material coming through the pipe. The fluid or gas velocity pressure also drops until it, plus the fluid or gas bouyant effect, exactly equals the float weight. The float then comes to rest in a dynamic equilibrium.

Additional increases in flow rate cause the float to rise higher in the tube, and decreases in flow cause the float to sink to a lower level. This means that every float position corresponds to one particular flow rate and no other for a specific gas, vapor, or liquid. The float then gives a reading on a calibration scale on the outside of the tube, and the flow rate can be determined by direct observation of the metering float.

Other methods of obtaining the varying area for the passage of fluid are a cylindrical tube with a taper hole inside, a cylindrical tube with a sideways discharge slot or a series of drilled holes, or a taper plug which can rise and fall in an orbit.

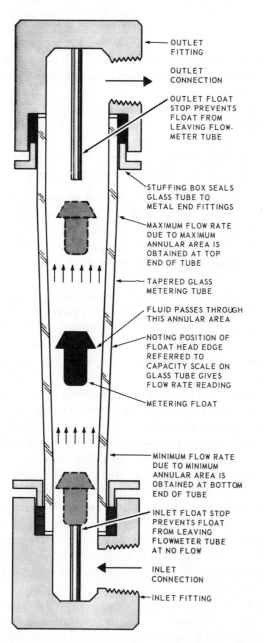

OUTLET
FITTING

OUTLET
CONNECTION

OUTLET FLOAT
STOP PREVENTS
FLOAT FROM
LEAVING FLOW-
METER TUBE

STUFFING BOX SEALS
GLASS TUBE TO
METAL END FITTINGS

MAXIMUM FLOW RATE
DUE TO MAXIMUM
ANNULAR AREA IS
OBTAINED AT TOP
END OF TUBE

TAPERED GLASS
METERING TUBE

FLUID PASSES THROUGH
THIS ANNULAR AREA

NOTING POSITION OF
FLOAT HEAD EDGE
REFERRED TO
CAPACITY SCALE ON
GLASS TUBE GIVES
FLOW RATE READING

METERING FLOAT

MINIMUM FLOW RATE
DUE TO MINIMUM
ANNULAR AREA IS
OBTAINED AT BOTTOM
END OF TUBE

INLET FLOAT STOP
PREVENTS FLOAT
FROM LEAVING
FLOWMETER TUBE
AT NO FLOW

INLET
CONNECTION

INLET FITTING

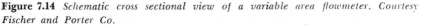

Figure 7.14 *Schematic cross sectional view of a variable area flowmeter. Courtesy Fischer and Porter Co.*

The basic equation for the area flow rate meter is identical to that for plates, nozzles, and venturi meters. This basic equation is:

$$Q = AC \sqrt{2gh} \qquad (7.14)$$

where Q = rate of the flowing fluid in cubic feet per second

C = experimentally determined flow coefficient which is dimensionless

A = flow area of the restriction in square feet

g = acceleration of gravity (usually the base value of 32.174 ft/sec² is used)

h = differential produced by the restriction at the rate of flow Q in feet of height of the flowing liquid

The flow rate varies directly as the area A in the equation. This type of meter permits measurement of flow rate without use of the square function in 7.14, because it determines the flow rate by measuring the free annular area available for flow around an obstruction. In plates, nozzles, and venturi meters, the flow rate varies as the square root of the head loss h.

There are some advantages obtained by measuring a factor that varies directly as the flow rate instead of as the square root of the flow rate. Variable area meters have almost linear graduation, and the calibration is almost linear. It is incorrect to say completely linear or completely uniform, because the area A is the inside tube diameter squared minus the float head diameter squared.

The variable head area flowmeter may be used equally well on liquids or gases, but corrections must be used when flowing gases of different densities through a meter having a fixed float weight. Each manufacturer has a set of curves for correction of liquid densities, both for volumetric and gravimetric units. When this type of meter is used for measuring the flow of various gases, a set of correction curves is provided to correct for gas specific gravity, gas pressure (both in pounds per square inch absolute and pounds per square inch gage) and gas temperatures in both degrees Fahrenheit and degrees Celsius. These corrections are used to provide the most accurate value for the flow, since each variable affects the flow through a meter designed for a specific float weight. The float weight is designed for specific conditions to give the true flow under the conditions of the scale calibration.

Variable area meters have the disadvantage that they must be mounted and held in a vertical position during use, because gravity is one of the factors in indicating the flow value. Vertical accelerations affect the apparent value of g. These two factors restrict the use of this type of meter. For example, they should not be placed on portable equipment when measurements are required while the equipment is in motion. The vertical glass tube is also subject to static pressure and any surge pressures in the system, and it is subject to breakage. Where there is doubt as to its ability to withstand the system pressure, an armored plate glass cover guard may be used.

While there is no theoretical limitation to the size of a variable area meter, the bulk and weight increase as the square of the pipe diameter. Therefore cost, weight, and handling problems become considerations when variable area meters are used for large pipe sizes.

Installation specifications should also be considered, because these meters replace a pipe section and must be mounted vertically even in a horizontal piping installation. However, there is minimal effect on flow distribution by upstream elbows, valves, and other bends as compared to variable head meters for which considerable lengths of straight pipe or straightening vanes must be used to provide uniform flow patterns for reproducible flow measurements.

Accuracy guarantees for a variable area meter are overall guarantees. This means that it is the guarantee of the degree of accuracy with which the final exhibiting device reflects the actual flow rate in the pipelines. Certifications of the accuracy of individual meters are available from the manufacturer. These certifications state the accuracy of all points on the scale and may be expressed as a percentage of instantaneous flow rate rather than of maximum flow, regardless of the fraction of maximum flow actually being measured.

Variable area meters are particularly suitable for small flows and can be used to measure accurately less than 0.1 cm^3/min of liquid and 1.0 cm^3/min of gas. However, for these small flow rates the indicating medium that appears to give the best results is a ball float instead of the tapered floats used in larger variable area meters. These floats may range in sizes from $\frac{1}{16}$ to $\frac{3}{8}$ in. diameter. The float material may be sapphire, stainless steel, or glass, and depends on the type of gas or liquid and the maximum capacities of the flow rate. These meters must be properly sized for the service for which they are intended. Correction and sizing graphs are available from the manufacturers of these devices, and are based on a knowledge of the actual density, viscosity, and temperature of the fluid to be measured.

When a variable area flow meter is on hand and it becomes desirable to evaluate its usefulness for a service other than that for which it was designed, it is possible to make the evaluation by using conversion equations designed for this purpose.

One set of equations for liquid conversion in equivalent capacity in gallons per minute of water consists of:

Volume rate:

$$\frac{(\text{gal/min}) \; (\rho) \; (2.65)}{\sqrt{(\rho_f - \rho) \; (\rho)}} \tag{7.15}$$

Weight rate:

$$\frac{(\text{lb/min}) \; (0.318)}{\sqrt{(\rho_f - \rho) \; (\rho)}} \tag{7.16}$$

Base or contract volume rate:

$$\frac{(\text{gal/min}_b) \; (\rho_b) \; (2.65)}{\sqrt{(\rho_f - \rho) \; (\rho)}} \tag{7.17}$$

A set of equations for gas conversion in equivalent capacity in gallons of water per minute consists of:

Standard volume rate:

$$\frac{(\text{SCFM}) \; (\rho_g \text{std}) \; (2.51)}{\sqrt{\rho_f \; (\rho_g \text{opt})}} \; \frac{530}{T\text{std}} \; \frac{P\text{std}}{14.7} \tag{7.18}$$

Operating volume rate:

$$\frac{(\text{OCFM}) \; (\rho_g \text{ std}) \; (2.51)}{\sqrt{\rho_f \; (\rho_g \text{ opt})}} \tag{7.19}$$

Weight rate:

$$\frac{(\text{lb/min}) \; (2.51)}{\sqrt{\rho_f \; (\rho_g \text{ opt})}} \tag{7.20}$$

where gal/min = maximum flow of liquid at metering conditions
gal/min$_b$ = maximum flow of liquid at base or contract condition
lb/min = maximum flow of fluid at metering condition
SCFM = maximum flow of gas referred to a base or standard condition in cubic feet per minute
OCFM = maximum flow of gas at operating conditions in cubic feet per minute
ρ_f = density of float in grams per cubic centimeter
ρ = density of flowing liquid at metering conditions in grams per cubic centimeter
ρ_b = density of flowing liquid at base or contract conditions in grams per cubic centimeter

ρ_g std = density of gas at 14.7 lb/in.^{2} absolute and 70°F or 14.4 lb/in.^{2} absolute and 60°F in pounds per cubic centimeter

ρ_g opt = density of gas at metering conditions in pounds per cubic foot

T std = temperature at base conditions in degrees Fahrenheit in absolute units (460 + °F)

P std = pressure at base condition in pounds per square inch absolute

The variable area flowmeters discussed so far have been visual indicating meters for pressures for which either straight glass tubes or armored glass tubes are suitable. However, for higher pressures metal tubes and either indicating, integrating, and or recording device readouts can be used.

Figure 7.15 is a schematic representation of an inductronic variable area flowmeter. The variable area flow meter is used as a primary detector to provide remote indication of the flow. A soft iron armature is attached to the variable area flowmeter float and, as the float changes position with respect to the rate of flow, the armature changes its position with respect to coils *a* and *b*. These two coils form two arms of an impedance bridge. As the soft iron armature moves with respect to these coils, an unbalance is set up in the bridge. This unbalanced voltage of the bridge is amplified and used to operate a servomotor which in turn operates a measuring meter in proportion to the flow. The servomotor may move a second soft iron armature with respect to a second set of coils *c* and *d* in the proper direction to balance the bridge. If dual soft iron armatures are used, the indicating or recording meter corresponds to the position of the primary detector and can be calibrated directly to the rate of flow.

Another method of achieving a direct readout is to attach a potentiometer to the end of the rod attached to the moving float, as shown in Figure 7.16. The rod moves the potentiometer arm along the resistance and gives an output voltage proportional to the float position. This output voltage is then calibrated in terms of the flow and can be remotely indicated or recorded. The use of transistor electronics to supply a constant voltage drop across the potentiometer makes this method simple, and as accurate as ac methods. This is possible because there is no drift in the transistor electronics, and the only disadvantage is the aging and wearing of the slidewire resistance of the potentiometer. The tapered tube can be made of metal to withstand any pressure for which the variable area meter is suitable.

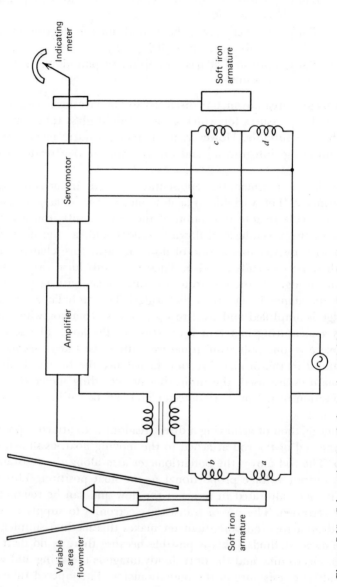

Figure 7.15 *Schematic of an inductronic variable area flowmeter.*

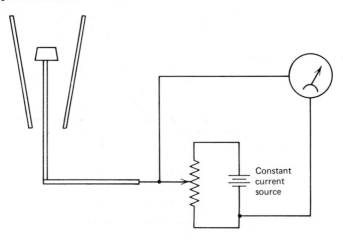

Figure 7.16 *Schematic of a potentiometric variable area flowmeter.*

7.3 ELECTROMAGNETIC FLOWMETERS

The electromagnetic flowmeter is strictly an electrical primary detector of the rate of flow. A typical system is shown in Figure 7.17.

In the electromagnetic flowmeter an emf is induced in the fluid by its motion through a magnetic field provided by an electromagnet. The dc magnetic field acts vertically through the pipe that carries the fluid. As shown, two electrodes are provided, one at each end of the horizontal diameter of the pipe. The emf induced in the liquid, according to Faraday's law, can be expressed by the equation

$$e = BDV \qquad (7.21)$$

where e = emf induced in the liquid
B = magnetic induction in webers per square meter
D = inside diameter of the pipe in meters
V = mean velocity of the fluid flow in meters per second

Therefore the induced emf is a function of the volume rate of flow output. An electronic indicating or digital voltmeter, or a self-balancing potentiometer can be used to measure the emf output. The fluid flowing in the pipe must have some conductivity, and some current must flow to operate the measuring instrument. The electromagnetic flowmeter shown in Figure 7.17 is limited to the measurement of fluids that have considerable conductivity. With liquids that have low conductivity, the elec-

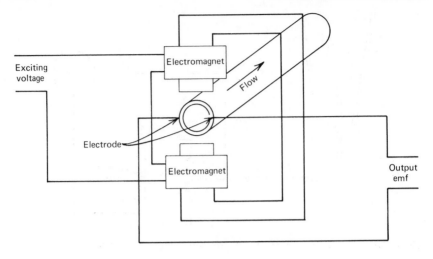

Figure 7.17 *Schematic of a typical electromagnetic flowmeter. Courtesy General Electric Co.*

trodes have a tendency to become polarized. Polarization effects can be overcome or minimized by the use of an ac field. However, when an ac field is used, the dielectric constant of the liquid influences the measurement because it has a shunting effect on the voltage developed. There is also an additional emf induced by the changing field in excess of that generated by the fluid flow in the electromagnetic flowmeter, as already discussed, and the induced emf is present when there is no fluid flow. This induced emf can be balanced out by adding an equal and opposite emf in series with the measuring electronic indicating or digital voltmeter.

The electromagnetic flowmeter is especially valuable in measuring liquid metals, corrosive fluids, and slurries. It is unaffected by viscosity, density, or turbulence. The transmitter can be installed in any position, the generated signal is linear with flow, there is an instantaneous response to flow changes, and range changes can be made easily.

As mentioned earlier, only flows of conductive fluids can be measured with electromagnetic flowmeters. Also, ac operated units must be operated at pressures and temperatures within the limitations of the transmitter, and the installation must be made so as to minimize or prevent stray 60 cycle pickup. Also, the line must be kept full of fluid, and the electrode tip kept clean.

Electromagnetic flowmeter head loss is the same as for a straight pipe of the same diameter and length, reverse flows can be measured, rangeability can be made as high as 30 to 1, and accuracy can be as good as 1% of full scale.

7.4 MASS FLOWMETERS

In some industrial applications it is desirable and essential to measure the mass flow rate or total mass flow of a gas or liquid. In such applications as dairies and other types of food processing industries, stringent sanitary standards covered by state and federal laws must be met. In other applications the gas or fluid may be corrosive, or the fluid may be a slurry.

Fixed volume or mass flowmeters, to be discussed in a later section, can be used in many applications, but these units have some type of piston or diaphragm inserted into the stream and are exposed to the flowing fluid. In some cases this is undesirable if the fluid is corrosive or contains contamination of any kind. Since these meters are discussed under positive displacement meters, this section is devoted to an electrothermal flowmeter of the boundary layer type in which the sensing and measuring elements are outside the pipe carrying the fluid and, as in the case of the electromagnetic flowmeter, do not obstruct the stream anymore than a pipe of the same diameter and length.

A smooth bore metal tube which seals the measuring elements from the stream is used as shown in Figure 7.18. A small heater element to introduce heat into the transducer is located at position c, an upstream temperature sensor is located at position a, and a downstream temperature sensor is located at position d. Often a temperature compensating element is located at position b. The two temperature sensing elements are insulated from the smooth bore metal tube by heat insulators e and f. The heavy outer wall of the pipe section acts as a heat shield and protects the measuring elements from physical damage. The type of temperature sensor depends on the application; thermistors, thermocouples, or resistance thermometers may be used.

The principle of operation is that heat is injected into the fluid through the walls of the tube and the boundary layer by the heater coil at position c. The upstream temperature sensor at a measures the initial temperature of the flowing stream, and the downstream temperature sensors measure the resultant temperature of the outside surface of the boundary layer. Small temperature differences across the boundary layer are sufficient for satisfactory operation.

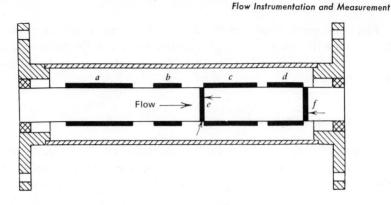

Figure 7.18 *Schematic of a boundary layer electrothermal mass flowmeter. Courtesy Neptune Meter Co.*

An equation for turbulent flow and one for laminar flow can be used to show a close approximation for the relation between heat flow Q, temperature difference ΔT, and mass flow F_{mass}.

For a turbulent flow $Q(T)$,

$$Q(T) = (A)\ \Delta T\ C^x\ K^y\ \frac{F^2}{\mu^a} \qquad (7.22)$$

where A = a design parameter
   C^x = specific heat
   K^y = thermal conductivity
   F^2 = mass flow rate
   μ^a = viscosity of the fluid

For a laminar flow $Q(L)$,

$$Q(L) = B\ \Delta T\ C^x\ K^y\ \frac{F^2}{\mu^a} \qquad (7.23)$$

where B = a design parameter

It should be noted that the values of the superscript letters of C, K, F, and μ are not the same for turbulent flow as they are for laminar flow. Also, these equations are valid only as long as the temperature gradient across the walls of the transducer element is negligibly small compared with the temperature across the boundary layer.

The two methods normally employed to measure the relation between heat transfer and mass flow rate are the variable temperature and the variable power techniques.

In the variable temperature method, the heater power is held constant and the temperature difference between the upstream and downstream temperature sensors is measured.

In the variable power method, the temperature differential Δt is held constant as the flow rate varies, and the heater power is measured. Heater power is equivalent to turbulent flow $Q(t)$.

In general, the response time is considered fast and may vary from a fraction of a second to a few seconds, depending on the rate of flow and the thickness of the section wall.

The measurement of flow rates over large ranges (total ranges of 1200 to 1 are possible) can be made with the boundary layer method by using a single transducer with the proper heating and sensing circuits. The measurement error is small for small changes in stream composition, and automatic compensation can be added, for significant composition changes, by monitoring the composition.

Axial Flow Mass Flowmeters. An axial flow mass flowmeter operates on the principle of conservation of angular momentum. Basically, an impeller is driven at a constant angular velocity, and this impeller creates an angular momentum in the fluid being measured. The rate of change in angular momentum in the fluid as it leaves the impeller is proportional to the velocity of the impeller and the mass rate of fluid flow. Some type of torque sensing device, such as a wheel, is located on the downstream side and generally adjacent to the impeller. This torque sensing device removes the angular momentum from the fluid at the same rate at which the impeller imparts it to the fluid. As long as the angular velocity of the impeller remains constant, the torque on the sensing device is proportional to the mass rate of fluid flow.

If a single impeller is used in the fluid stream, fluid flow measurement is possible in one direction only. To measure flow in both directions a second impeller, located on the downstream side of the torque sensor and driven in the direction opposite that of the upstream impeller, is required. The torque sensor must also be capable of sensing torque in either direction. The block diagram shown in Figure 7.19 is for the single-impeller system.

In this system a regulated motor generator power supply drives the impeller at a constant frequency. The sensing wheel, which is on the downstream side of the impeller, is spring restrained and magnetically

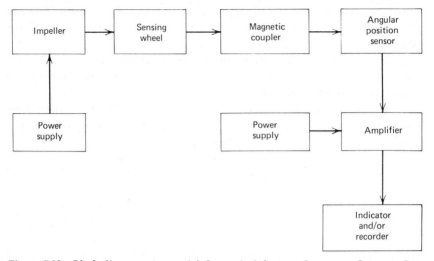

Figure 7.19 *Block diagram of an axial flow principle mass flowmeter. Courtesy General Electric Co.*

coupled to an angular position detector. A stationary disc is used to reduce the coupling between the impeller and sensing wheel for zero flow. The output voltage of the angular position detector is proportional to the mass flow rate and feeds into a servoamplifier, a ratemeter, and an integrator-counter.

The single-impeller design has the limitations of measuring flow in only one direction, in accuracy at low flow rates where extraneous torques are an appreciable percentage of the signal, and inability to measure rapid changes in flow. It is relatively simple in comparison to other methods of measuring mass flow and is adequate for unidirectional flow when there are no rapid changes in the flow.

Turbine Flowmeters. The construction features of a turbine volumetric flowmeter are shown in Figure 7.20*a*. As shown, the turbine rotor is held between two sets of concentric cylinders which serve to guide the flow and to position the rotors in the pipe mounting as pictured in Figure 7.20*b*. Figure 7.20*b* also gives some relative idea as to sizes of turbine flowmeters using the same type of construction features.

As the turbine rotor revolves, each vane generates a pulse and represents a unit volume for flow totalization. These meters generate a digital electrical signal output which is detected by a flowmeter or tachometer pickup coil designed specifically for the Barton Series 7000 turbine

Figure 7.20 *Turbine flowmeters. (a) Construction features of a turbine volumetric flowmeter. (b) Turbine mounting in a flow section. Courtesy ITT Barton.*

meters illustrated. The total number of rotor revolutions or output pulses is related to the total throughput or volume of flow. The frequency of the pulses generated is directly proportional to the flow rate of the material being monitored or measured. The pulses generated in the pickup coil are of sine wave form and can be transmitted electrically over great distances, if necessary, to a variety of readout devices for computing, indicating, recording, controlling, or automation.

A pickup coil is shown mounted on the top of the turbine flowmeter shown in Figure 7.21a. This pickup coil may feed into a direct reading totalizer with the construction features shown in Figure 7.21b. This unit gives both a total flow and a flow rate indication at any time. Other units that may receive the pickup coil output are shown in Figure 7.21c. The unit on the left is a flow rate indicator with control current output. The next unit is a flow totalizer, and the two units on the right are direct reading totalizers.

Turbine flowmeters can be used for measuring flows ranging from 0.003 to 15,000 gal/min as standard liquid flowmeters, 1790 to 73,500 bbl/h as pipeline flowmeters, and 20 to 9000 ft³/min as gas flowmeters. Standard and pipeline meter flows are dependent on the viscosity of the liquid being measured, and gas meters on the density of the gas being measured.

In performance, standard liquid turbine flowmeters have a linearity of ±0.25% of reading to normal maximum rated flow, and ±0.4% of reading to extended maximum from rates above the normal maximum. These meters have approximately twice the accuracy over any 6 to 1 range downward from rated maximum flow, and repeatability of ±0.02% of reading in the rated linear flow range. Repeatability in the nonlinear low flow range is ±0.25% of reading.

Turbine gas flowmeters have a linearity as high as 1% of reading for any 10 to 1 range from the maximum downward. Below the 10 to 1 from maximum, the linearity is 0.1% of the full scale reading.

The operating temperature range is from −450 to 1000°F, and the pressure rating is as high as 50,000 lb/in². The limits on pressure are established only by the housing fittings, ASA flanges, and other accessories.

7.5 POSITIVE DISPLACEMENT FLOWMETERS

Positive displacement flowmeters basically capture and release a fixed volume of fluid by some type of pumping action. This action can be performed by a piston, a rotary vane, a diaphragm, or peristaltic pumping. These meters normally count the total number of cycles that occur, and indicate or register an integrated flow volume. These flow devices can also be considered volumetric pumps delivering a fixed volume per piston, vane, diaphragm, or peristaltic stroke or movement. The flow rate is determined by the frequency of the cycle.

Positive displacement meters are available from $\frac{1}{2}$ to 16 in. sizes, are usually simple to install, can have as high as 0.1% accuracy, usually have close mechanical tolerances, have a rangeability of 5 to 1 for fluids and 100 to 1 for gases, do not require electric power or an air supply, and can be damaged by exceeding their capacity. Although many positive displacement meters do not basically require electrical power or an air supply, some make use of electrical drive motors to deliver given volumes, and either electrical, pneumatic, or hydraulic power, or a combination, for control purposes.

Piston Pumps.   Metering pumps are used to inject an exact amount of fluid into a flow line or a collecting vessel. The piston pump is generally a reciprocating type, which means that a piston or plunger delivers a fixed volume on each stroke. The basic pump is shown schematically in Figure 7.22. In the simple single-action unit shown, the fluid is drawn into the piston cavity through an inlet valve as the piston is moved back in its cavity, and is discharged into the flow line or vessel as the piston moves forward and fills the cavity. As the piston moves back and forth, it delivers a fixed volume in a pulsating flow.

Piston meters are usually used to deliver controlled volumes at very high pressures. Where the pulsating flow cannot be tolerated, it can be evened out by means of reservoirs or special pumping arrangements. An electric motor, air, or steam is used to drive these pumps.

In piston pumps the amount of flow or volume delivered per stroke can be changed by varying the length of the piston stroke, and total volume, as a function of time, can be controlled by changing the pumping speed. The stroke adjustment can be varied manually or automatically as required by the application.

Nutating Piston Pumps.   The nutating piston pump shown in cross section in Figure 7.23 is probably the most widely used piston type pump in use today, at least in the United States. The nutating pump may also be termed a disc pump, as will be quite evident when its action is studied.

The piston is the only moving part in the measuring chamber. The action of this piston resembles the action of a top when it has passed its peak speed and starts to wobble or nutate just before it loses speed and goes out of control. In the nutating piston meter, the motion of the disc piston *A* is controlled by the shaft *B* as it moves around the tapered cam *C*. This cam keeps the lower face of the piston in contact with the bottom of the measuring chamber on one side of the pump, and keeps the upper face of the piston in contact with the top of the measuring chamber on the opposite side. As shown, the piston is positioned so that the lower side of the disc is in contact with the bottom of the measuring

(a)

(b)

(c)

Figure 7.21 *Turbine flowmeter sensing and measuring devices.* (a) *Pickup coil mounted on the top section of a turbine flowmeter section.* (b) *Construction features of a direct reading turbine flowmeter totalizer.* (c) *Left to right, flow indicator with control current output, flow totalizer and flow rate indicator, and two direct reading totalizers.* *Courtesy ITT Barton.*

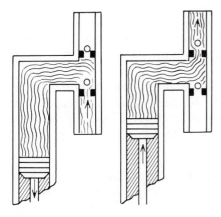

Figure 7.22 *Cross sectional view of a typical single-piston flowmeter. Courtesy National Instrument Co.*

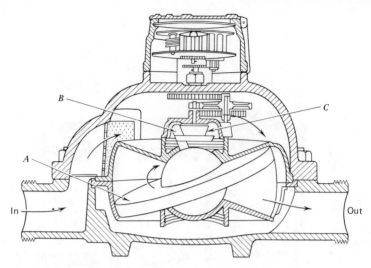

Figure 7.23 *Cross sectional view of a typical nutating piston flowmeter. Courtesy Hersey Products, Inc. Ametrol Division.*

chamber on the left-hand side, while the upper side of the disc is in contact with the top of the measuring chamber on the right-hand side. This method of pumping produces a smooth and continuous flow with no pulsations as the separate compartments of the measuring chamber are successively filled and emptied. The measuring chamber is sealed off into separate compartments, and each compartment holds a definite volume. The seal is maintained by the fluid between the piston and the measuring chamber wall as a result of capillary action. This capillary action minimizes leakage or slippage and provides accuracy at low rates on well made meters of this type.

Nutating piston meters are designed for the rate of flow of the liquid to be measured and for nominal line pressures. Selection of a meter should be based on the flow rate, line pressure, and allowable pressure drop for the intended application.

From the figure it can be seen that the measuring chamber and the piston are completely enclosed by the liquid, so that variations in line pressure should not distort the chamber and affect its accuracy. As a result, the measuring device must be brought out through a seal such as a stuffing box.

Rotary Pumps.   Rotary pumps are capable of furnishing smooth, pulsation-free flows at pressures up to the 1500 lb/in.^{2} range. This type of pump is also suitable for high viscosity fluids. Volumetric efficiencies up to 90% can be obtained by maintaining close tolerances between the mating surfaces.

Smooth flow is obtained by having more than one vane in action, so that some flow is maintained on a continuous basis. The flow in this type of pump is controlled by valves for internally bypassing some of the fluid.

In action, there are four vanes, and each vane carries a fixed volume as it rotates around a center shaft. While one vane cavity is filled under the line pressure, the backflow is sealed off by the next succeeding vane. Under normal operating conditions one vane discharges its volume into the discharge line at the same time the feed line fills the cavity of the receiving vane. In this flowmeter the line pressure keeps the vanes in motion, and no external electrical or pneumatic source of power is required. The seal between the vanes and the measuring cavity is maintained by capillary action; the fluid being measured acts as the sealant in the same manner as in the nutating piston meter.

Diaphragm Pump. Diaphragm pumps use a membrane which is activated by some external means such as a push rod, or by means of air or hydraulic pressure. A cross section view is shown in Figure 7.24.

For pressures up to 125 lb/in.^{2} Teflon or other limp membranes or diaphragms are used. For higher pressures up to 2500 lb/in.^{2} use is made of small metallic diaphragms. The material used depends on the type of fluid, the temperature, and the pressure. The membrane or diaphragm is in direct contact with the fluid and the driving mechanism. In other words, it is a partition that seals in the fluid and separates it from the mechanical or hydraulic drive.

These pumps are controlled by varying the drive motor speed, or by varying the stroke of the diaphragm. The diaphragm pump is comparable in action to the piston pump in which the diaphragm is used for the pumping action instead of the piston.

Peristaltic Pump. When it is necessary to pump fluids that cannot contact the pump or metering components, the peristalic pump is quite useful. The fluid is usually pumped through a plastic tube and does not come in physical contact with the mechanism generating the measured flow. The pumping mechanism can be rollers, cam operated fingers, or a rotor on an eccentric shaft. In each case the fluid is pushed through a given length of pliable tubing, capable of withstanding the system pressure, by a squeezing motion. These pumps can handle measured flows

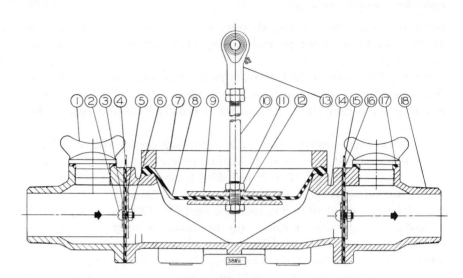

CROSS SECTION

KEY NO.	DESCRIPTION
1	Clean-Out Plug - Cast Iron
2	1/4" -20x1/2" Machine Screw - Stainless Steel
3	Lower Weight - Stainless Steel
4	Flapper Valve - Neoprene & Nylon
5	Upper Weight - Stainless Steel
6	1/4" -20 Hex Nut - Stainless Steel
7	Clamping Ring - Cast Iron
8	Diaphragm - Natural Rubber & Fabric
9	Clamping Plate - Cast Iron
10	Pump Rod - Cold Rolled Steel
11	1/2" -20 Hex Thin Nut - Steel
12	1/2" Thread Seal Washer - Steel With Rubber Insert
13	Bearing Rod End - Steel, Cadmium Plated
14	Pump Body - Cast Iron
15	Gasket - Lexide
16	Valve Seat - Stainless Steel
17	Gasket - BUNA-N
18	Suction & Discharge Fitting - Cast Iron

Figure 7.24 *Cross sectional view of a typical diaphragm flowmeter. Courtesy ITT Marlow.*

from a fraction of a milliter per minute to 40 gal/min. A typical roller type peristaltic pump is shown in Figure 7.25.

The flexible tubing is used to withstand corrosion and for volatile liquids or liquids requiring a high degree of cleanliness. One area of use is in the medical profession for pumping blood during analytical tests.

Control of flow in the peristaltic pump is effected by changing the speed of the drive motor. Flow control is accurate to better than the normal $\pm 1\%$ for industrial applications.

7.6 OPEN CHANNEL FLOWMETERS

Open channel flowmeters can handle from 4 to 1,000,000 gal/min and are normally weir or Parshall flume types. This flowmeter is relatively inexpensive, and usually the only problem is that it can trap entrained matter. One of the largest uses is in flow-measurements for agriculture irrigation and industrial wastes.

Weirs. The weir is shown in cross section in Figure 7.26. These flow-meters make use of V notches, rectangular notches, and trapezoidal notches. Both 60° and 90° V notches are used. They have ranges from 20 to 1 to 60 to 1.

The weir is essentially a dam with a notched opening in the top through which the liquid flows. It is one of the simplest and most accurate devices for measuring the flow of water under proper conditions. Its accuracy is approximately 2%.

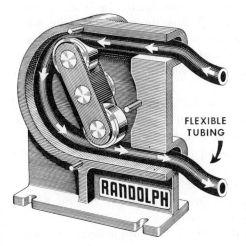

FLEXIBLE
TUBING

Figure 7.25 *Cross sectional view of a typical roller peristaltic pump flowmeter. Courtesy Randolph Co.*

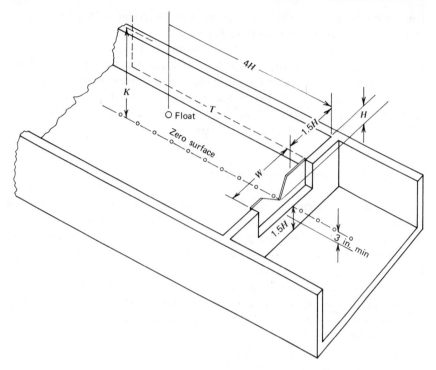

Figure 7.26 *Cross sectional view of a typical weir. Courtesy Penn Meter Co.*

The rate of flow is measured by simply measuring the head of water above the lowest point of the weir opening through which the water flows. This height is measured by means of a float, installed in a box, called a stilling well, which is a part of the total structure. The float is so located that it is not disturbed by the velocity of the flow, or by any turbulence in the stream.

The V notch offers the widest range for a single size, since the small opening in the V can accommodate small flows and the top portion larger flows. It also offers the greatest head loss because of its shape.

The rectangular notch is the oldest type and probably the most common because of its simplicity, ease of construction, and accuracy.

The trapezoid notch, also known as the Cipolletti, is designed so that the trapezoidal side slopes produce a flow correction so that the flow is proportional to the length of the weir crest.

A weir has to be cleaned periodically if the liquid being measured contains any entrained material. Deposited materials produce errors in flow rate just as deposited materials create errors in flow rate measurements with concentric, orifice plates.

Since the weir is a liquid head measuring device, a generalized equation is:

$$Q = KWH^n \qquad (7.24)$$

where Q = flow rate in cubic feet per second
  K = a constant depending on the type of notch and the size of the weir
  W = width of notch throat for the rectangle or trapezoid and of the crest for the V notch
  H = head in feet
  n = an exponent depending on the type of weir

In Figure 7.25 a rectangular notch is shown backed by a V notch to demonstrate both types. The zero measuring surface is shown for the base of the V notch but, if only the rectangular notch is used, its bottom edge then becomes the zero surface for the float at no flow.

The curves in Figure 7.27 show weir capacities for a 30° V notch, a 60° V notch, and a 90° V notch with values for K and the ideal location of the transmitter T that is attached to the float. To obtain accurate flows the upstream side of the V notch must be sharp and maintained free of nicks, burrs, or rounded edges.

Flumes. Only Parshall flumes are considered here, although Palmer-Bowlus, standing wave, and venturi flumes are available. A plan view of a Parshall flume is shown in Figure 7.28 with a table showing the dimensional relationships.

Parshall flumes are self-cleaning and operate with a small loss of head or channel grade. The loss is about one-fourth of the loss occurring in weirs. These flumes are recommended for use where sand, grit, or other heavy solids are present in the stream whose flow is being measured.

Parshall flumes are designed for use under conditions in which flow velocities are moderate, and they should be located so they are not affected by control structures, bends in channel alignment, or other devices that cause eddies, waves, or uneven flow. Flume capacities are shown in Figure 7.29. The flows are indicated as cubic feet per second, gallons per minute, and million gallons per day, as a function of head and W.

7.7 SUMMARY

Flow rate measurements can be made for very small flows, medium flows, and very large flows. The accuracy of measurement and the application

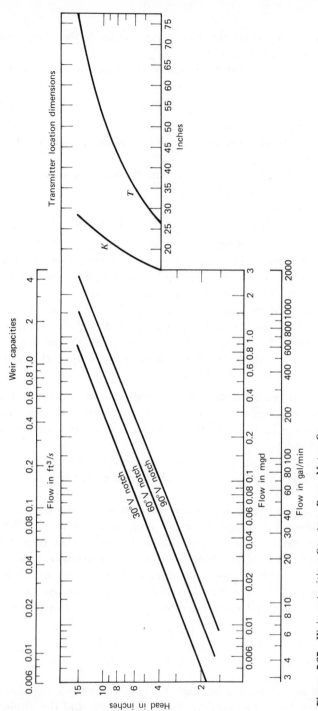

Figure 7.27 *Weir capacities. Courtesy Penn Meter Co.*

Parshall flume dimensions in inches

W	A	B	C	D	E	F	G	K	N
3	18.38	18.0	7	10.19	24.0	6.0	12.0	1.0	2.25
6	24.44	24.0	15.5	15.63	24.0	12.0	24.0	3.0	4.50
9	34.63	34.0	15.0	22.63	30.0	12.0	18.0	3.0	4.50
12	54.0	52.88	24.0	33.25	36.0	24.0	36.0	3.0	9.0
18	57.0	55.88	30.0	40.38	36.0	24.0	36.0	3.0	9.0
24	60.0	58.88	36.0	47.50	36.0	24.0	36.0	3.0	9.0
36	66.0	64.75	48.0	61.88	36.0	24.0	36.0	3.0	9.0
48	72.0	70.63	60.0	76.25	36.0	24.0	36.0	3.0	9.0
60	78.0	76.50	72.0	90.63	36.0	24.0	36.0	3.0	9.0
72	84.0	82.38	84.0	105.0	36.0	24.0	36.0	3.0	9.0
84	90.0	88.25	96.0	119.38	36.0	24.0	36.0	3.0	9.0
96	96.0	94.13	108.0	133.75	36.0	24.0	36.0	3.0	9.0

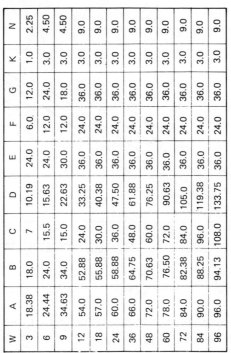

Plan view

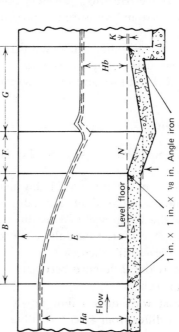

1 in. × 1 in. × 1/8 in. Angle iron

Figure 7.28 *Plan view of a typical Parshall flume. Courtesy Penn Meter Co.*

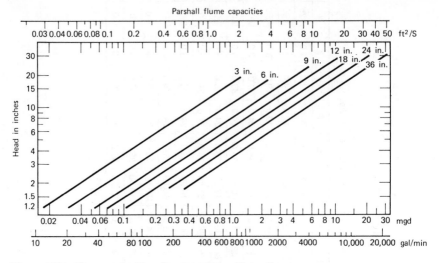

Figure 7.29   *Flume capacities. Courtesy Penn Meter Co.*

ultimately determine the type of flowmeter and its cost. It has also been shown that the principles of pressure, mechanical motion, and temperature can be converted into equivalent flow rates, and that the flow rates can be controlled and measured to produce better and consistent quality products.

Table 7.1 lists the different flow instruments discussed, the media for which they are best suited, the volume they can accommodate, their relative cost, their head loss, and their comparative accuracy.

Review Questions

7.1   What four factors determine the method used in the measurement of flow?

7.2   What is meant by a head flowmeter? How many types of head flowmeters are available for industrial applications?

7.3   Name the advantages of a venturi tube. What are its disadvantages?

7.4   Does the venturi tube follow the theory and equations derivable from Bernoulli's theorem?

7.5   What is the difference between a standard venturi and an insert venturi tube?

7.6   In what ways does a flow nozzle resemble a venturi tube? How does it differ from a venturi tube?

7.7   Why are orifice plates so popular for flow measurements? Give their advantages and disadvantages.

7.8   Name the features required in an orifice plate when consistent measuring results are to be obtained.

7.9   Under what conditions would a segmented or eccentric orifice be substituted for a concentric orifice?

7.10   How are orifice plates mounted in pipelines and what precautions must be observed?

7.11   Why are vena contracta taps used? Are they always the same distance apart? Why or why not?

7.12   How can the area of an orifice be eliminated from equations used in the sizing of an orifice?

7.13   To make the most accurate flow measurements with an orifice plate, what additional factors, not shown in the basic equations, must be considered?

7.14   Name the specifications that should be furnished to an orifice plate supplier so that one can obtain the proper orifice to make an accurate flow measurement.

7.15   Where is a pitot tube useful and why is it not considered an industrial instrument? How does the Annubar element overcome this limitation?

7.16   Why are range tubes used in mechanical flowmeters? What are their advantages?

7.17   How can an evenly graduated metering scale be provided for square root measuring sensors such as orifice plates?

7.18   What function does the bell serve in the Ledoux bell meter?

7.19   Give the three major parts of an electronic integrator.

7.20   Explain the operating principle of an electronic integrator.

7.21   Name the major types of transducers used in flow measurements involving head meters.

7.22   What are the advantages of using transducers for measuring flow? What are their disadvantages or limitations in flow measurements?

7.23   Explain the float action of a variable area flowmeter. What does the term variable area mean in terms of a flow measurement?

7.24   What are the disadvantages of the majority of variable area flowmeters?

7.25   What does an accuracy figure on a variable area meter include? Is this true for other types of flowmeters as well? Explain.

7.26   Is an inductronic or a potentiometric type variable area flowmeter simpler to operate? Why have potentiometric flowmeters been less popular until recently?

7.27 Under what conditions can an electromagnetic flowmeter be used?

7.28 Why choose an electromagnetic flowmeter for a flow application?

7.29 Under what conditions would you choose a mass flowmeter of the type shown in Figure 7.18 for a flow application? Name its chief advanatages.

7.30 Under what conditions would you choose an axial flow mass flowmeter? Give the advantages and disadvantages of this type of instrument.

7.31 What is meant by the term reciprocating pump for flow measurements? What is its main disadvantage for smooth flows? How can the pulsations be minimized?

7.32 Describe the action of a nutating piston pump. How are the seals maintained?

7.33 What is the advantage of a rotary pump for controlled flow measurement?

7.34 What disadvantages do all piston, rotary, and diaphragm pumps have in corrosive and sanitary flow measuring applications?

7.35 How is the volume delivery of a diaphragm pump controlled?

7.36 How is accuracy built into piston, rotary vane, and diaphragm flow measuring devices?

7.37 Under what operating conditions would you choose a peristaltic flow metering pump?

7.38 What type of operating conditions would lead you to choose an open channel weir for flow measurements? What advantages does a V notch offer?

7.39 What advantage does a flume offer over a weir for slurry flow measurements?

7.40 Give the advantages and disadvantages of a weir as compared to a flume for general high flow, large volume, open channel flow measurements.

Problems

7.1 A flow measurement is required for a water flow of 6000 ft³/min. Find the β factor, assuming the pipe has an inside diameter of 8 in., the orifice is made of stainless steel with an area factor for the thermal expansion of 1 for the temperature range to be used, and the differential pressure does not exceed 10 lb/in.² for full range.

7.2 A variable area flowmeter capable of measuring a gas flow of 100 ft³/min at 76°F for a gas density of 0.175 is available to measure a water flow. If the operating temperature is 68°F and the operating pressure does not exceed 25 lb/in.^{2}, what is the maximum water flow the float is capable of reading?

7.3 Calculate the value of the Reynolds number in volumetric units if a pipe is 4 in. in diameter and the fluid is water.

Bibliography

Aronson, Milton H., *Flow Measurement and Control Handbook*, Vol. 2, Instruments Publishing Co., Inc. 1964.

Bristol Co., *Differential Pressure Flow Meters*, Engineering Data Bulletin F1607, Waterbury, Conn., 1957.

Buzzard, William, *Handbook Flow Meter Sizing*, No. 10B9000, Fischer and Porter Co., Hatboro, Pa., 1963.

Ellison Instruments Co., *Anatomy of an Underground Storage Field*, Profile of a breakthrough, Gas Reprint, Boulder, Colo., December 1968.

Emerson Electric, *Design Specifications, DS5530*, Model 5530 Indicating Integrator, Brooks Instrument Division, Hatfield, Pa., 1965.

Fischer and Porter, *Analog and Digital Magnetic Flowmetering Systems*, Catalog 10D 1400, 1964; *An Introduction to Flowrator Meters*, Catalog 10-A-10, Hatboro, Pa., April 1953; *The F & P Tri-Flat Variable Area Flow Meter Handbook*, Supplement 1, 1955; *The F & P Variable-Area Flow Meter Handbook*, Catalog 10A90, Hatboro, Pa., 1952.

Hasting-Raydist, Inc., *HF Series, Mass Flowmeter*, Specification Sheet No 506, February 1964; *LF Series, Mass Flowmeter*, Specification Sheet, No. 505-A, Hampton Va., May 1964.

Honeywell, Inc., *Fundamentals of Instrumentation for the Industries*, G00003-0010M, Ft. Washington, Pa., January 1958.

Olive, Theodore R., and Danatos, Steven, Guide to Process Instrument Elements, *Chemical Engineering*, June 1957.

Richards, E. Curtis, *Principles and Operation of Differential Fluid Meters*, Brown Instrument Co., Division of Honeywell, Inc., Ft. Washington, Pa., 1947.

CHAPTER 8

Automatic Measurement and Control Concepts and Systems

Automatic measurement and control can be partial, complete, or some intermediate compromise. The control always involves some preset or predetermined value designated a set point. The system is designed to measure and control economically any deviations of the variable within the span and range of the controls, so that the predetermined value is maintained within the limits chosen for producing a quality product.

Automatic measurement and control systems operate on an error principle and have a definite limit of error span. The measurement system detects the error and produces a measurement to indicate manual control is required, or the meter movement automatically initiates an action, through a control system, to correct the error. This is normally accomplished by some form of feedback, anticipatory, or feed forward system. The accuracy required of the control system, as well as the speed of response for corrective action for the particular application, determine both the type of system and the cost. The range is from a simple manual on-off switch to a complex computer control system.

8.1 COMMON BASIC CHARACTERISTICS OF AUTOMATIC MEASUREMENT AND CONTROL SYSTEMS

To understand automatic control and measurement systems, there are certain basic characteristics that are common to most automatic measurement and control systems. We cover the basic characteristics of measuring devices, basic process characteristics, and the basic characteristics of automatic control for the industrial application.

8.2 BASIC CHARACTERISTICS OF MEASURING DEVICES

A measuring device has to sense or detect some physical parameter used in the industrial process such as pressure, temperature, level, flow, resistance, voltage, power, or motion. The measuring device must be capable of faithfully and accurately detecting any changes that occur in the detected parameter. For control purposes the measuring instrument movement either generates a warning signal to indicate the need for a manual change, or activates an actual control device. Both accuracy and speed characteristics must be considered, and each is discussed separately.

Accuracy.   Accuracy in measurement is influenced by static error, dynamic error, reproducibility, and dead zone.

We define *static error* as the deviation of the measuring device indication or recording from the true value of the measured variable. There is static error in all measuring devices, and large static errors are undesirable. However, in some cases large static error may not be detrimental to automatic control, while in other cases it is practically disastrous.

Static error does not interfere when it is more important to hold a variable at a constant and reproducible value instead of at an exact value. This may be true when products are processed in the same plant under exactly the same conditions.

Large static errors are disastrous if the control parameters are established in a pilot plant, the product processed or manufactured in several places, and the entire lot of parts assembled into a final product with a random parts choice during the final assembly. Close tolerances to an exact calibration value are required in such operations.

Accuracy is normally expressed as a percent of the instrument range or full scale value, with the value ranging plus and minus.

Dynamic error is the error that occurs within the instrument as it measures a source or load change occurring in the variable. To illustrate an instrument response involving dynamic error, we will assume that it has no static error. In the curves shown in Figure 8.1 at 0 min, the actual temperature and measured temperature are the same. The instrument measures the actual temperature. The temperature of the process gradually increases as a function of time. The measured temperature immediately falls behind or lags the actual temperature as it rises. This lag continues until the maximum dynamic error is reached. It then follows with this maximum error until the process temperature levels off at a new value. As the temperature rise decreases, the measuring instrument dynamic error decreases, and at the end of 10 min, as shown, the error is again at zero.

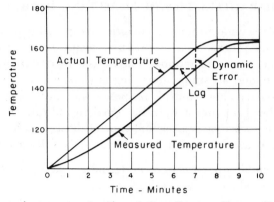

Figure 8.1 *Dynamic error as a function of time. Courtesy Honeywell, Inc.*

Dynamic error is present in all instruments, because they deal with a transfer of energy that takes time. It is independent of static error and is an additional error when static error is present. The time delay occurs under dynamic conditions and must be considered when setting up an automatic measuring and control system.

If both the static error and dynamic error of the measuring instrument are known and are reproducible every time the instrument is used, the instrument can still perform a useful function.

Reproducibility means that the measured value can be closely matched each time the value of the variable is measured. The degree of reproducibility is measured by how closely the same value of a variable can be measured at different times. In some respects reproducibility is more important than accuracy in automatic control, because it is a dynamic condition which exists in all cases in which automatic measurement and control are most useful in industrial processes.

Not all changes are gradual. Both step changes and sinusoidal changes occur. A step change is seldom met in actual industrial practices, primarily because it is a theoretical condition which involves an instantaneous change of a variable to a new value. Step changes are used in analytical evaluations for design and system performance predictions. Step changes can be simulated by rapid flow and pressure changes.

However, sinusoidal changes usually occur because cycling takes place in automatic control. As shown in Figure 8.2, when a corrective action is taken to cause a change in a variable about a set point at which the variable is to be held, a sinusoidal change occurs. In actual practice, when more heat is called for by the control device, the temperature

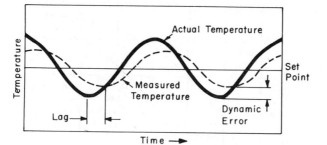

Figure 8.2 *Sinusoidal cycling as a function of time. Courtesy Honeywell, Inc.*

overshoots the set point and the lag of the measuring instrument in reaching the set point values increases the amount of overshoot. When the measuring instrument value reaches the set point value, it actuates the control device to lower the temperature. The dynamic error causes it to lag the actual temperature change back to the set point value, and the actual temperature then undershoots the set point value until the measuring instrument instructs the controller to increase the heat input. This continues through each heating and cooling cycle, so that both the actual temperature and the measured value vary sinusoidally throughout the process. It should be noted that the values cross each other twice during each cycle, but never on the set point value, and that the measured value otherwise never reaches the actual value of the variable.

The dynamic error prevents the indicator or recorder from showing the true conditions at the process. If there is a process lag, which is discussed later, it will add to the total system error.

In addition to static and dynamic error of the measuring instrument, there is a *dead zone* of some finite size where the variable can change without being detected, and a *dead time* before the sensor can possibly detect a change in the variable.

The dead zone effect creates an initial delay, or lag, which reduces the speed of response of the instrument. As an example, the dead zone of an on-off controller is the zone that lies between the on and off set point values. Essentially, it is the span during which there is no control, or the span between the on control and the off control. The dead zone may also be considered a neutral zone where some variable change occurs and the change must exceed the established value for control action to be initiated.

Dead time effects are illustrated in Figure 8.3. Dead time does not cause any change in the process reaction characteristics. It simply delays

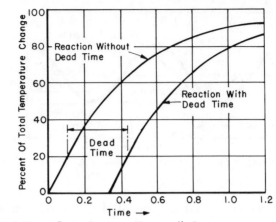

Figure 8.3 *Dead time effects. Courtesy Honeywell, Inc.*

the reaction. Dead time probably introduces more difficulty in automatic control than a lag occurring at any other point during a control cycle. During this period a controller is useless, because it cannot initiate corrective action until after the change has occurred.

Speed of Response. A complete instantaneous response to a change in a variable is an ideal that has not been achieved in any physical system. While the response may start immediately, it takes time to complete its effect on the system.

The time required for a response to have an effect on the system, in a general sense, is called *lag*. The preferred engineering term is *delay in response*. It is the falling behind, or retardation, of one physical condition with respect to another physical condition to which it is related, as discussed under dynamic error and dead time.

The faster the response of the primary detector, the more rapidly a measurement and corrective action can occur.

The atomic age has developed equipment capable of delivering pulses of energy that last only a very short time. This time can be measured in nanoseconds (10^{-9} s). Electronic measuring equipment has been developed that can measure and display these rapidly generated signals. Such equipment is becoming available for general use, on an industrial basis, and is already widely used in research and development activities, radar detection, and nuclear energy applications.

In the use of radioactive sources in industrial applications, discussed in Chapter 10, responses of the order of microseconds are acceptable for starting corrective action.

For dynamic measurements of strain, acceleration, and motion, responses in the millisecond time span are being accomplished, and servomechanisms are capable of making corrective action in this same type of time span. These fast response units serve only a small part of our total process and manufacturing establishments, because they are not needed in all operations.

As discussed in Chapter 5, the response of thermocouples now appears to be quite slow, yet they have filled and still fill an important role in many industrial processes. In many cases they are adequate to produce a perfectly good product that man needs for his existence or comfort. To illustrate the difference in response time as a percent of the total temperature change for thermal detectors, the curves in Figure 8.4 for a thermal radiation detector, in Figure 8.5 for two types of thermocouples in a still liquid, in Figure 8.6 for thermometer bulbs in a moving liquid, and in Figure 8.7 for different thermocouple arrangements in air tell a significant story about the wide range of response.

These curves indicate that there are several factors that can affect the speed of response of a primary detector. The *mass* of the detector that transmits the change signal (such as the size of a thermocouple measuring junction), the *length* of the capillary of a filled system thermometer, the *mass* of a protecting well, the *insulating air space* between a sensor and the protective well, the effect of the *velocity* of the moving medium being measured, and the *kind of medium* being measured. While the figures shown are for temperature applications, the conclusions apply generally to all measuring systems in some degree.

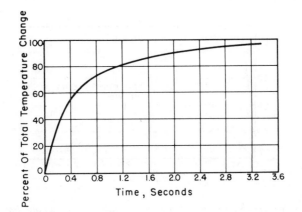

Figure 8.4 *Speed of response of a thermal radiation detector. Courtesy Honeywell, Inc.*

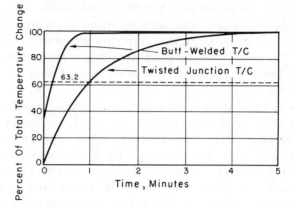

Figure 8.5 *Speed of response for two types of thermocouples in a still liquid. Courtesy Honeywell, Inc.*

Dynamic Response of First Order Type Instruments.

First order type instruments include elementary temperature measuring devices such as thermometers using a bulb or other primary element, which is mainly thermocapacitance, surrounded by a heat transfer film, which is primarily resistance. A pressure gage may have a lightweight bellows or diaphragm enclosing a given volume, and a pressure inlet made by a small tube that provides resistance to fluid flow.

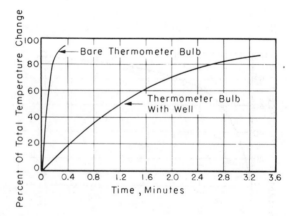

Figure 8.6 *Speed of response for thermometer bulbs in a moving liquid. Courtesy Honeywell, Inc.*

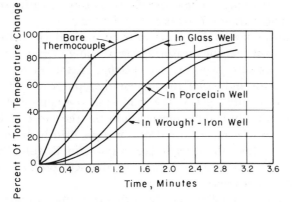

Figure 8.7 *Effects of wells on the speed of response of thermocouples compared to a bare thermocouple. Courtesy Honeywell, Inc.*

Ideally, the dynamic response of first order type instruments to a step change can be represented by the equation

$$T \frac{d\theta}{dt} + \theta = \theta_F \tag{8.1}$$

where θ = value indicated by the instrument
   θ_F = final steady value
   t = time
   T = a time constant, RC seconds or minutes in an equivalent resistance circuit

For given initial conditions this particular linear first order differential equation has the solution shown in the equation

$$\frac{\theta}{\theta_F} = 1 - e^{-t/T} \tag{8.2}$$

This equation represents a single exponential response so that, if plotted in dimensionless ratios, time t/T represents a change in time constant T and alters the time scale. Under these circumstances, as the time constant T, which is the measuring lag, becomes larger, the response, on maintaining the same shape, becomes proportionately slower. Following the normal nomenclature for industrial instrumentation, the time constant T is the time required to indicate 63.2% of the complete change. In electronic instrumentation and in other special cases, the measuring lag can be specified by the time required to attain 90, 95, or

99% of the full change. These specifications can be easily converted to obtain the time constant T by use of tables for e^{-x}.

The measuring lag for first order instruments is given by time constant T. It is significant that the measuring lag of many thermometers, thermocouples, and pressure gages may be specified in this manner with accuracy sufficient for industrial instrumentation purposes. The time constant T is numerically equal to the product of resistance and capacitance in electrical units. In a thermometer bulb the resistance to heat flow is given in degrees of temperature drop for each British thermal unit per minute of heat flow; the capacitance is given in British thermal units per degree of temperature rise, and so the RC product for thermal resistance and capacitance is in minutes.

The dynamic error following a step change is not as important as the dynamic error during a linear or sinusoidal change, since a measured variable does not ordinarily change in this manner.

The dynamic response of these instruments to a linear change can be represented by the equation

$$T \frac{d\theta}{dt} + \theta = Kt \tag{8.3}$$

where K is the rate of change of the true value of the measured variable. In the last term Kt shows that the actual measured variable changes linearly with time. One particular solution for a set of given initial conditions is shown in the equation

$$\frac{\theta}{KT} = \frac{t}{T} - 1 + e^{-t/T} \tag{8.4}$$

If these values are plotted using dimensionless quantities, it is apparent that after the transient has disappeared the instrument lags behind by essentially a constant time and a constant value. The transient normally disappears after a time equal to about three constants, since by then 95% of the lag is normally attained.

The lag of the instrument is given directly by the time constant T, and we now see why this particular value was chosen. As an example, for a thermometer having a time constant of 1 min, it will ultimately lag behind a linear change by 1 min. The dynamic error is given in this case by the value KT. With a thermometer having a 1 min lag and a linear change of 5° of temperature per minute, the dynamic error is 5°. By this we also see that the dynamic error increases directly with the rate of change of the measured variable. When a small dynamic error is desired or necessary (we may call this high fidelity), either the instrument must have a small lag or the measured quantity must change very slowly.

Dynamic response to a sinusoidal change can be evaluated in the equation

$$T \frac{d\theta}{dt} + \theta = A \sin \omega t \tag{8.5}$$

where θ = value indicated by the instrument
t = time
T = time constant, RC seconds or minutes
A = amplitude of cycle of measured variable
ω = angular frequency of cycle

By omitting the transient terms, a solution with given initial steady state conditions can be represented by the equation

$$\left(\frac{\theta}{A} \right)_{ss} = \frac{1}{\sqrt{1 + (\omega t)^2}} \sin (\omega t - \phi) \tag{8.6}$$

This equation shows that the instrument lags the measured variable by a geometric angle, and also that the amplitude is reduced or attenuated. The complete equation yields little information, so that the lag and dynamic error may be treated separately.

If the instrument produces a sine wave, which is displaced by the lag angle θ varying from 0 toward 90 geometrical degrees, the phase lag in degrees is not as convenient as the lag of the instrument expressed in time units. The lag of the instrument in time units is expressed by the equation

$$\text{Lag} = \frac{1}{\omega} \arctan \omega T \tag{8.7}$$

As an example, suppose that a thermometer with a time constant of 1 min measures a temperature cycling with a 10 min period. The lag is then:

$$\text{Lag} = \frac{10}{2\pi} \arctan \left(\frac{2\pi}{10} \times 1.0 \right) = 0.89 \text{ min} \tag{8.8}$$

When the same thermometer is used, the amplitude indicated by the instrument is only 85% of the actual value. If the measuring lag of the thermometer increased to 5 min, which is entirely possible, the amplitude drops to 31%. If the measuring lag is 1 min, but the cycle increases to a

2 min period, the amplitude drops to 31%. Under such conditions the dynamic error of the instrument is so large that the instrument readings are meaningless. For the instrument to indicate a true value within 10%, the number $\omega\ T$ must be about 0.5 or less.

Dynamic Response of Second Order Type Instruments. Many instruments have additional lag because of the existence of a mass that must be accelerated, or because more than one element of fluid capacity is involved. These are second order type instruments. For example, a thermometer bulb is surrounded by a metal or ceramic tube for the purpose of protection. The primary element can be assumed to be mainly thermocapacity, the air depth mainly resistance, the wall mainly thermocapacitance, and the heat transfer film resistance. A pressure gage may have a similar construction involving separate resistances and capacitances.

Most mechanical instruments, such as pressure gages and manometers, have some small but significant mass which must move during instrument operation. A pressure gage has a mass in a system, a spring, and viscous damping means. This viscous damping is often nothing but the air damping and viscous friction of the moving members. Liquid manometers also have the same components, the damping being caused by viscous fluid friction.

These second order instruments possess a dynamic response to a step change which may be ideally described by means of a second order differential equation:

$$a_2 \frac{d^2\theta}{dt^2} + a_1 \frac{d\theta}{dt} + a_0\theta = a_0\theta_F \tag{8.9}$$

where θ = value indicated by the instrument
            θ_F = final steady value
            t = time
   a_2, a_1, a_0 = constants

The constants in the previous equation must be derived by an analysis of each instrument. For instruments similar to the temperature instruments, the constants are combinations of the resistances and capacitances. For the pressure gage the constants include the mass, the damping, and the spring gradient.

For instruments having rotating instead of linear motion, mass becomes moment of inertia. For the fluid manometer the constants include the length of the column and the damping factor. For galvanometer electric instruments, the constants become those for moments of inertia, electrical damping, and spring gradient.

The solutions to the differential equation are well known and involve three cases: an oscillatory condition in which the roots of the auxiliary equation are conjugate complex with negative real parts; critical damping in which the roots are negative, real, and equal; and an overdamping condition in which the roots are negative, real, and unequal.

8.3 BASIC PROCESS CHARACTERISTICS

In the selection of automatic control equipment, there are two basic effects, exhibited by every process, which must be considered. These are *load changes* caused by altered conditions in the process, and *process lag*. Process lag is the delay in time required by the process variable to reach a new value after a load change occurs. Process lag is caused by capacitance, resistance, and dead time. One or more of these process characteristics may be involved.

Load. The load of a process is considered the total amount of corrective control element required at any one time to maintain the established process level. Since the load change can be in either direction, that is, it can call for corrective action either above or below the selected set point, the control element must be capable of furnishing the entire span of corrective action. For example, if a factory is to be maintained at a constant temperature, the load can call for either heating or cooling. This means that both a heating plant and an air conditioning unit are needed as corrective control elements, and each unit is rated at a capacity capable of maintaining the temperature under ambient temperature extremes of the area in which the factory is located.

Load changes can occur gradually over a period of time, or they may change quite rapidly. The electrical power generating industry is faced with both types of changes on a daily basis. From midnight to 6:00 A.M. every morning, there is a relatively small demand for power and few large demand changes occur. From 6 A.M. to 8A.M. there is a growing demand as most people wake up, use lights, and turn on electric stoves or other electrical appliances. Between 7 A.M. and 9 A.M. the load increases, as businesses open and shops use power for machining, welding, and other activities. From 11:30 A.M. to noon the requirements suddenly drop off for the lunch period. From 12:30 P.M. to 1:00 P.M. the demand increases, and from 4:30 P.M. to 5:00 P.M. the commercial demand again drops. During the winter the demand for power increases from 6:00 P.M. to 10:00 P.M., as cooking, lighting, and television sets consume electricity. During the summer the demand usually peaks during the afternoon, when air conditioning units have their largest loads. Without fast acting controls and reserve generating capacities, electrical power utilities would face some real difficulties. If the demand drops sharply and the control action is too slow, there will be a power surge that will pop light bulbs, burn out ballasts, speed up or burn out motors, and blow out overvoltage protective circuit breakers or other devices. However, if the demand suddenly increases, and there is an inadequate reserve to meet the increase, the voltage drops, lights dim, and motors slow down. Electric power companies operate on a set point for voltage and frequency, with high and low levels designed to maintain the majority of electrical equipment depending on them.

Process Lag. How much time does a power utility have to meet a sudden demand for more or less power? It should be prepared to meet a load change within seconds. Power distribution centers are usually set up on a maximum 20 s cycle. If trouble develops, the controls automatically reset every 20 s until the difficulties are corrected. In a few rare cases there have been widespread power failures because the automatic control shut down the equipment before corrective action could be carried out.

All systems are not as large, complicated, and fast acting as an electrical power system. Where large volumes of materials have to be heated, cooled, or maintained at a given temperature, pressure, or flow rate, there is a definite process lag in response to a load change. This change takes time which is controlled by the volume, the density of the material, and the capacitance of the system. Two systems may have exactly the same volume and the same material density, but completely different capacitances. This means that in defining a process for control features the type of capacitance must be specified.

To more clearly illustrate differences in capacitance and capacity and to differentiate clearly between them, we use Figure 8.8. Both tanks have the same volume and therefore the same capacity of 128 ft^3. This capacity is the measure of the ability of a process to contain or hold energy or material. In this case the liquid volume for both processes is the same. However, there may be a difference in the densities of the liquids, so that there may be a distinct difference in the total weight of the volumes. Also, if a flow tap were placed at the base of each of the volumes, there would be a difference in the pressure due to the difference in the head of the two volumes.

There is also a distinct difference in the capacitance of the two volumes when referred to the reference volume per foot of height. Here we define the *capacitance* as a measure of the ability of the process to hold

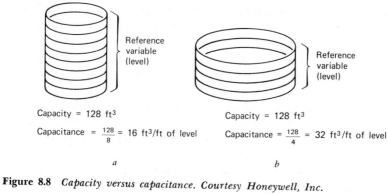

Capacity = 128 ft³

Capacitance = $\frac{128}{8}$ = 16 ft³/ft of level

Capacity = 128 ft³

Capacitance = $\frac{128}{4}$ = 32 ft³/ft of level

a

b

Figure 8.8 *Capacity versus capacitance. Courtesy Honeywell, Inc.*

a quantity of energy or material per unit quantity of some reference material. The capacitance is measured in units of quantity divided by the reference variable. In the body of *A* in Figure 8.8, the capacitance for a level reference variable is 16 ft³/ft of level and the body of *B* has a capacitance of 32 ft³/ft of level. This shows that, while *A* has twice the liquid level capacity of *B* with the same volume capacity, *B* has twice the liquid volume capacitance of *A* with respect to foot of level. This comparison emphasizes that care must be exercised in analyzing and defining the capacity and capacitance of a process for which a measurement and control system is being selected.

To further demonstrate that there can be other differences in capacitance, let us assume that the liquid in *A* has a density that requires 100 Btu of heat to increase its temperature 1°F and that the liquid in *B* requires only 50 Btu to increase its temperature 1°F. The thermal capacitance per degree Fahrenheit of *A* is then twice as great as that for *B*, which is the reverse of the volume-to-level capacitance ratios.

A large process capacitance, whether a continuous or a batch system is involved, has a tendency to keep the controlled variable constant in spite of load changes. Conversely, a large capacitance make it more difficult to change to a new value. However, a small process capacitance makes it more difficult to control the variable at a constant value, but makes it easier to change to a new value. Since capacitance of a process has these characteristics, it becomes a major factor in the analysis of any process.

In a process in which flow is involved, whether it is liquid, solid, or electrical flow, we find the second basic type of process lag defined as *resistance*. Resistance is measured in units of potential required to cause or produce a unit change in flow. This resistance may be to the flow of heat through a solid or air, or flow of electricity through a wire or a semiconductor. The same principle is involved, because a given potential is required to cause the flow. In the case of temperature, there is a thermal resistance which is actually the change in temperature that occurs per unit of heat flow, usually expressed as British thermal units per second. In electrical flow the electrical resistance is usually expressed in ohms per foot or voltage drop per foot.

Resistance to heat flow is often caused by gas or liquid films forming on the surface of a thermal conductor. These film surfaces require a larger control element or thermal potential source to accomplish the required thermal transfer. The time required to overcome this resistance causes a process lag which must be considered in the analysis of the system and its response for measurement and control of the process.

In addition to the process lags just discussed, dead time must also be considered in the process reaction, just as it is considered for the measuring instrument. This is especially true in continuous processes in which it is necessary to transfer energy by means of a fluid flow through a distance at a given velocity. It takes a finite time for the fluid entering the process to flow through the process, and it takes the process time to absorb the energy. During this dead time the control element is helpless, because it cannot take corrective action until the change has occurred, and it takes almost the same length of time for the control to take corrective action unless some means are provided to prevent the delay. These are discussed later, after we examine the basic characteristics of automatic control.

First Order System Response.   A first order system can be described by one factor *T, the time constant,* just as first order instruments can be described by one constant *T*. A basic equation can be written as

$$\frac{\text{Output}}{\text{Input}} = \frac{1}{Ts + 1} \tag{8.10}$$

This basic equation describes a single-capacitance and a single-resistance type of first order system. This type of expression is often referred to as a transfer function. Actually, it is simply the Laplace transform of the differential equation describing the system. Any two systems that have the same transfer function have the same response with respect to time. That is, any single-capacitance, single-resistance system has this transfer function.

Several techniques have been developed and can be used for evaluating the time constant. The methods most commonly used are: step analysis, frequency of response, ramp input, correlation, and statistical analogies. The best method in any given instance is dictated by both the objective of the test and the equipment or tools available. Techniques such as the step analysis method require only simple equipment and paper calculations, while others require computer evaluation.

To describe a process a mathematical expression must relate output to input and show how one varies with respect to the other with the passing of time. Classically, this expression is a differential equation differentiated with respect to time. For example, the equation

$$A - B = C \frac{dH}{dt} \tag{8.11}$$

says that, if the outflow of a volumetric rate B is subtracted from an inlet rate A, the difference equals the area of the tank bottom C in head dH during a change in time dt. Stated in differential terms,

$$\text{Rate } A - \text{rate } B \times \text{time } dt = \text{volume change } dV \tag{8.12}$$

Using Ohm's law as a basic equation $(E = IR)$, we can use a hydraulic analog for flow through a linear resistance as an example. If B, the outflow rate, is equivalent to an electrical current produced by a difference of potential across a linear resistance, we can rewrite the equation as

$$A = B(CsR + 1) \tag{8.13}$$

where $H = BR$

and $\frac{d}{dt} = s$

We can place this in the general output-input equation:

$$\frac{\text{Output}}{\text{Input}} = \frac{B}{A} = \frac{B}{B(CsR + 1)} = \frac{1}{CsR + 1} \tag{8.14}$$

where $T = RC$, so that our basic equation is the same as equation 8.10:

$$\frac{\text{Output}}{\text{Input}} = \frac{1}{Ts + 1} \tag{8.15}$$

The time constants so far discussed are useful in creating an analog of a system and in selecting the type of controller and optimum control settings necessary for successful operation of the process.

Step Analysis of a Single–Time Constant System. The time constant of a single–time constant process is determined simply by finding the time required to reach 63.2% of the final change in the output following a step change. A step change in input to a single–time constant process can be a sudden increase in flow rate, a sudden decrease in flow rate, or a change in input voltage which produces a change in the output response. The equation for such a response curve is:

$$1 - e^{-t/T} \qquad (8.16)$$

This equation is quite useful in plotting the percent incomplete $e^{-t/T}$ on logarithmic paper. Such a curve represents the percent of the total rise that has not taken place at each point in time. Thus the line starts at 100% and drops downscale.

If the plot is a pure exponential on a logarithmic scale, the curve will be straight. The reason for emphasizing this point is that, when the percent incomplete curve is straight on logarithmic paper, it verifies the fact that the process has a single time constant.

Time constant T is the time required for a 63.2% rise to final value or completion, which is the same as 36.8% incompletion, or 36.8 on the percent incomplete scale.

Step Analysis of a Two–Time Constant System. Assuming a two–time constant process in which all feedback or corrective action is eliminated, a step change is applied and the resultant change in measured variable is recorded. The magnitude of the step change is not critical, however, the steps should be large enough to produce the necessary data but not large enough to disturb the process operation or cause the process to exceed its normal or linear limits. In other words, the largest allowable upset is the best for evaluation purposes. If the upset must be kept small, the recorder can be recalibrated to expand the reading. The resulting reaction curve can be used to evaluate the first and second order time constants using the percent incomplete method.

The reaction curve or typical second order, two–time constant system is used to calculate and tabulate the percent incompletion of the process (100% minus percent completion of the process). If any dead time is known to exist in the system, it should be subtracted from the time coordinate of each percent incomplete point. The dead time, which is also called the transportation time, is revealed by the reaction curve, and is the time it takes for a change in variable to "travel" to the measurement point. For example, the dead time in a heating process may be the time required for the heated fluid to reach a measuring element located some distance downstream from the heating element. Under these circumstances calculations can be made from a reaction curve whose time axis is shifted to eliminate dead time.

The percent incomplete curve is plotted on semilogarithmic graph paper. Percent incompletion is plotted on the vertical axis, and time on the horizontal axis. At time zero no change has taken place, and the reaction is 100% incomplete.

To determine the time constant, three curves must be completed. In order to determine the first time constant T_1, the reaction curve is extended to the zero time basis for the linear portion. The intercept of this extended linear portion will exceed 100%. The first time constant T_1 is the time it takes the reaction curve to reach 36.8% of the linear extension value in excess of 100%. This is shown in Figure 8.9.

8.4 BASIC CHARACTERISTICS OF AUTOMATIC CONTROL

Automatic control indicates that the control occurs as a corrective measure in response to a signal, and that it is accomplished without human intervention to effect the control action. In other words the action occurs automatically.

One basic characteristic is the manner in which the controller acts to restore the controlled variable to the desired value. This is functionally the mode of control. The common modes of control consist of (1) two position, (2) floating, (3) proportional, (4) proportional plus reset, and (5) rate. A control system may employ one or more of these modes in its industrial function.

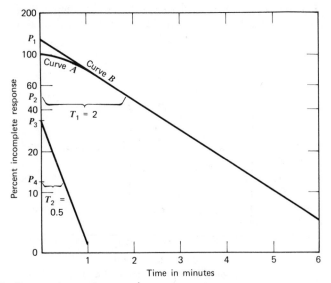

Figure 8.9 *Percent incomplete curve for a two–time constant system step analysis.*

Two Position Control. In Chapter 5 the two-position controller for temperature control was discussed. In its basic concept a two-position controller can only operate when the control element is turned fully on or fully off. There is no intermediate position of control between the two extremes on and off.

In the case of the on-off temperature controller, the controller either turned the heat source on or turned it off. If the controlled element is a valve, the controller either fully opens the valve or fully closes it. A valve on-off control is shown graphically in Figure 8.10.

This figure illustrates the neutral or dead zone during which there is no intermediate control in the system. As shown, the neutral zone has been designed at 1% of the instrument range. When the controlled variable is $\frac{1}{2}$% below the set point, the valve opens, and when it is $\frac{1}{2}$% above the set point, the valve closes. From 49.5 to 50.5% the valve remains open, and from 50.5 to 49.5% the valve remains closed.

Two-position control is simple and is quite popular for processes with slow reaction rates for which it can furnish adequate regulation. Two-position control is best suited to a process with minimum transfer rate in which the two extreme positions can be adjusted to permit an input just slightly above and slightly below the requirements for normal operation. Some load change can be handled by two-position control, provided the changes are neither fast nor of extensive magnitude. Changes in load cause the controlled variable to cycle at a new rate and magnitude, depending on the rate and direction of the load change. The cycling at each new load assumes a different average value, depending on

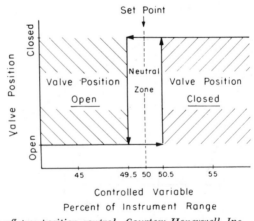

Figure 8.10 *On-off two-position control. Courtesy Honeywell, Inc.*

the direction of the load change. When the load change is too large or occurs too often or too rapidly, the control is unstable. The system may supply either too much or too little control agent, and it may go into a rapid cycle change commonly referred to as *hunting*. Hunting is a continuous cycling from on to off to on, with the controlled variable continuously cycling above and below the normal set point condition.

In industrial applications the two-position control is well suited to the control of ovens and furnaces in which the main losses, after loading, are changes in ambient temperature and radiation losses.

Floating Control. When load changes are large and/or rapid, it is necessary to provide a better type of control than the two-position on-off type. The first refinement that can be added to supply intermediate positioning of a control element is called *floating control*. Floating control moves the final control element at a constant speed, in either direction, whenever the controlled variable changes a predetermined value from the set point. In contrast to a two-position controller, which changes the position of the final control element from on to off, the floating controller changes its speed from on to off.

In a floating control system the final control element does not move while the controlled variable remains within the neutral zone. When the controlled variable moves outside the neutral zone, the final control element moves in the appropriate direction to correct the controlled variable. This movement continues until the controlled variable moves into the neutral zone, or until the full final control element input is actuated or until it is completely off, depending on the correction needed.

In floating control the final control element moves more slowly than in two-position on-off control, because intermediate positioning is desirable for the application. Floating control has the advantage of counteracting gradual load changes by the gradual changing of the final control element. This minimizes cycling but does not eliminate it entirely.

Care must be used in applying floating control. If the control element moves too rapidly, two-position control results. If the control element moves too slowly, the control system is not capable of keeping pace with a sudden change. Ideally, the control element should be chosen to change at a rate fast enough to keep pace with the most rapid rate at which load changes can occur.

Floating control is not designed for use in processes having a significant lag. If used in such a process, excessive cycling takes place, often beyond that occurring in two-position control. Also, it was not designed for use in processes in which load changes occur rapidly, even for small changes. Other types of control should be used on these types of processes.

Proportional Control. In processes requiring a smoother control than can be provided by the simpler two-position on-off or the floating controller, *proportional control* can be used.

Proportional control provides a fixed linear relationship between the value of the controlled variable and the amount of control exercised by the final control element. A proportional controller moves the final control element to a definite position for each value of the controlled variable.

To illustrate the action of a proportional controller, let us examine Figure 8.11. The valve chosen is for a range of action to provide a control element to a process for the temperature range from 100 to 200°F. The set point is 150°F for a control band of ±50°F. When the controlled variable is at 100°F or less, the valve is wide open. When the temperature is between 100 and 200°F, the controller moves to a position that is proportional to the value of the controlled variable. It is a definite percentage of the amount of control element passed by the valve position and is represented by the proportional position action line shown. At 125°F the valve is 75% open, at 150°F the valve is 50% open, at 175°F the valve is 25% open, and at 200°F or more the valve is fully closed.

As shown, the control covers the range from 100 to 200°F, so we have proportional control over a 100°F span and this is known as the proportional band. To express this value as a percentage of the full scale range of the controller, we can see that the scale is from 50 to 250°F, a span of 200°F. While we have a 200°F range on the controller, we have control only over 100°F, so the proportional band is 50% of the full scale range. The majority of proportional controllers are designed with adjustable proportional bands. Normally, the width of the proportional band determines the amount of valve motion for any given change in the controlled variable. This means that, the larger the proportional band, the

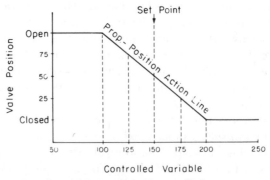

Figure 8.11 *Proportional control. Courtesy Honeywell, Inc.*

smaller the change in valve position for any given change in the controlled value. Conversely, it also means that, the smaller the proportional band, the greater the change in valve position for a given change in the controlled variable.

While the proportional controller gives smoother action than the simpler modes of control, it should be noted that there is a definite or fixed relationship between the value of the controlled variable and the valve position. This can be a distinct disadvantage whenever there is a load change.

As discussed earlier, when a load change occurs, it requires more or less control element to maintain the predetermined set point value of the controlled variable. The proportional controller cannot change the valve position. As shown in Figure 8.10, it can move the control valve to one, and only one, position for any given value of the controlled variable regardless of the process load. This characteristic of the proportional controller results in *offset*. Offset is a sustained deviation of the controlled variable from the set point value as long as the load change is maintained. Offset can be overcome by a *manual reset* adjustment provision on most proportional controllers by shifting the proportional band about the set point. However, *this means that we do not have an automatic control for load changes with proportional control when offset occurs.*

The effects of manual reset are shown graphically in Figure 8.12. The same proportional band is maintained for each of the three curves. Curve *A* is for normal load operation, curve *B* is for a lighter load operation, and curve *C* is for a larger load, with the shift of the proportional band in the direction of the load change. If the proportional band is wider when the load change occurs, the offset will be greater. With a wide proportional band even small load changes lead to offset.

Proportional control was designed to reduce cycling below that of two-position control. When used in the right applications, it is an excellent control mode. The proportional controller performs best on processes that have a large capacitance, a relatively slow reaction rate, and relatively small process lag and dead time. All these characteristics promote stability and make a narrow proportional band possible. This results in fast corrective action and a minimum of cycling.

Another type of proportional control is available only on electric controllers. This is a time proportioning control in which the final control element is either on or off as in two-position control, but the ratio of the on time to the off time for each cycle varies as the controlled variable moves above or below the set point.

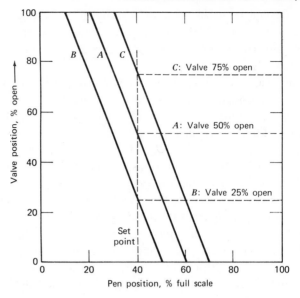

Figure 8.12 *Manual reset. Courtesy Honeywell, Inc.*

As an example, let us assume that the controller has a full cycle time of 10 s and a proportional band of 20°. At the set point the controller turns the final control element on for 5 s, and then off for 5 s. If the temperature drops 10°, the final control element is on for the full 10 s cycle. However, if the temperature is 5° above the set point, the final control element is on for 2.5 s and off for 7.5 s during the 10 s cycle. At 10° above the set point the final control element is off for the full 10 s cycle.

The time proportioning controller has the same disadvantages as the proportional position controller with respect to load changes and offset.

Proportional-plus-Reset Control. To have an automatic system using proportional control, when there are load changes, the reset must be automatic. In general the control mode is the same for automatic reset in proportional control, except that the reset is automatic instead of manual for the proportional band adjustment. In addition to the proportional band adjustment, the proportional-plus-reset control has a reset rate adjustment which determines the rate at which the proportional band is shifted to meet load change conditions.

For a given reset rate setting, the reset is proportional to the amount of deviation of the controlled variable. Reset is also proportional to the duration of the deviation. It is cumulative as long as there is a deviation. As soon as the controlled variable returns to the set point, the reset returns to zero.

Proportional-plus-reset controllers position the final control element with respect to the location of the controlled variable within the proportional band and with respect to the time duration and amplitude of the controlled variable from the set point.

A graphic illustration of how a proportional-plus-reset controller changes the position of the control valve when there is a sustained change in process load is shown in Figure 8.13. The control action is shown in its component parts. The proportional action is shown in *A,* the reset action in *B,* and the proportional-plus-reset control in *C.*

We now have an automatic control action which can be used with small process capacitance, a fast reaction rate, and large load changes, providing there is no large dead time or excessive transfer lags. Large dead times or excessive process lags can still cause overshoot and excessive cycling. These characteristics of a process require another control known as rate action.

Rate Action. Rate action exists only in combination with proportional or proportional-plus-reset action. When a deviation occurs, rate action basically provides for an initially large overcorrection. It causes the final control element to provide more correction than it would on the initial signal from proportional or proportional-plus-reset action. After the large initial change, the controller begins to remove its effect, leaving the proportional or proportional-plus-reset responses to position the

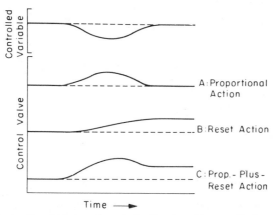

Figure 8.13 *Proportional-plus-reset control. Courtesy Honeywell, Inc.*

the final control element. This results in an early stage extra supply of control agent designed to counteract or overcome the unfavorable effect of process lag. Essentially, it is a temporary overcorrection proportional to the amount of controlled variable deviation from the set point.

Proportional-plus-Rate Control. Proportional-plus-rate control is illustrated in Figure 8.14. The control action is broken up into its components similar to Figure 8.13. The proportional response is shown in *A*, the rate response in *B*, and the proportional-plus-rate response in *C*. By comparing the graphs of Figures 8.13 and 8.14, it is clear that the controlled variable is the same, there is no change in the proportional action of *A*, the offset remains fixed in Figure 8.13 calling for reset action in *B*, while the rate action of *B* in Figure 8.14 overcorrects first in one direction and then in the other; the proportional-plus-rate action of *C* in Figure 8.14 shows the overall result of the controlled variable being brought back to the set point. In the same time span the proportional-plus-reset action shows the controlled variable still above the set point.

Rate action can be adjusted to provide for the lag in the process and is usually a time adjustment to advance the effect of proportional action on the final control element. This is shown graphically in Figure 8.15. The rate time is determined by subtracting the time required for a certain movement of the final control element called for by the combined effect of proportional-plus-rate action from the time required for the same movement due to the effect of proportional action alone.

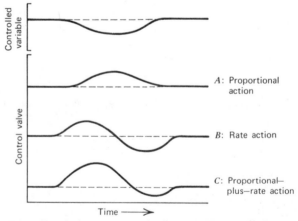

Figure 8.14 *Proportional-plus-rate control. Courtesy Honeywell, Inc.*

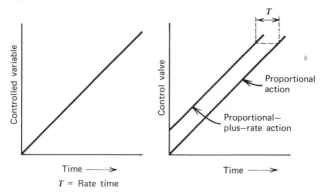

Figure 8.15 *Rate action. Courtesy Honeywell, Inc.*

Proportional-plus-Reset-plus Rate. Processes with a large dead time (usually considered to be longer than 2 min) and/or large process or transfer lag are sometimes difficult to control with either proportional-plus-reset or proportional-plus-rate actions. The proportional band must be set exceptionally wide and the reset time quite slow to prevent excessive cycling. When load changes occur, a wide deviation takes place and it takes a long time for the controlled variable to return to the set point. The addition of rate action to the proportional-plus-reset mode often solves the control problem.

The addition of rate action is particularly useful during startup of batch processes. The initial large correction brings the controlled variable up to the set point very quickly, reduces overshoot, and reduces both the time and magnitude of cycling. The variable lines out at the set point sooner than it would with the proportional-plus-reset mode alone.

In conclusion it can be seen that each mode of control is applicable to processes having certain combinations of basic characteristics. In choosing a control mode for automatic control, the best choice is the simplest mode that will accomplish the control necessary to produce a marketable product at a profit. This choice will be the one that gives the best results and will ultimately be the most economical. A too complicated control, if not absolutely necessary, is expensive and may give poor control, while a too simple system may make adequate control difficult or impossible.

The principal characteristics of the process required for the different modes of control are summarized in Figure 8.16. Extensive experience is

Mode of Control	Process Reaction Rate	Transfer Lag or Dead Time	Process Load Changes
Two-position	Slow	Slight	Small and slow
Proportional	Slow or moderate	Small or moderate	Small
Proportional-plus-rate action	Slow or moderate	Unlimited	Small
Proportional-Plus-reset action	Unlimited	Small or moderate	Slow, but any amount
Proportional-plus-reset-Plus-rate action	Unlimited	Unlimited	Unlimited

Figure 8.16 *Control modes and process characteristics. Courtesy Honeywell, Inc.*

required in making the best choice of control mode for a specific application. This choice may well be a general control mode which can be used on more than one operation. The batch process industries are masters of choosing control functions. Batch control is well planned; it is not a rare occurrence.

8.5 COMPUTER CONTROL SYSTEMS

Computers serve the industrial field in three areas. One is the purely business area, for inventory, payrolls, taxes, and other high volume non-producing but necessary business paper handling. The second area is design, research, and development of new products and processes. In the industrial process and manufacturing functions, the third area is the operation and control of the process. Here extreme care must be duly exercised to be certain that a system concept has been really evaluated so that management will have a profitable success story. True system concept evaluation is not always true, because an on-line computer system can go wrong as a result of improper hardware, control program software insufficiencies, and negative personnel approaches. Any one of the three can spell trouble and most probably disaster to such a venture.

On-line computers can be designed to do a competent job, but this requires a *thorough systems analysis* so that *the computer is programmed to do a specific area of work under a specific set of circumstances.* It has to be a *closed loop system,* and *there cannot be any ambiguity in its instructions.* It should be permitted to do its work without operator interference, such as an operator attempting to override or outguess it by manual control. If such a condition is considered desirable, the program logic must be prepared to accommodate it.

When a computer is to control several process variables or operational functions on a time shared basis, the program has to be prepared to cover such items as (1) proper control in the event any one of the control devices in a functional area fails, (2) the overload that may be imposed on associated controllers if one controller is removed for repair or calibration, and (3) the effects of altering a memory (stored information). In fact, the program logic must be prepared to cover any contingency that might occur in the closed loop system. Otherwise there will be control problems that will cause loss of time and product.

A computer is a *tool* which faithfully and accurately carries out the functions for which it has been programmed, assuming that the proper computer has been chosen, the system has been adequately analyzed, and the program fits the analysis. As stated earlier, computers are normally used only when a large number of comparisons is needed to make a decision, or when a large number of control points must be scanned, recorded, and evaluated rapidly to maintain proper control characteristics. An example of process control computer usage is the control and operation of blast furnaces such as the one shown in Figure 1.4b at a United States Steel plant. Figure 8.17 graphically shows the functions and path of the process flow which allows the operator to "see" that the blast furnace's automatic charging system is performing according to its programmed instructions. The communications center shown in Figure 8.18 is the nerve center for the complex instrumentation that operates Pittsburgh's largest iron maker. On the Graphic Panel 70 instruments and controls are displayed to the operator. Figure 8.19 shows a different role for the computer in business applications. This is the Dearborn Message Center of the Ford Motor Company. The "wall" of clocks and computers shown is the center of an operation that receives and transmits in excess of 25,000 messages daily using private line facilities for 165 stations, plus commercial teletype (TWX and Telex), for worldwide communication on sales, replacement parts, and manufactured parts ready for delivery.

8.6 SUMMARY

The simplest mode of control that will obtain the desired results for a particular process application can be considered the best. To make this decision, an engineer must evaluate all the errors in the process and the

Figure 8.17 *System flow display and control center. Courtesy U.S. Steel Corp.*

control system, the variables to be controlled, the uncontrollable variables and their effects, the instrument calibration techniques, the system calibration techniques, and the compensation techniques to be applied.

Process and manufacturing industries must develop, train, and use experienced engineers, good engineering practices, reliable and well trained operating personnel, good maintenance practices, and reliable equipment to succeed in the selection and operation of adequate measurement and control systems.

Review Questions

8.1   Give the main functions of a sensor in a measuring system.

8.2   What characteristics require high priority in the selection of a sensor?

8.3   What is static error? When is it important? How can it be eliminated;

Figure 8.18 *Blast furnace control center. Courtesy U.S. Steel Corp.*

8.4 How can you determine the magnitude of a dynamic error? Is dynamic error always a detriment?

8.5 Why is reproducibility important in a measurement system? Is reproducibility more desirable than accuracy? Explain.

8.6 What is sinusoidal variation? When does it occur?

8.7 Explain the difference between dead zone and dead time.

8.8 What factors affect speed of response to effect a control action? Why is fast response normally considered highly desirable? Is it always necessary?

8.9 What is a first order response and what types of instruments have such characteristics?

8.10 What is the relationship between the time constant T and the response in a measuring system? What is the time constant T in a first order instrument?

Figure 8.19 *Dearborn message center. Courtesy Ford Motor Co.*

8.11   How long does it take a transient to disappear in a normal system and approximately what percentage of the lag is attained at this time?

8.12   When is a measurement system considered to have high fidelity?

8.13   What types of instruments are considered to be second order? What characteristics result in this consideration?

8.14   What instruments exhibit a moment of inertia? What is the linear equivalent of inertia?

8.15   What two basic effects must be considered in the selection of automatic control equipment for any process?

8.16   What is the load in a process measurement? What load requirements must be considered to maintain a constant temperature?

8.17   How does process lag compare with instrument lag? Which is more important in a control system?

8.18   Explain the difference between capacitance and capacity of a system.

8.19   What common factor is shared by first order instruments and first order processes?

8.20   What causes resistance to heat flow reaching a thermal conductor?

8.21 How can the time constant of a process be evaluated? Explain why the time constant is considered important.

8.22 Explain how the time constant of a single–time constant system can be found. Is the same technique applicable to a two–time constant system? Explain.

8.23 Explain the meaning of two-position control. When is there really no control in such a system?

8.24 What causes hunting in an automatic control system? How can it be overcome?

8.25 What is the difference between two-position and floating control?

8.26 What function is supplied by proportional control? What is the limitation of proportional control?

8.27 What causes offset? Explain how it can be overcome.

8.28 What is the difference between normal proportional control and time proportioning control?

8.29 When can rate action be used? Why is rate control desirable?

8.30 Where can proportional-plus-reset-plus-rate control be used and why?

8.31 What is the best mode of control to select for an automatic control system?

8.32 What does a good engineer do about the error in a process and control system for which he must provide automatic control?

8.33 How can the overall error for a system having a combination of errors such as dead zone, lag, and dynamic response be calculated?

8.34 Why are instruments designed with a specific range and span of measurement?

8.35 Why is calibration of instruments and instrument systems important if reproducibility is considered more valuable than accuracy in automatically controlled processes?

8.36 With all the possible errors that can occur in a process and control system, how is it possible to obtain accurate measurements with good automatic control?

Bibliography

Anderson, N. A., Step-Analysis Method of Finding Time Constant, *Instruments and Control Systems*, Vol. 36, No. 11, November 1963, p. 130.

Considine, D. M.. *Process Instruments and Control Handbook*, McGraw-Hill, New York, 1957.

Edmunds, Bill, Practical Dynamic Analysis, *Instruments and Control Systems,* Vol. 36, No. 6, June 1963, p. 127.

Honeywell, Inc., *Fundamentals of Instrumentation for the Industries,* G00003-00 10M, Minneapolis, January 1958.

Page, C., Error Analysis and Compensation with DDA, *Instruments and Control Systems,* Vol. 37, No. 7, July 1964, p. 141.

Stein, Peter K., Error analysis, *Instruments and Control Systems,* Vol. 37, No. 6, June 1964, p. 137.

Towner, W. C., Practical Dynamic Analysis, *Instruments and Control Systems,* Vol. 36, No. 11, November 1963, p. 138.

CHAPTER 9

Analytical Industrial Instrumentation

Customers quite often submit stringent specifications to a manufacturer for his product. Such a customer may be the city, state, or federal government. Other customers, such as utility companies, automotive manufacturers, and heavy equipment manufacturers, insist on guaranteed qualities in the generators, motors, metals, plastics, and other products they buy for use in their processes or end products.

The supplying manufacturer must not only meet these specifications, but must also produce evidence that he has met them. Such specifications may require evidence that the product will withstand certain forces related to stress, strain, vibration, acceleration, or shock. Other requirements may specify the measurement of speed, sound, pH, viscosity or density. This means that the manufacturer must either make these measurements at his own facility or have them made, so that he can sell his product. It also means that while the product is being made he has to have an in-process capability for checking that the specifications are being met and a means of final inspection of certain qualities before the product, whether a basic raw material or a complicated assembly, is shipped to the customer.

In this chapter we discuss some of the analytical instruments needed to ensure that the manufacturer has a product that can pass the final inspection, but we reserve nondestructive types of inspection for Chapter 11 because they have applications to both in-process and final inspections and represent a large enough group to be covered separately.

Analytical techniques and instruments to be discussed include strain gages for stress and strain, accelerometers for acceleration, vibration, and shock, linear differential variable transformers for displacement, spectrophotometers and densitometers for density measurements, pH meters for

323

acidity and alkalinity, viscometers for viscosity measurements, relative humidity and dew point units for moisture measurements, and chromatographs for gaseous and liquid mixture analyses.

9.1 STRESS AND STRAIN

Stresses and strains are generated in component parts, subsystems, and systems by weight, temperature, pressure, vibration, or displacement forces.

One of the more popular methods of making these measurements is by means of strain gages. Other methods include stress coat, photoelastics, and a technique called holography.

Metal Strain Gages.   When a load is imposed on any material object, the object expands or contracts or is subjected to shear. If a grid of wire or foil with selected resistive characteristics is securely bonded to the object, it will theoretically stretch or be compressed exactly as the surface to which it is attached. The metallic resistance strain gage is based on the principle that, when a conductor is subjected to a tensile or compressive strain, it exhibits a change in resistance. The magnitude of the change, related to the original resistance, is proportional to the magnitude of the applied strain. This strain is defined as:

$$\text{Strain} = \frac{\text{Change in length}}{\text{Original length}} \tag{9.1}$$

or in symbolic form,

$$E = \frac{\Delta L}{L} \tag{9.2}$$

The unit strain E as determined by strain gages is expressed in microinches per inch.

In the application of strain gages, use is made of a constant of proportionality known as the gage factor which is normally designated G, E, or K. This constant ranges from a value of 2 to 4 for most commonly used strain gage alloys. The gage factor is based on the change in resistance that occurs in the total resistance as related to the change of length of the conductor with respect to its unit length. In equation form:

$$\text{GF} = \frac{\Delta R/R}{\Delta L/L} \tag{9.3}$$

The gage factor and the K factor are considered the same when the input strain is the strain in the speciment to which the gage is attached. To be technically correct the K factor is the transfer ratio or sensitivity of a strain gage that relates mechanical strain input to unit resistance change output. The K factor and the gage factor are not the same when related to the strain in the gage itself.

The transfer ratio K is affected by temperature and, when it is considered synonymous with gage factor, it becomes an uncontrolled variable which contributes to inaccurate strain measurements.

To obtain relatively accurate measurements with metal strain gages, several factors have to be considered. These are the chemical composition of the metal alloy, the heat treatment of the metal, the effects of temperature, the gage size and configuration, the adhesive used in bonding, the cure cycle of the adhesive, the type of gage backing, and the zero shift of the gage with respect to any temperature compensation. When semiconductor gages are considered, a new set of rules applies, and these are discussed next so that a more complete picture can be formed. The discussion of semiconductor gage characteristics is followed by a description of types of gages and their applications.

Semiconductor Strain Gages. Semiconductor strain gages operate on a piezoresistive principle. This is defined as the change in electrical resistivity with applied strain. While all materials show this effect to some degree, in certain semiconductors it is quite large, and an appreciable change occurs with an applied stress. The resistance change occurs under all conditions of static and dynamic strain.

In a semiconductor the resistivity ρ is inversely proportional to the product of the number of charge carriers Ni, and their average mobilities μ_{ave}. In equation form the resistivity may be expressed as:

$$\rho = \frac{1}{eNi\ \mu_{\mathrm{ave}}} \tag{9.4}$$

where e = electronic charge

The effect of an applied stress is to change both the numbers of carriers and their average mobility. The magnitude and the sign of the change depend on the particular semiconductor material, its carrier concentration, and its crystallographic orientation with respect to the applied stress. In a simple tension or compression, when the current through the gage is along the stress axis, the relative change in resistivity $\Delta\rho/\rho_0$ can be expressed as:

$$\frac{\Delta\rho}{\rho_0} = \pi_L\sigma \tag{9.5}$$

where π_L = longitudinal piezoresistive coefficient

σ = stress

For other stress systems similar equations can be developed with different values of piezoresistive coefficients.

As previously discussed, metal wire and foil gages have a gage factor of 2 to 4. Semiconductor strain gages can have gage factors ranging from 45 to 200. The gage factor for a semiconductor strain gages can be expressed as:

$$\text{GF} = \frac{\Delta R}{R_0 E} = 1 + 2v + \pi_L Y \qquad (9.6)$$

where v = Poisson's ratio

Y = Young's modulus

The first two terms represent the change in resistance due to dimensional changes, while the last term represents the change in resistivity with strain.

The larger the gage factor, the higher the material resistance change that gives a higher output and resolution of the strain. When semiconductors are considered for stress measurements, certain characteristics should be considered. In semiconductors with a relatively high number of carriers (in the order of 10^{20} carriers/cm^3), the gage factor is essentially independent of both temperature and strain. Gages made from such materials require few, if any, correction factors to obtain highly accurate data. Manufacturers code their gages for these highly desirable characteristics.

When semiconductor materials contain less than 10^{17} carriers/cm^3, the gage factor becomes quite temperature and strain dependent. Corrections must be applied to obtain accurate data. Under these circumstances the gage factor assumes the equation form:

$$\text{GF} = \frac{T_0}{T}(\text{GF})_0 + C\left(\frac{T_0}{T}\right)^2 E \qquad (9.7)$$

where GF_0 = ambient temperature zero strain gage factor

T_0 = ambient temperature

T = gage operating temperature in degrees Fahrenheit

For such gages appropriate corrections for both temperature and strain dependence are required. These corrections are discussed later.

Gage Forms and Construction. The metallic strain gage active element consists essentially of a conductive grid, supported on some type of backing matrix. The conductor may be a fine wire, or it may be die cut or etched out of metallic foil. Semiductor gages are formed from semiconductor materials which are doped with appropriate carriers to produce the desired characteristics. Both P and N type crystals are used in gage construction.

In the case of a wire conductor, the grid may be planar, formed by weaving the wire around pins, or wraparound, formed by coiling the assembly. The foil gage is made by photographing a reduced image of the desired grid form onto a chemically treated metallic form and then etching away the surplus metal, or by cutting with a very precise die set. These three forms of metal gages are shown diagrammatically in Figure 9.1.

Usually, the wire diameter or foil thickness is 0.001 in. or less, so some support is required for handling. The most common matrix for wire girds is a fine, thin, rag-pulp paper tissue covering both sides of the grid and impregnated with a nitrocellulose cement. Because of the nature of

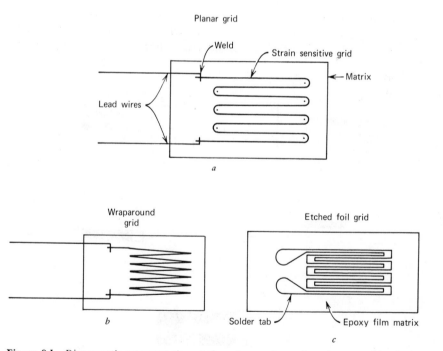

Figure 9.1 *Diagramatic representation of three forms of metal strain gages. (a) Planar grid. (b) Wrap around grid. (c) Etched foil grid.*

its construction, the wraparound gage uses a heavier grade of paper, making a thicker assembly. Foil gages may have a paper matrix, but the more common support is a thin film of epoxy resin. Both types of gages may also be encapsulated in a paper matrix impregnated with a phenol-formaldehyde resin (Bakelite) and cured to a hard shell.

The grid wire used in wire gages is so fine that special equipment must be used to handle it. In order that connections to the measurement system may be made, a comparatively heavy lead wire is welded to the grid wire and permitted to protrude from the matrix, as shown in Figure 9.1. The foil gage is designed to have enlarged tabs at the end of the grid to which wires or conductor ribbons may be soldered. Some types of gages have ribbons welded onto the tabs.

Semiconductor gages may be constructed as shown in Figure 9.2. A single-unit, encapsulated, ruggedized gage is shown diagrammatically in *a*, a dual-element, encapsulated, ruggedized gage in *b*, a bare, ruggedized gage in *c*, a bare, plated, and soldered lead gage in *d*, and a U gage in *e*. These gages are very small compared to most wire and foil gages. Actual semiconductor gage lengths are in thousandths of an inch.

Comparison of Gage Types.   The strain gage is cemented to the specimen being tested, and senses the strain field in a direction parallel to the grid of the gage. This is considered a bonded strain gage. Every

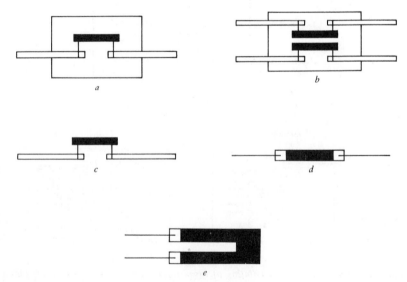

Figure 9.2   *Diagramatic representation of three forms of semiconductor strain gages.*

strain field has an associated second field at right angles. This may be the Poisson effect in the case of a simple uniaxial field, or may be comparatively unrelated in the case of a complex stress pattern. Whichever the case, the second field acts on those parts of the gage grid that do not lie parallel to the primary field. This gives rise to a change in resistance compounded from the effects of both fields, and is controlled by the geometry of the grid. This is known as the *transverse strain effect*. One way of minimizing this effect is to decrease the resistance of those parts of the grid that are affected by the transverse field, so as to make the strain-induced change in resistance negligible. Within a given gage geometry, resistance of the transverse portions can be reduced only by increasing the cross section. This is practically impossible for a wire gage, but can be easily accomplished in a foil gage by widening the end loops, as shown on the etched foil grid in Figure 9.1c.

A mounted gage senses the strain in the plane of the grid, which is separated from the mounting surface by the cement and the matrix. When purely axial strains are being sensed, this presents no problem, since the strain path lies on the surface and is transmitted through the cement and the matrix to the gage. However, when the specimen is under a bending stress, the strain at a surface is defined by the distance of the surface from the neutral axis and the radius of curvature of the axis. As shown in Figure 9.3, the gage grid lies farther from the neutral axis than the specimen surface. This means the gage senses a strain larger in magnitude than that occurring on the specimen surface. This is known as the *offset error*. The offset error has little effect when the bend radius is large or when the distance from the neutral plane is substantial, but becomes critical when either or both of these factors approach the dimensions of the offset between the surface plane and the gage plane.

Offset error may be minimized by placing the gage as close to the specimen surface as possible. In general, the center plane of a wire grid gage is approximately 0.004 in. above the specimen surface, and the center plane of a foil gage is approximately 0.0025 in. above. This means that a foil gage has a substantially lower offset error than a wire gage.

In actual applications offset error may be completely eliminated by calibrating the installation to obtain the applied load/indicated strain characteristic.

Construction of the gage grid and methods of manufacturing the base alloy grid material make it much easier to control the properties of a foil than a wire. As a result, there is a greater degree of reproducibility among gages and better stability of the individual element. These fea-

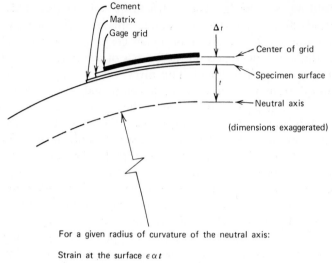

For a given radius of curvature of the neutral axis:

Strain at the surface $\epsilon \propto t$

Strain at the gage grid $\epsilon' \propto t = \Delta t$

When $\Delta t \ll t \quad \epsilon' - \epsilon$

When $\Delta t - t \quad \epsilon' \neq \epsilon$

$$\epsilon' \sim \epsilon = \Delta \epsilon = \text{offset error}$$

Figure 9.3 *Offset error in strain gages. Courtesy Wm. T. Bean, Inc.*

tures make the foil gage a good first choice for most purposes for which a metail strain gage can be used. Where higher gage factors are a necessity, the semiconductor gage is a natural choice and will become more useful as better control of its characteristics is developed.

Strain gages are attached to a surface in different orientations, so that the strain being measured is in the direction of the stress in the part under analysis and the plane of the gage grid. Measurements can be made for torsion, shear stresses, compression, material yield points, and axial, biaxial, and triaxial strain. The gages used in these applications are not selective enough to measure just one stress or strain, without any other contributions, because there are associated fields at right angles to the direction of the strain, which enter into strain measurement. Extreme care has to be exercised so that associated stresses and strains are minimized. Factors that contribute a voltage or current, such as temperature, can be compensated for by using a dummy unit exposed to the same temperature but connected so that the temperature effects in the bonded gage and the dummy gage cancel each other. Special gage arrangements are also used to obtain the most meaningful stress or strain measurements in analytical studies in which the exact stress or strain direction is being established.

Strain gages can also be used to obtain dynamic strain information. A simple dynamic measuring circuit is shown in Figure 9.8.

Strain Gage Circuits. Measurement of strain with a resistance strain gage requires a very accurate measurement of a very small change in resistance. The most commonly used electrical circuit for such an application is the Wheatstone bridge. A schematic representation is shown in Figure 9.4. Using a constant supply voltage, potential across the output load is zero for a balanced bridge but has a predictable and measurable amplitude when one arm of the bridge changes in resistance by a known amount. Since it is possible to relate the resistance change ratio to the applied strain, the output voltage can also be related to strain. The relationship between voltage and strain can be kept linear by proper choice of operational limits. The output voltage is in the order of a few millivolts, which makes it necessary to use high sensitivity detectors and amplifiers to obtain suitable signal levels for recording. With a carefully designed system, the signal-to-noise ratio can be controlled to give good,

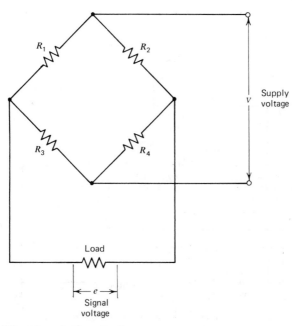

Figure 9.4 *Wheatstone bridge circuit.*

reliable signals. A schematic circuit system is shown in Figure 9.5 for a dc excited gage.

Multiple Gage Bridges. While only one active gage is shown in Figure 9.5, one, two, three, or four gages can be active, or use can be made of unstrained gages usually called dummy gages to obtain compensation for temperature and other effects.

Gages can be used in either half bridge or full bridge configurations. Each type of circuit has advantages and disadvantages. Figure 9.6 illustrates the use of two active gage circuits, and Figure 9.7 illustrates the use of four active gage circuits. These circuits are used primarily to make static strain measurements.

In this circuit a strain gage is connected in series with a power supply and a fixed or *ballast* resistor. The circuit is capacitively coupled to an amplifier, with the capacitor presenting infinite impedance to steady dc voltage but transmitting pulsating dc signals generated by alternating strains. The output e of the circuit in Figure 9.8 can be written in equation form as:

$$de = \frac{(Rg + R)\, dRg - (Rg\, dRg)}{(Rg + R)^2}\, V \tag{9.8}$$

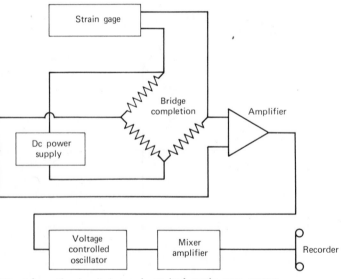

Figure 9.5 *Schematic circuit for a dc excited strain gage system.*

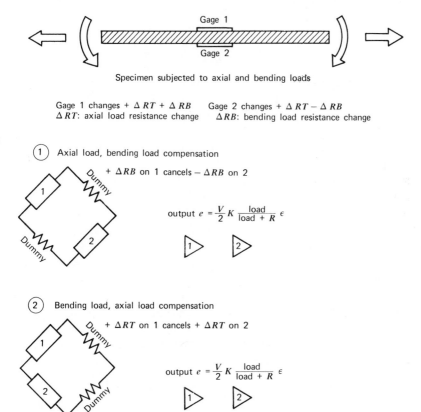

Gage 1

Gage 2

Specimen subjected to axial and bending loads

Gage 1 changes $+ \Delta RT + \Delta RB$ Gage 2 changes $+ \Delta RT - \Delta RB$
ΔRT: axial load resistance change ΔRB: bending load resistance change

① Axial load, bending load compensation

$+ \Delta RB$ on 1 cancels $- \Delta RB$ on 2

output $e = \dfrac{V}{2} K \dfrac{\text{load}}{\text{load} + R} \epsilon$

② Bending load, axial load compensation

$+ \Delta RT$ on 1 cancels $+ \Delta RT$ on 2

output $e = \dfrac{V}{2} K \dfrac{\text{load}}{\text{load} + R} \epsilon$

 Planes of gages must be equidistant from neutral axis. If not, output will be changed by an amount equal to the strain present at the plane of symmetry of the gages.

Dummies should have same resistance as gage elements. They may be resistors or matched gages. Dummy gages may be mounted on coupon or on specimen as shown. The latter will increase the output by a factor of $(1 + \mu)$.

Figure 9.6 *Two active strain gage circuits. Courtesy Wm. T. Bean, Inc.*

Multiplying and dividing the right-hand member of the equation by Rg gives

$$de = \frac{VRRg}{(Rg + R)^2}\frac{dRg}{Rg}\qquad(9.9)$$

By definition of gage factor,

$$\frac{dRg}{Rg} = \text{GF }ds\qquad(9,10)$$

so that equation 9.9 can be rewritten as

$$de = \frac{VRRg}{(Rg + R)^2}\text{ GF }ds\qquad(9.11)$$

where ds = distance moved in inches

Equation 9.11 is the most useful form in calculating the equivalent output for a given strain input when a constant voltage power supply is used. For simple dynamic circuits the basic equation shows that the gage output e is increased by making R large relative to Rg. As a rule, a compromise is made in the size of the ballast resistor R, since a large value of R tends to maintain a constant circuit current regardless of the gage change dRg but also requires a large input voltage V. R is usually limited to from two to four times the magnitude of Rg to maintain V within reasonable limits.

Types and Sizes of Strain Gages.   Strain gages can be uniaxial, meaning that strain is measured in one direction, biaxial for measuring strain in two directions or triaxial for measuring strain in three directions. The gages may be mounted at right angles (90°), at 60°, or at 45° to each other. Figure 9.9 shows several arrangements that are commercially available to strain gage users. Special sizes and shapes can be obtained, although many shapes and sizes are available as standard items. Gages can be purchased with compensating features to enable the user to obtain the most reliable measurements for his application. A strain gage user normally sets up specifications and orders gages to meet a specific application, or procures gages that require the least correction if none meet his exact needs.

Strain Gage Calibration.   There are several ways to calibrate strain gages and strain gage measuring systems. One method is to actually place known loads on the gages and measure the output. In many applications this method is virtually impossible, because the part or system cannot be exercised to obtain calibration values. In many cases, if one were to cali-

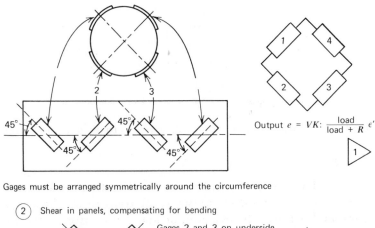

(1) Torsion only. Compensating for axial and bending loads

Output $e = VK: \dfrac{\text{load}}{\text{load} + R} \epsilon'$

$\triangleright$ 1

Gages must be arranged symmetrically around the circumference

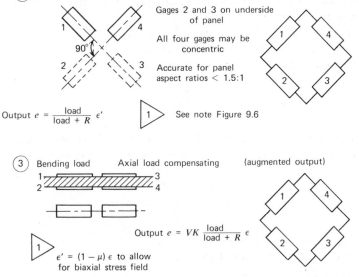

(2) Shear in panels, compensating for bending

Gages 2 and 3 on underside of panel

All four gages may be concentric

Accurate for panel aspect ratios < 1.5:1

Output $e = \dfrac{\text{load}}{\text{load} + R} \epsilon'$

$\triangleright$ 1   See note Figure 9.6

(3) Bending load   Axial load compensating   (augmented output)

Output $e = VK \dfrac{\text{load}}{\text{load} + R} \epsilon$

$\triangleright$ 1

$\epsilon' = (1 - \mu) \epsilon$ to allow for biaxial stress field

Figure 9.7 *Four active strain gage circuits. Courtesy Wm. T. Bean, Inc.*

brate the gages to eliminate all the errors that have been discussed, a full time mathematician or a computer would be required. However, for the majority of applications the gages and/or the whole strain measuring system can be calibrated electrically for satisfactory measurement results.

Errors exist in both methods as a result of (1) the temperature coefficient of the materials used, (2) sensitivity changes with temperature

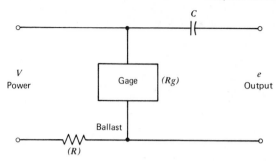

Figure 9.8 *Simple dynamic circuit for strain gages. Courtesy B&F Instruments.*

variations and methods of correction, (3) calibration errors caused by series or shunt resistances, (4) resistance of transmission lines and the configuration used, namely, one arm, or four arms active, in the gage or transducer bridge circuit.

For all practical purposes, when use is made of a half bridge configuration, the following equation gives a good R_{cal} value which can be applied with confidence for parallel calibration resistor circuits:

$$R_{cal} = \frac{Rg}{\epsilon K} - Rg \tag{9.12}$$

When the signal leads are of sufficient length to have some significant resistance, equation 9.12 can be modified by adding the resistance of the leads:

$$R_{cal} = \frac{R_T}{\epsilon K} - R_T \tag{9.13}$$

where R_{cal} = calibration resistance in ohms
  R_T = total resistance of gage and leads in ohms
  Rg = gage resistance
  K = gage factor
  ϵ = strain value in microinches per inch

The plot in Figure 9.10 shows the R_{cal} values for a 350 and 120 Ω gages under different strain conditions.

The error correction, in percent, for lead length when twisted shielded pair conductors are used is shown in Figure 9.11. An analysis of this curve shows that the smaller the resistance of the gage the greater error a given lead resistance produces.

Wheatstone Bridge Circuit Analysis. Assume that in the bridge circuit shown in Figure 9.12 the bridge is balanced and $R_1 = R_2 = R_4$. According to Ohm's law:

$$Ead = I_4R_4 \quad \text{or} \quad I_4 = \frac{Ead}{R_4} \tag{9.14}$$

$$Edc = I_3R_3 \quad \text{or} \quad I_3 = \frac{Edc}{R_3} \tag{9.15}$$

$$Eab = I_1R_1 \quad \text{or} \quad I_1 = \frac{Eab}{R_1} \tag{9.16}$$

$$Ebd = I_2R_2 \quad \text{or} \quad I_2 = \frac{Edb}{R_2} \tag{9.17}$$

For an open circuit from b to d,

$$I_4 = I_3 \quad \frac{Ead}{R_4} = \frac{Edc}{R_3} \tag{9.18}$$

$$I_1 = I_2 \quad \frac{Eab}{R_1} = \frac{Ebd}{R_2} \tag{9.19}$$

However, if $R_1 = R_2 = R_3 = R_4 = 350 \ \Omega$ when the system is balanced and a measuring instrument load R_L is 150 Ω, we can demonstrate how the strain gage equations actually work. Assume that R_2 is an actual strain gage with a gage factor of 2.04 and is subjected to a strain of 1200 μin. per inch. The gage is excited with a dc input of 22.5 V. If the load is a microammeter coil, how much current will flow through its coil when the strain is applied if the bridge was balanced before the strain was applied?

Since we no longer have an open circuit with the microammeter as a load, our open circuit equations are not valid. Using the circuit shown in Figure 9.13, the output current and voltage of a strain gage bridge are calculated by using Thevenin's theorem.

The impedance Z_0 of the circuit as measured across points a and b, where all arms and R_L are considered simple resistive elements, is:

$$Z_0 = R_L + \frac{(R_1 + R_3)(R_2 + R_4)}{R_1 + R_2 + R_3 + R_4} \tag{9.20}$$

In the event that all the bridge components are not simple resistors, Z_0 must be the *true impedance* across a and b, and not the same as calculated by equation 9.20.

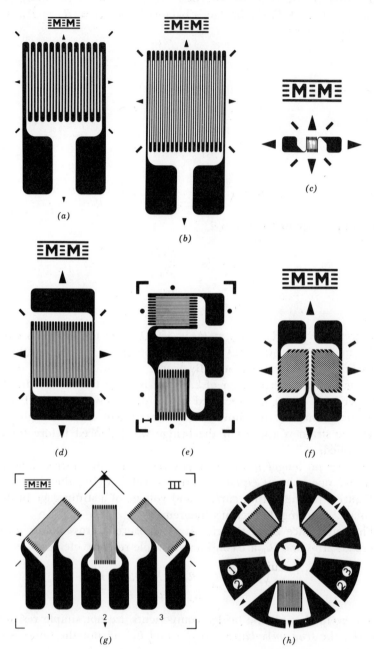

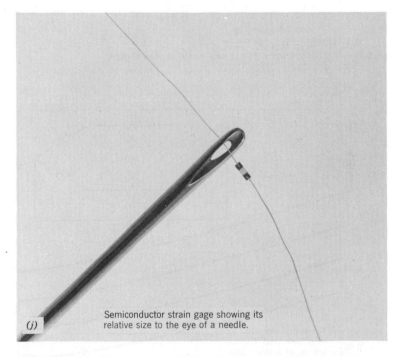

(i)   Weldable strain gage designed for high temperature and adverse environmental conditions.

(j)   Semiconductor strain gage showing its relative size to the eye of a needle.

Figure 9.9 *Typical types of commercial strain gages. (a) Single-element gage with large tabs for easy soldering. (b) Single-element strain gage. (c) Strain gage for minimum length applications when width is available. (d) Strain gage for minimum width applications. (e) Two-gage rosette for simple tension or compression type load cells. (f) Two-gage rosette for torque measurement. (g) Three-gage rosette for applications where approximate direction of strain is known. (h) Three-gage delta rosette for applications where direction of strain is unknown. Courtesy Micro-Measurements. (i) Courtesy Microdot, Inc. (j) Courtesy Kulite Semiconductor Products, Inc.*

339

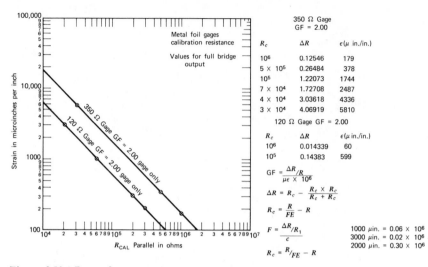

Figure 9.10 R_{cal} values for a 350 Ω and a 120 Ω strain gage.

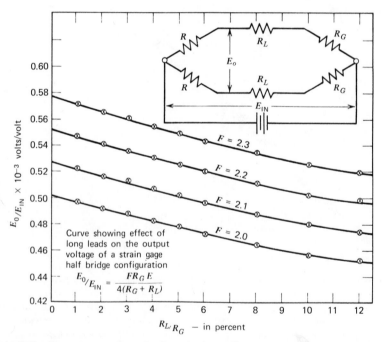

Figure 9.11 Percent error curve for long leads on the output voltage of a strain gage half bridge configuration.

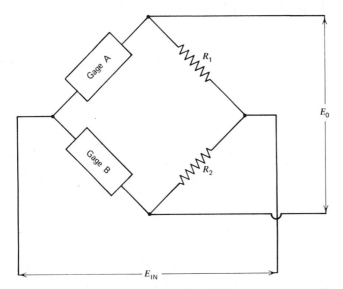

Figure 9.12 *Wheatstone bridge circuit for a half bridge strain gage configuration.*

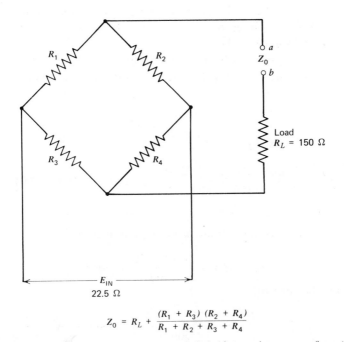

$$Z_0 = R_L + \frac{(R_1 + R_3)\,(R_2 + R_4)}{R_1 + R_2 + R_3 + R_4}$$

Figure 9.13 *Wheatsone bridge circuit for a half bridge strain gage configuration with a load.*

When the points a and b are shorted to close the circuit, Thevenin's theorem gives us:

$$I_L = \frac{dE_0}{Z_0} \qquad (9.21)$$

and

$$E_L = I_L R_L = \frac{dE_0 R_L}{Z_0} \qquad (9.22)$$

where I_L = load current
E_L = voltage drop across the load

Under normal strain gage measurement conditions in which there is no inductance or capacitance and $R_1 = R_2 = R_3 = R_4$,

$$Z_0 = R_L + R_1 \qquad (9.23)$$

and

$$E_L = \frac{R_L}{R_L + R_1} dE_0 \qquad (9.24)$$

This shows that the reduction in output voltage from the open circuit to the load condition is determined by the ratio $R_L/(R_L + R_1)$. This relationship of open circuit to load is a convenient way of determining the suitability of an amplifier or galvanometer to measure the strain generated in a strain gage bridge circuit.

With this background information we can now solve our problem for current through the bridge as a loaded circuit. We can write our basic equation as

$$dE_0 = \frac{R_1 R_2}{R_1 + R_2} \times I \text{ GF } ds \qquad (9.25)$$

since $R_1 = R_2 = 350 \ \Omega$,

$$dE_0 = \frac{350 \times 350}{350 + 350} \times \frac{22.5}{350 + 350} \times 1200 \times 10^{-6} \times 2.04$$

$$= 13.75 \times 10^{-3} \text{ V}$$

The impedance Z_0 of the circuit, using equation 9.20, is

$$Z_0 = 150 + \frac{700 \times 700}{1400} = 500 \ \Omega$$

Solving for the current,

$$I_L = \frac{dE_0}{Z_0} = \frac{13.75 \times 10^{-3}}{500} = 0.0000275 \text{ A}$$

$$= 27.5 \times 10^{-6} \text{ A or } 27.5 \ \mu\text{A}$$

This illustration demonstrates several earlier statements which may not have been fully appreciated, namely, that we are dealing with very small currents and voltages, that the gage factor is important, and that improper application of a strain gage involves large errors. It is also evident that very small strain measurements are extremely difficult to obtain. In practice, strain levels of less than 250 μin. are considered in the same magnitude as instrument noise when metallic gages are used. This is the area where the semiconductor strain gage can make one of its better contributions.

Holography. Holography is a special type of three-dimensional image recording on a photographic emulsion without the use of optical lenses. The object to be photographed is illuminated with coherent laser light. The photographic plate is positioned to detect the reflected waves from the illuminated object. A reference beam of coherent light is reflected simultaneously from a mirror toward the photographic plate to set up an interference pattern with the reflected light from the object. The net result is that the interference pattern is recorded on the photographic emulsion. When the photographic plate is developed, it shows only the interferogram generated by the interference pattern.

When this interference pattern on the photographic plate is illuminated by directing the reference coherent beam, or one of similar coherent light, onto the back of the plate, the reflected waves from the object that had reached the photographic emulsion are brought out and the observer views the object or objects through the plate in their original three-dimensional configuration. The advantage is that, as the angle of observation changes, the observer sees objects in the background that were not visible in a straight line view. By this shifting of the observation angle, a true three-dimensional effect is obtained. The observations can be made by the human eye or a camera. Pictures can be reproduced for each angle at which the camera was placed for an observation.

To measure stress, strain, or vibration, a grid can be placed on the object and the grid coordinates can be recorded photographically under the different conditions to be studied by the holography technique. These patterns can then be reconstructed as pictures for comparison to evaluate the stresses, strains, or vibrations occurring within the object at the points of measurement.

This technique is not a good industrial measuring medium in its present state of development, but there is wide interest in it for future industrial applications in areas not adequately covered by other techniques.

Photoplastics. Photoplastics, like holography, operates on an interference principle. The object to be studied must be covered with a plastic sheath, or the object may be a plastic model which can be stressed or strained to produce an interference or fringe change. Color blindness in an operator cannot be tolerated for this type of technique.

After the object has been covered with the plastic sheath, it is stressed and then illuminated with polarized light and observed. As the stressing occurs, the plastic exhibits changes in color. The number and complexity of these color changes can be related to the stress, and is a good indication of the exact stress points.

The advantage of the technique is that it gives the exact points of maximum stress, but it can be used only when there is sufficient time to apply the stress coat or make a model. This may take from days to months, and as a result it is used only when other techniques are inadequate. At best this technique is used on only a very small number of objects requiring stress and strain evaluation in industrial applications.

9.2 TRANSDUCERS

Transducers for measurement of physical parameters such as pressure and temperature have been discussed in earlier chapters, but it is well to remember that a transducer is defined as a device that converts energy from one form to another. In instrumentation terms a transducer should exhibit the following characteristics.

1. It must accurately measure the magnitude of the physical phenomenon.

2. It should accurately reproduce the physical event in relation to time. Ideally, there should be no time lag.

3. It must accurately reproduce the entire frequency range of the physical phenomenon without change or degradation in any portion of the spectrum being measured.

4. It must produce accurate data under extreme environments of humidity, temperature, shock, or vibration.

5. It must be capable of supplying an output signal that is compatible with signal conditioning equipment without modification of the original event characteristics.

6. It must be ruggedly constructed and simple enough to operate so that it can be handled by inexperienced personnel without being damaged and without having its signal characteristics affected. Needless to

say, there are few transducers that possess all six characteristics, because users continually stretch the state of the art to obtain more sensitivity and more accuracy for faster occurring phenomena. This means that a compromise has to be made in ruggedness, simplicity, speed of response, and accuracy if a measurement is to be made.

Transducers are used to interpret physical energy in terms of equivalent electrical current or voltage. Much of the physical energy we measure is related to mechanical motion or forces such as displacement, acceleration, vibration, and pressure.

The relationship of these forces to the performance of equipment under simulation testing forms a sound basis for design evaluation and improvement.

We discuss only three types of transducers, piezoresistive, piezoelectric, and inductive. However, the operating principles of other type transducers involve mutual inductance, the Hall effect, transformer-rectifier techniques, transformer bridge circuitry, electromagnetic and force balancing, vibrating reeds, attenuation thermocouple comparative bridges, potentiometers, magnetic amplifiers, sum and difference squaring, electronic multiplication, $E \times I$ plus integration, saturable cores, resonant circuits, variable reluctance, electron spin resonance, nuclear magnetic resonance, electron paramagnetic resonance, electrical differentiation, photoelectric emitters and detectors, coulomb transfer by relay, electrokinetics, modulation bridges, and capacitance. Both the number and variety of transducers will continue to grow as new materials are fabricated and conversion techniques are developed.

9.3 PIEZORESISTIVE TRANSDUCERS

Strain gages are used in commercially available transducers for transforming pressure, acceleration, load, force, and deflections to electrical outputs proportional to the physical input. They are useful for both static and dynamic applications within the limits of their frequency response and linear proportionality, and in many applications are one of the better methods of obtaining the needed measurement. Piezoresistive transducers generally use special purpose foil gages. The frequency response characteristic is comparatively low because of the mechanical limitations of the mechanisms used, particularly in more sensitive units. The use of semiconductor gages in these transducers greatly increases the output voltage available and the frequency response due to the increased gage factor of this type of strain gage. Commercially available units are wired as four-arm bridges and are treated as such in wiring and powering. There are also temperature limitations imposed by the materials used to bond the gages to the mechanisms used.

9.4 PIEZOELECTRIC ACCELEROMETERS

Certain solid state materials which are electrically responsive to mechanical force are used as transduction devices in shock, vibration, and pressure pickups. Pressure applications were discussed in Chapter 3 under pressure, so we limit our discussion here to those used for acceleration, shock, and vibration. Piezoelectric materials generate an electrical emf when subjected to mechanical motion or force, and *do not* generate an output under static force conditions. This statement is emphasized, because these types of materials are useful only in dynamic measurements. Piezoresistive materials in a bridge configuration are suitable for static measurements, as discussed in Section 9.3. Piezoelectric materials have a linear elastic range in which they produce an emf proportional to the mechanical stress they receive.

Piezoelectric action described in electrical terms of charge q, capacitance c, and open circuit voltage e is shown in Figure 9.14. When a mechanical force F is exerted on the piezoelectric crystal, a charge q is generated. Through incorporation of a mass m in direct contact with the crystal, we have the essential components of an acceleration transducer. By applying a varying acceleration to the mass–crystal assembly, the

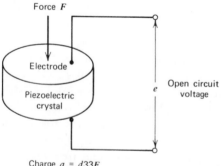

Charge $q = d33F$

$$e = \frac{q}{C} = \frac{d33F}{C}$$

where $d33$ is piezoelectric constant and
C is crystal capacity

Figure 9.14 *Electrical representation of piezoelectric action. Courtesy Consolidated Electrodynamics Corp., Data Instruments Division.*

crystal experiences a varying force $(F = Ma)$ generating a varying charge q across the crystal:

$$q = dF = dMa \qquad (9.26)$$

where d is the piezoelectric constant and the electrical charge q is directly proportional to the acceleration experienced by the transducer. Because of the capacitance c of the crystal, a voltage e is developed across the electrodes:

$$e = \frac{q}{c} = \frac{dF}{c} \qquad (9.27)$$

A typical linear seismic piezoelectric accelerometer is shown in Figure 9.15. This transducer utilizes a piezoelectric element in such a way that an electric charge is produced which is proportional to the applied acceleration. Ideal seismic piezoelectric pickups can be represented by the elements shown in Figure 9.15b. A mass is supported on a linear spring fastened to the frame of the instrument. The piezoelectric crystal that produces the electrical charge acts as the spring. Viscous damping between the mass and the frame is represented by the dash pot C. In Figure 9.15c the frame is given an acceleration upward to a displacement U, which produces a compression on the spring equal to S, the displacement of the mass relative to the spring.

For frequencies far below the resonant frequency of the mass and spring, the displacement is directly proportional to the acceleration of the frame and is independent of frequency.

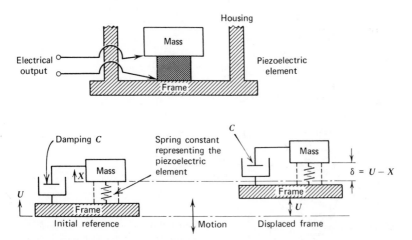

Figure 9.15 *Linear seismic piezoelectric accelerometer. Courtesy Gulton Industries, Inc.*

Accelerometer Forms and Construction. A compression accelerometer, in its simplest form, consists of a piezoelectric material disc and a mass placed on a frame, as shown in Figure 9.16. Motion in the direction indicated causes compressive forces to act on the piezoelectric element, producing an electrical output proportional to the acceleration. In this unit the mass is cemented to the piezoelectric element with a conductive material, and in turn the piezoelectric element is cemented to the frame. The components must be cemented firmly to avoid being separated from each other by the applied acceleration. Figure 9.17 shows a compression accelerometer in which the mass is held in place by means of a stud extending from the frame through the piezoelectric material.

Figure 9.18 shows a typical piezoelectric seismic system operating on a bending principle. The construction shown consists of a frame to which a flat mass loaded cantilever is attached. Acceleration causes this cantilever to bend. The magnitude of the deflection depends on the stiffness of the cantilever and the mass loading. At frequencies well below the resonant frequency, the bending strain is proportional to the applied

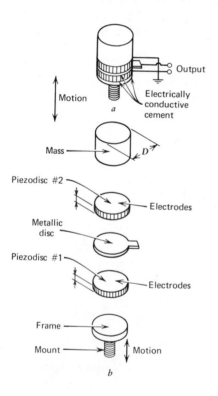

Figure 9.16 *Compression piezoelectric accelerometer without housing. Courtesy Gulton Industries, Inc.*

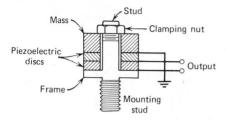

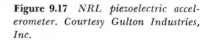

Figure 9.17 *NRL piezoelectric accelerometer. Courtesy Gulton Industries, Inc.*

acceleration of the frame. A strip of piezoelectric ceramic, having electrodes on both the top and bottom, is bonded to the centilever. Upward curvature of the cantilever causes compression in the length of the ceramic, and downward curvature causes tension. The resulting strain in the piezoelectric strip generates an electrical output voltage proportional to the applied acceleration.

Figure 9.19 shows a dual-output bender type accelerometer. This accelerometer is supported at its center and has a mass attached to both ends of the metal strip. A piezoelectric strip is bonded to each side of the metal bar between the masses. The metallic support for the element is connected to the output ground. The two outer piezoelectric strips are silvered and are connected together to the high side of the output. The two ceramic strips are oriented so as to generate electric charges of the same polarity when one is in compression and the other is in tension, so that the generated charges are additive. When acceleration is applied to

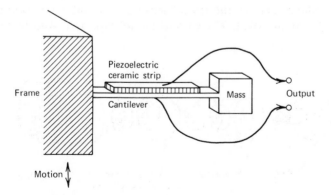

Figure 9.18 *Typical piezoelectric seismic element used in a bending principle accelerometer. Courtesy Gulton Industries, Inc.*

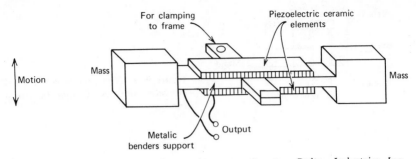

Figure 9.19   *A dual-output bender accelerometer. Courtesy Gulton Industries, Inc.*

the frame, compressive stresses on one ceramic are produced simultaneously with tensile stresses on the other.

Another feature of this type of accelerometer is that the outputs can be connected in series instead of in parallel. This doubles the sensitivity and reduces the capacitance by a factor of 4.

A low impedance accelerometer is shown in Figure 9.20, and an equivalent circuit in Figure 9.21. This accelerometer contains a miniature source follower built into the accelerometer case. The accelerometer sensing element uses compliant rod construction to provide maximum isolation from the case. The sensing crystals are electrically isolated from the case to prevent ground loops. Solid state components are used in the circuits, and the circuits are constructed in a welded module and hard potted to withstand the rugged conditions under which accelerometers are used.

In the circuit diagram of Figure 9.21 the source follower A provides a high impedance load to the crystal sensing unit and a low output impedance of less than 150 Ω. The gain of the follower is 1. Additional ampli-

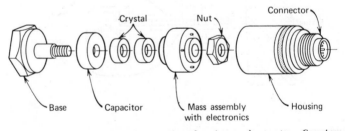

Figure 9.20   *A typical low impedance piezoelectric accelerometer. Courtesy Consolidated Electro-dynamics, Data Instruments Division.*

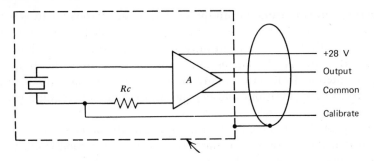

Figure 9.21 *Equivalent circuit for low impedance accelerometer. Courtesy Consolidated Electro-dynamic Corp.*

fier stages can be added to meet low signal range requirements. The use of field effect transistors in the source follower provides high input impedance and low noise over a temperature range of −320 to 300°F. The output impedance is resistive, and the circuit sensitivity is equivalent to the accelerometer sensitivity. A temperature response curve for a typical low impedance accelerometer is shown in Figure 9.22.

It should be noted that each piezoelectric material has a characteristic temperature called the *Curie point.* When a piezoelectric element is heated above its Curie point, it is permanently damaged and completely loses its piezoelectric activity. At elevated temperatures resistivity is reduced, and permanent changes in the piezoelectric and dielectric constants occur. This emphasizes the fact that accelerometers should be used at temperatures substantially below the Curie point of the piezoelectric element.

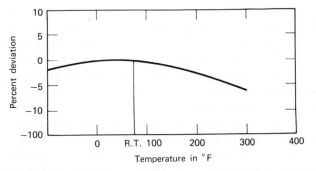

Figure 9.22 *Typical low impedance accelerometer temperature response. Courtesy Consolidated Electro-dynamics Corp.*

Accelerometer Mounting Techniques. For an accelerometer to generate accurate and useful data, it must be properly coupled to the test specimen. The method of attachment must not introduce any distortion. This requires that the accelerometer mounting be rigid over the frequency range of interest. In practice, many mounting methods are suitable for a wide variety of applications and depend on the practical requirements of the test system.

When possible, the best method is to mount the accelerometer with a stud, so that the entire base of the accelerometer is in good contact with the test object. Care should be taken to ensure a flush mate to a smooth, flat surface. Normally, a mounting torque of 18 to 25 in. lb is adequate. A good technique with all studs, when frequencies are above 5000 Hz or when shock accelerations are to be measured, is to use a drop of light oil between the mating surfaces. When electrical isolation is desirable to prevent ground loops, insulated mounting studs are particularly useful.

The torque used to mount an accelerometer is important and should be controlled. Excessive torque can damage the pickup by stripping the threads or breaking the mounting studs. Insufficient torque results in a poor mounting contact and causes large errors in high frequency vibration measurements. Some compression types of accelerometers have sensitivities that are a function of mounting torque. The same type of mounting and mounting torque should be used in each application to obtain reproducible and accurate data.

For surfaces that cannot be drilled and tapped to accept the normal threaded studs of accelerometers, use can be made of cementing studs. Cementing studs are used to prevent contamination of the accelerometer mounting surfaces. They also facilitate the removal of the transducer. The efficiency of this technique depends entirely on the adhesive used. A thorough investigation is recommended for each particular application.

Another technique of mounting accelerometers to test structures is the use of pressure sensitive (double-backed)) tape. Such mountings have an effective mounting frequency lower than the resonance frequency available with a solid mounting stud or a cemented installation. Pressure sensitive adhesive mounting methods should be evaluated, particularly for measuring vibration over a few hundred hertz, and they should never be used for shock measurements.

Performing an acceleration measurement can and does affect some systems being measured and changes the nature of the data obtained. There are two reasons for this effect. (1) The fixture required to couple the accelerometer to a somewhat flexible structure may introduce local stiffening in the mounting area and change the structural response. (2)

The added mass of the transducer may change the system characteristics. These effects may be reduced or eliminated by choosing the smallest and lightest accelerometer capable of making the measurement.

Directional Sensitivity. The majority of piezoelectric accelerometers are of the uniaxial type for measuring acceleration along a single axis. This axis is perpendicular to the plane of the mounting surface. Accelerometers are also made as biaxial and triaxial units.

A triaxial accelerometer contains three uniaxial seismic units, and a biaxial unit contains two. In the triaxial unit the three uniaxial units are orthogonal to each other, and each unit produces an independent output. Precision manufacturing of a single housing for the three units produces a more compact triaxial accelerometer and ensures more precise orthogonality than one can obtain using three individual uniaxial accelerometers. In the simultaneous measurement of acceleration components along three mutually perpendicular (orthogonal) directions, unless the tranverse sensitivity of each seismic system is relatively low compared with the maximum sensitivity, erroneous indications are obtained regardless of whether three uniaxial pickups or a single triaxial pickup is used.

Let us consider a triaxial accelerometer containing three identical elements. Assume that the transverse response is 10% of the value in the direction of maximum sensitivity. If there is an acceleration of 10 g in a transverse direction, the orthogonal elements will produce an output equivalent to 1 g. Thus an error of 10% may occur.

Using Figure 9.23 we consider the true sensitive axis of the accelerometer to be on a 10° angle. If we accelerate the accelerometer along the indicated sensitive axis, it can be shown that we are using only the sine of 80°, or 98.5% of the true sensitivity. Now, accelerating the transducer perpendicular to the sensitive axis, we obtain a transverse sensitivity of the cosine of 80° or 17.3%. This indicates that the direction of applied acceleration is important in most accelerometers.

It is possible to make an accelerometer that is independent of direction. Such a unit is known as an omnidirectional accelerometer, and its voltage output is dependent only on the magnitude of the applied acceleration. Generally this type of accelerometer is fabricated only for special applications.

Physical Characteristics. Commercially available accelerometers are usually cylindrical in shape. They are available with both attached and detachable mounting studs, or with bondable mounting surfaces on the base of the cylinder. A coaxial cable connector is provided at either the top or the side of the sensor. Most accelerometer pickup assemblies are

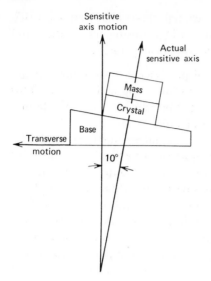

Sensitive
axis motion

Actual
sensitive axis

Mass

Crystal

Base

Transverse
motion

10°

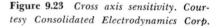

Figure 9.23 *Cross axis sensitivity. Courtesy Consolidated Electrodynamics Corp.*

relatively light in weight. They range from $\frac{1}{2}$ to 60 g. In general, the larger the accelerometer, the higher its sensitivity and the lower its resonant frequency. The higher the resonant frequency, the limit being determined by the combination of mass and piezoelectric element, the lower the capacitance or sensitivity and the more difficult it is to provide mechanical damping. The highest fundamental resonant frequency may be above 100,000 Hz.

The amplification factor of an accelerometer is defined as the ratio of the voltage sensitivity at its resonant frequency to the voltage sensitivity in the band of frequencies in which the sensitivity is independent of frequency. This ratio depends on the amount of damping in the seismic system, and it decreases with increased damping. While some pickups with resonant frequencies below 20,000 Hz use silicon oil damping, most piezoelectric pickups are essentially undamped and have amplification ratios ranging from 5 to 50.

Piezoelectric Voltage Amplifier Systems. To serve any useful purpose the output signal of an accelerometer has to be measured. In order to make such measurements a cable and signal conditioning have to be added to the pickup. These additions present system considerations.

In our earlier discussion it was shown that a piezoelectric accelerometer produces a charge q proportional to the acceleration applied. The ratio of the charge to the acceleration can be defined as the charge sensitivity S_Q:

$$q = S_Q a \tag{9.28}$$

or

$$S_Q = \frac{q \text{ (in picocoulombs)}}{a \text{ (in } g)} \tag{9.29}$$

If a follower is used, the measured voltage e_0 (in millivolts) is equal to $e_i A$, where e_i is equal to the charge produced divided by the capacitance and A is the follower gain:

$$e_i = \frac{q}{c_t} = \frac{S_Q a}{C_t} \tag{9.30}$$

$$C_t = C_a + C_b + C_i \tag{9.31}$$

where e_i is in volts, q is in picocoulombs (1 pC $= 1 \times 10^{-12}$ C), C_t is in picofarads (1 pF $= 1 \times 10^{-12}$ F), and a is acceleration in g.
Since we normally measure e_0 in millivolts,

$$e_0 = \frac{1000 S_Q A}{C_t} \times a \tag{9.32}$$

and the system sensitivity in M_v/g becomes

$$\frac{e_0}{a} = S_E = \frac{1000 S_Q A}{C_t} \tag{9.33}$$

The total system capacitance C_t includes the cable capacitance, so that there is a maximum cable length that can be used for a maximum frequency with a maximum voltage. Source followers are essentially cathode followers and are used for impedance matching functions. In general, source followers solve most of the problems associated with high impedance voltage amplifiers.

The maximum length of signal cable is dependent on the capacitance of the cable, the frequency and the voltage. Nomographs of a particular source follower can be used to an advantage to determine the frequency and output voltage for different values of cable length, or the maximum output voltage and/or frequency for a given cable length. There is always a problem of matching with a follower. This problem can be minimized by use of a charge amplifier system.

Charge Amplifier Systems. A charge amplifier can be described by considering it an operational amplifier with capacitive feedback. A block diagram schematic for a typical unit is shown in Figure 9.24. The output voltage e_{out} that results from a charge signal input e_{in} is returned to the input circuit through the feedback capacitor C_f, in a direction that maintains the input circuit voltage at or near zero. This means that the net charge input is stored in the feedback capacitor, producing a potential difference across it equal to the value of the charge divided by the value of the capacitance. This potential difference determines the relationship of the output signal voltage magnitude to the input charge signal magnitude. The transfer characteristic of the amplifier is dependent on the value of the feedback capacitor.

In equation form it can be shown that

$$\frac{e_{out}}{e_{in}} = \frac{Z_f}{Z_{in}} \tag{9.34}$$

If the impedance of the amplifier is large and if the gain is large,

$$Z_f = \frac{1}{2\pi f C_f} \tag{9.35}$$

and

$$Z_{in} = \frac{1}{2\pi f (C_a + C_b)} \tag{9.36}$$

where C_a = transducer capacitance
C_b = cable capacitance

Therefore

$$\frac{e_{out}}{e_{in}} = \frac{2\pi f (C_a + C_b)}{2\pi f C_f} \tag{9.37}$$

or

$$e_{out} = \frac{e_{in}(C_a + C_b)}{C_f} = \frac{q}{C_f} \tag{9.38}$$

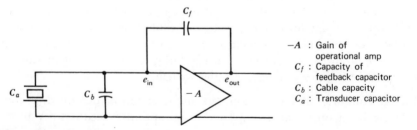

$-A$: Gain of operational amp
C_f : Capacity of feedback capacitor
C_b : Cable capacity
C_a : Transducer capacitor

Figure 9.24 *Schematic block diagram of a typical charge amplifier. Courtesy Columbia Research Corp.*

This simply means that e_{out} is proportional to the charge q produced by the accelerometer. The amount of charge present is not affected by the cable capacity. Since the amplifier detects charge rather than voltage, the system is completely independent of shunt capacitance and of changes in shunt capacitance. This is important because it means that: (1) cable length can be ignored in system output level calculations, and (2) the system temperature characteristics are dependent only on the charge-versus-temperature characteristics of the transducer and are not affected by the changes in capacitance of the transducer or the cable.

The major advantages of the charge amplifier in an accelerometer measurement system are (1) the greatly lowered dynamic input impedance (less than 1 MΩ); this materially reduces the effects of humidity and cable connector contamination and reduces the noise due to pickup from radiofrequency fields; (2) flat low frequency response to 10 Hz or less, which can be maintained without high impedance; (3) the fact that any length of cable up to 5000 ft or more can be used between the transducer and amplifier without a reduction in output.

A charge amplifier in an accelerometer circuit offers the same cabling advantage as a self-balancing potentiometer offers in a thermocouple temperature measuring system.

A typical charge measuring system provides a voltage corresponding to the actual vibratory motion of the accelerometer at the amplifier output. The amplitude of this signal is a function of the transducer charge sensitivity, the actual vibration amplitude, and the amplifier gain setting. It is not affected by cable length. The wave form of the signal is controlled only by the amplifier frequency response. No RC time constant consideration is required.

Charge amplifier system temperature response is a function of the charge characteristics of the accelerometer. Compensation of a given accelerometer charge amplifier system can be accomplished by means of a series capacitor to flatten the charge-versus-temperature curve. However, the gain of the charge amplifier would then change as a function of the source parallel capacitance which includes the accelerometer and cable capacitances. This is not desirable and eliminates one of the major advantages of using a charge amplifier, namely, that cable capacitance does not affect the amount of charge present.

Accelerometer Calibration. To produce useful test measurement data, accelerometers require calibration just as any other type of sensor.

The basic sensitivity of an accelerometer is determined by mounting it on a vibration table and setting the peak-to-peak displacement at 0.040 in. at 100 Hz. The displacement is measured optically, and the frequency is set and continuously monitored by means of a frequency counter.

The acceleration level is present in this calibration technique. However, it can be calculated to a calibration accuracy of $\pm 1\%$ using the following equation:

$$a = \frac{2\, d_0 f^2}{19.58} = 0.1023\, d_0 f^2 \qquad (9.39)$$

where d_0 = peak displacement
    f = frequency
    a = resultant acceleration in peak g

The basic charge sensitivity can be calculated from the following equation:

$$S_Q = \frac{\sqrt{2eC_t}}{A_a} \qquad (9.40)$$

where S_Q = charge sensitivity p coulomb per gravitation unit
    e = system output in volts rms
    a = acceleration level in peak g
    A = follower gain
    C_t = total capacitance, of the accelerometer plus the cable plus the follower input capacitance, in picofarads

The basic open circuit voltage sensitivity can then be calculated by the equation

$$S_{E_0} = \frac{1000 S_Q}{C_a} \qquad (9.41)$$

where S_{E_0} = open circuit voltage sensitivity
    C_a = accelerometer capacitance in picofarads

Frequency response is best determined by a back-to-back method using a calibrated shaker. The drive of the table is set for the calibration by using an NBS certified standard, and the response is read on a working standard. The NBS certified standard is replaced by the accelerometer being calibrated. The data from the standards is averaged, and the working standard is used as the drive control for the calibration. NBS calibration accuracies are: 1% for frequencies up to 900 Hz; 2% for frequencies over 900 Hz.

Cross axis sensitivity or traverse response is determined by mounting the accelerometer on the sensitive axis and recording the output at a given frequency with a given displacement. The accelerometer is then mounted perpendicular to the sensitive axis and rotated $360°$ while vibrating at the same g level to determine the maximum output.

The percent cross axis sensitivity is the maximum output perpendicular to the sensitive axis divided by the output on the sensitive axis at a fixed vibration level F_0 times 100. In equation form:

$$\text{Percent } S_{ca} = \frac{\text{Output } \perp \text{ to sensitive axis}}{\text{Output on the sensitive axis}} \times 100 \text{ at } F_0 \qquad (9.42)$$

where percent S_{ca} = percent cross axis sensitivity
$\qquad F_0$ = a fixed frequency

Temperature response is determined by recording the accelerometer output at 100 Hz with a 2 *g* drive level at room temperature and at the temperature extremes of the accelerometer specifications. Response is then calculated with respect to the room temperature reference sensitivity.

Accelerometer capacitance is measured with a precision capacitance bridge as required.

The *natural frequency* is determined by mounting the accelerometer on a plate with 10 times the effective mass of the accelerometer. This system is then suspended on a rubber band, and the impedance method is used to establish the natural frequency.

A variable high frequency oscillator is used to drive the accelerometer. The accelerometer current is held constant as the frequency is increased. A point is observed where the output across a resistor is maximum along with a 90° phase shift. This point is the natural frequency of the accelerometer plate system and is likewise the frequency of minimum impedance.

9.5 LINEAR VARIABLE DIFFERENTIAL TRANSFORMERS (LVDT)

There are many mechanical quantities in modern scientific and industrial operations that can be measured directly in terms of distance or linear displacement. This displacement, or distance, can be converted to electrical responses for measurement, remote indication, remote control, computation, and/or amplification. There is a choice of methods by which the conversion can be made, as indicated in the introduction of section 9.2. Many considerations must be weighed and tradeoffs evaluated in order to determine the best transducer for a particular application. The choice of an ideal electromechanical transducer generally is a compromise with respect to environment, linearity, reliability, and ruggedness.

Principle of Operation. The LVDT is an electromechanical transducer which produces an electrical output proportional to the displacement of a separate movable core. The physical layout of an LVDT is shown in Figure 9.25. Three coils are equally spaced on a cylindrical coil form. A rod shaped magnetic core is positioned axially inside this coil assembly and provides a path for a magnetic flux linking the coils.

By energizing the center or the primary coil with an alternating current, voltage is induced in the two outer coils. In the transformer configuration the outer or secondary coils are connected in series opposition so that the two induced voltages are opposite in phase. Thus the net output of the transformer is the difference of these two voltages. For one central position of the rod shaped magnetic core the output voltage is zero. This is the null position or balance point.

By moving the core from the balance point, the voltage induced in the coil toward which the core is moved increases, and the voltage in the opposite coil decreases. This produces a differential voltage from the transformer. By proper design this voltage can be made to vary linearly with change in core position. Motion of the core in the opposite direction beyond the null position produces a similar linear voltage characteristic with the phase shifted 180°. A continuous plot of voltage output versus core position within the linear limits of the transducer is shown in Figure 9.26. The curve appears as a straight line through the origin or null point if opposite algebraic signs + and − are used to indicate the opposite phases. By plotting voltages as a positive quantity with disregard to phase, the curve appears as a V as shown in Figure 9.27.

In practice, the output voltage does not quite reach a zero value at the null position because of a small residual voltage that does not cancel. This means that the bottom of the V curve when sufficiently magnified is not sharp but appears slightly rounded and does not reach the zero value. In good LVDT units the null voltage output is small enough to be insignificant for most applications in which they are used.

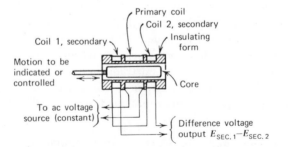

Figure 9.25 *Operation of the LVDT. Courtesy Schaevitz Engineering Co.*

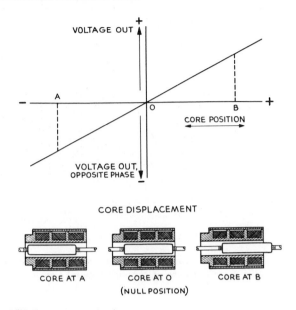

CORE DISPLACEMENT

CORE AT A   CORE AT O   CORE AT B
                 (NULL POSITION)

Figure 9.26 *LVDT output and phase as a function of core position, linear graph. Courtesy Schaevitz Engineering Co.*

Form and Application. An LVDT mounted in a cantilever spring arrangement for use as an accelerometer is shown in Figure 9.33a. As shown, the total mass of the resilient member plus any attached mass determines the position of A with respect to B. The center position is established as the zero acceleration value. Since an accelerating force may be a variable quantity, this same LVDT can be used to measure the rate of acceleration.

In general, the unit is cylindrical in shape because the coils are wound as parallel wires around a cylinder used to house the magnetic core. The whole unit can be easily mounted as shown in Figure 9.28.

The LVDT can be mounted on various types of mechanical devices such as the Bourdon tube shown in Figure 9.28. The proper mechanical arrangement can be made to read gage, absolute, or differential pressure, or derived quantities such as vacuum or flow. Each of these forms of measurement is fundamentally based on the distortion of a resilient member, so that the relative position of the movable point with respect to a fixed reference is a measure of the pressure. This positional measure is in turn translated into an electrical voltage with the LVDT.

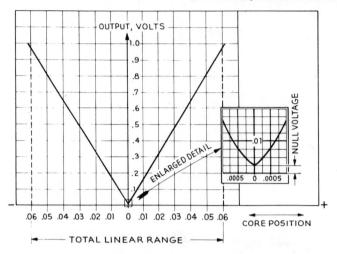

Figure 9.27 *V graph. Absolute magnitude of LVDT output voltage as a function of core position. (Inset is a magnified view of the null region.) Courtesy Schaevitz Engineering Co.*

The velocity of a moving or vibrating object can be measured by two methods using an LVDT. In each method the body of the LVDT is fixed and the core of the LVDT is attached to the object.

In the first method the motion of the core produces an output in which the carrier frequency is modulated in accordance with the motion. The carrier frequency is eliminated by a detector circuit, leaving only

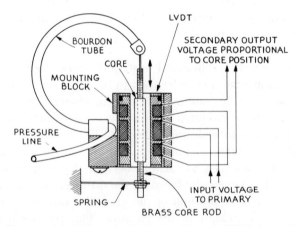

Figure 9.28 *Pressure transducer, showing operation of LVDT by a Bourdon tube. Courtesy Schaevitz Engineering Co.*

the modulation frequency, the signal amplitude of which is proportional to the displacement. By passing this signal through a differentiating network, an output proportional to the velocity is obtained.

In the second method the primary is energized with dc instead of ac and the two secondary coils are connected in series adding instead of series opposition. As long as the core is stationary, there is no output. When the core moves, an output voltage is produced and its instantaneous magnitude is proportional to the core velocity.

Frequency Response. The small mass of the core, the absence of friction, and negligible actuating force make the LVDT well suited to follow rapid changes in displacement without lag. This makes the LVDT useful in accelerometers and vibration pickup systems.

When a sinusoidal or other oscillating movement is applied to the core, the output voltage is varied accordingly so that the modulation envelope accurately portrays the waveform, frequency, and amplitude of the applied motion. In such an arrangement, the carrier frequency should be at least 10 times the highest frequency component of the acceleration or vibration to be measured. A higher ratio is preferred.

LVDT units with a straight movable magnetic core can be designed for operation at any ac frequency in the range from below 60 Hz up to the radiofrequency of 1 MHz and beyond. In normal applications LVDT units are available in the 60 to 20,000 Hz range. There are few requirements at frequencies above 20 kHz.

The curve shown in Figure 9.29*a* is for an LVDT output voltage modulated by a vibrating core as observed on an oscilloscope. This modulating vibration cannot be measured directly as an LVDT voltage output, but it can be readily rectified and filtered to produce the alternating voltage waveform shown in Figure 9.29*b*.

The output voltage of an LVDT is directly proportional to the primary voltage. Unless the LVDT is being used in a computer application where primary voltage (or current) is one of the variables, the excitation voltage should be closely regulated and monitored. Constant voltage regulation should be used on the source of the excitation unit. When the excitation voltage cannot be precisely regulated, it should be monitored with a meter and readjusted as necessary.

When a choice of output impedances of an oscillator is available, the lowest output impedance that provides the required input voltage minimizes any possible voltage variation due to any slight variation in LVDT primary impedance with different core positions.

Output voltage also varies as a function of frequency. The percentage change in output is always less than the percentage change that causes

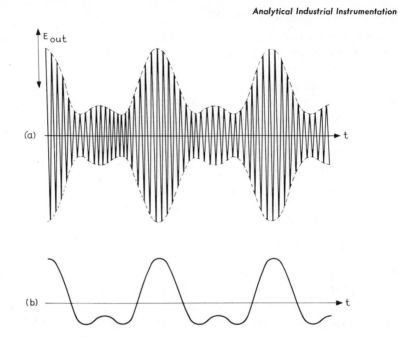

Figure 9.29 *LVDT waveshapes. (a) Oscilloscope trace of LVDT output voltage modulated by the vibrating core. (b) Voltage corresponding to vibration waveform, from the rectifier and filter circuit. Courtesy Schaevitz Engineering Co.*

it, and in some frequency ranges the variation in output is very small or nonexistent as frequency is varied. Since the frequency of some oscillators may vary appreciably with fluctuations in line voltage, it is important to use well regulated input power to minimize any changes in voltage or oscillator frequency.

It should also be noted that the LVDT should be calibrated with the same input voltage values at which the measurements are made.

Mounting Techniques.   A common method of mounting an LVDT is by means of a two-piece split block. This block is bored to fit the outside diameter of the LVDT, and screws are used to clamp the two block sections together around the LVDT. A single-piece C block open on the side of the bore may also be used.

Transformers, especially those that are not mounted in magnetic cases or shields, are preferably mounted in nonmetallic blocks or **mounting** structures. If metal parts must be used, whether the transformer is shielded or unshielded, it should be assembled so that such adjacent metal is distributed symmetrically in both directions from the center of the transformer. This minimizes the possibility of adverse effects such as deviation from linearity and increased residual voltage at the null position of the core. This type of mounting makes it difficult to adjust the LVDT in its holder to obtain a good null position in the application setup.

Magnetic material may be used in mounting the transformer if absolutely necessary, but in such cases special methods may be required to produce a low null voltage. In some cases a special transformer design and/or special core dimensions may be required to maintain precise linearity. A magnetically shielded transformer should always be specified for mounting in iron or other magnetic material.

The linearity of response of the LVDT should always be checked when using metallic mounts. This check can be made in the actual setup by measuring the transformer output as a function of the core position when the core is given a series of equal steps over the linear range of operation.

In normal operation the moving core should not slide against or make direct contact with the inside surface of the standard LVDT coil form. Although in some applications in which random or sliding contact may be difficult to avoid, such contact may not prove objectionable.

Standard LVDT cores are threaded internally for assembly on non-magnetic screws or threaded core rods. These are usually designed and produced by the user in the special form required for each particular application. A core rod may be a threaded integral extension of some specialized moving part, or a simple element such as a screw, stud, or screw eye for attachment to an internal actuating element or support. In many applications the core is completely supported and guided clear of the transformer by the core rod or rods.

When it is necessary or desirable to support and guide the core directly inside the transformer, a tubular bearing, usually of nonconductive material, may be fitted full length into the transformer. Usually this core guide fits the core diameter closely enough to prevent any tilting effects. Occasionally, the overall mechanical arrangement may render a close fit undesirable.

Sensitivity. The rated sensitivity of an LVDT is usually stated in terms of voltage (millivolts or volts) output per thousandths of an inch core displacement per volt input (commonly written mV out/0.001 in./volt in). If the input voltage has a constant specified value, the sensitivity may simply be described in millivolts (or volts) output per 0.001 in. core displacement. Since voltage sensitivity varies with frequency, except in some designs over a limited frequency range as previously stated, the frequency should be stated when specifying sensitivity. The actual output voltage for a given core displacement is determined by multiplying the sensitivity by the displacement in thousandths of an inch, and then multiplying this product by the input voltage.

Another method of specifying the sensitivity of an LVDT, preferred by some users, is to specify the output voltage produced at the rated input voltage with the core positioned at one end of the rated linear range. This is more simply stated as the nominal range output.

A typical curve of changing output voltage as the excitation frequency is varied is shown in Figure 9.30. This is an example of the relationship between frequency and sensitivity.

The sensitivity and output of an LVDT can often be increased by raising the primary voltage above the rated value, depending on frequency and ambient temperature. The input voltage should be limited by the maximum internal winding temperature that can be tolerated by the insulation specifications for a particular type of LVDT. Such increased excitation, for LVDTs with which it can be used, does not cause

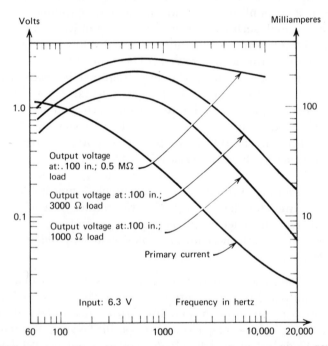

Figure 9.30 *Output volts and primary current versus frequency for an LVDT. Courtesy Schaevitz Engineering Co.*

excessive distortion, since the magnetic core is not easily saturated because of the relatively high reluctance of the magnetic path.

The output voltage is closely proportional to input voltage, except for very slight and usually unimportant deviations at low and very high output levels.

Either voltage or power output can be increased beyond the published ratings for a particular LVDT by connecting a capacitor across the load. Such an arrangement should be verified experimentally in order to avoid chance adverse effects on performance, particularly effects on linearity of response, and sensitivity to frequency variation.

The output voltage variation of an LVDT is stepless. As a result, the effective resolution depends entirely on the minimum voltage or current increment that can be sensed by the associated electrical system.

The LVDT output is readily resolved to within 0.1% of the full range output by a suitable null balance indicating or servo system.

The LVDT usually demonstrates little or no disturbing sensitivity to small displacements of the core in any direction perpendicular to the transformer axis, providing the core remains parallel to the axis. This means the core need not be accurately centered within the transformer, but should be accurately restrained from incidentally tilting or assuming an angular position when it is actuated. The sensitivity to traverse displacement may vary from about 1% to less than 0.1% of the useful sensitivity measured parallel to the axis, depending on the specific LVDT design.

Linearity and Linear Range. The output voltage of an LVDT is a highly linear function of core displacement within a certain range of motion. As shown in Figure 9.26, a graph of the output voltage versus core displacement for the core range chosen is essentially a straight line. Beyond this range the curve starts to deviate from a straight line.

The linear range of any LVDT varies to some degree with frequency. It is a convenient practice to identify each type of LVDT with a figure corresponding to a single nominal linear displacement range measured plus or minus from the null core position. The nominal, or rated, linear displacement range specified by the manufacturers is usually a conservative value for any frequency at which the transformer is rated. When the transformer is used with the correct core provided for the specified frequency, the actual linear displacement range always equals or exceeds the nominal range.

The degree of linearity within the linear range is defined as the maximum deviation of the output curve from the best fit straight line passing through the origin. This is expressed as a percentage of the output at nominal range.

Many standard LVDTs have a linearity of ±0.5% or better. Special design and/or selection or by using a selected portion of the nominal range make better linearities possible. Linearities of ±0.1% or better have been attained in some types of LVDTs.

The linearity and linear range of an LVDT are most commonly specified for a 0.5 MΩ load, since most grid input amplifiers and vacuum tube voltmeters have input resistances of 0.5 MΩ or higher. Unlike the potentiometer, the LVDT may be connected to a wide variety of load impedances from infinity down to an impedance of the same order as the differential secondary impedance of the LVDT. The load may be given any value in this range, with only a small effect on linearity or linear range, as shown in Figure 9.31.

Effects Due to Stray Magnetic Fields. If an LVDT is operated in the presence of stray magnetic fields of considerable magnitude, such as the fields frequently produced by power transformers and motors, an unwanted voltage component may be added to the output.

In order to determine the presence of such extraneous voltage, the core should be placed at one end of the operating range, away from null, and the LVDT disconnected from any voltage source. The LVDT output is then measured with the adjacent magnetic field source in operation, and then with the source disconnected. As many measurements are needed as there are sources suspected. A sensitive vacuum tube voltmeter is useful for this purpose.

One way to avoid extraneous voltage output is to disassemble the transformer and the core, and wedge the core inside the transformer at the approximate limit of the rated displacement range. The LVDT is then moved through different angles and locations to determine the position producing the minimum induced voltage, or a value of induced voltage that does not adversely affect the operation.

Calibration of Linear Variable Differential Transformers. Calibration of an LVDT should be made as nearly as possible under conditions of actual use, including the same mountings and environment and the same excitation voltage and frequency.

The linearity of the LVDT output allows the calibration to be made by establishing two points as far apart as possible, but within the range in which it is to be actually used. Two accurately known external reference positions or quantities are required.

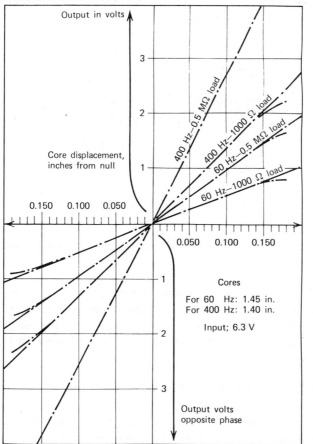

Figure 9.31 *Output voltage at different frequencies and different loadings. Courtesy Schaevitz Engineering Co.*

In applications in which the LVDT is to be used on both sides of null, the calibration points should span the entire range to be used. In this case a minimum of three calibration points is required. One at null and one at each end of the range.

The precise response of the LVDT to linear core motion can be measured independently of installation, if desired, by use of an LVDT micrometer stand, as shown in Figure 9.32. A nonmagnetic and preferably nonconductive threaded core rod extends from the micrometer feed to position and support the core. The core should be within the transformer, and the LVDT clamped at least 1 in. or more away from any

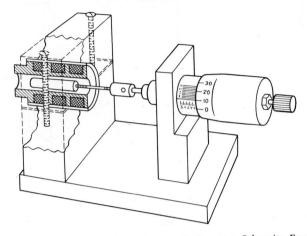

Figure 9.32 *LVDT calibration micrometer stand. Courtesy Schaevitz Engineering Co.*

part of the micrometer or micrometer feed mount. The core should be aligned axially when necessary by means of an insulated bearing inside the transformer, or by means of a second core rod and spring or bearing support at the opposite end of the transformer, as shown in Figure 9.33.

9.6 OPTICAL ELECTRONIC CHEMICAL ANALYSES

Optical electronic chemical analytical instruments such as spectrometers, spectrophotometers, and colorimeters find extensive use in chemical, drug, medical, and petroleum industries for continuous flow and element analyses.

Spectrometry is based on the absorption or the transmission of portions of the electromagnetic spectrum. Absorption spectrometry is based on the observation and comparison of absorption spectra. Each chemical compound has its own absorption spectrum which produces a curve showing the amount of light absorbed for each wavelength. The spectrometer must be capable of furnishing a wavelength spectrum with a controllable bandwidth narrow enough to be useful. Practical monochromators produce a group of wavelengths rather than a single wavelength. A monochromator in a spectrometer is an optical system and a prism which transmits light of one color when a wide spectrum light source is focused on the prism. The prism may be made of a variety of materials, but the material chosen depends upon the spectrum to be used, that is, the visible, the infrared, or the ultraviolet.

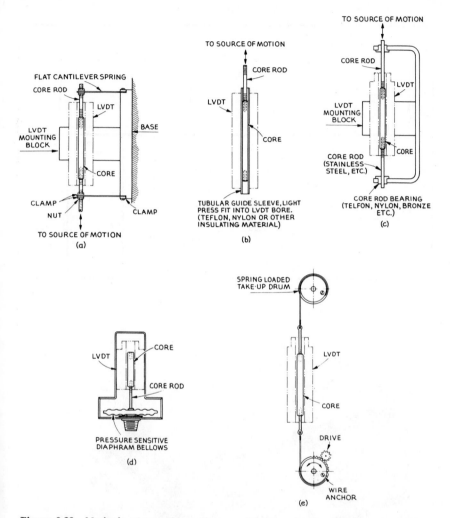

Figure 9.33 *Methods of assembling the movable core. (a) Cantilever spring support (no friction). (b) Core guide or bearing. (c) Core rod bearings. (d) Integral core rod extension from the moving element. (e) Cable suspension. Courtesy Schaevitz Engineering Co.*

371

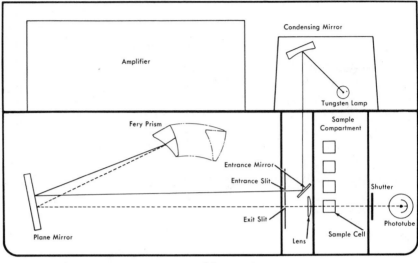

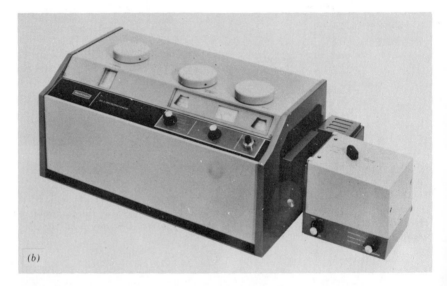

Figure 9.34 *Spectrometer. (a) Schematic of a spectrophotometer. (b) Model DU spectrophotometer. Courtesy Beckman Instruments.*

In transmittance spectrometry the transmittance T is the ratio of the radiant energy P_s transmitted by the sample to the radiant energy P_i incident on the sample. Both radiant energies must be obtained at the same wavelength and with the same spectral slit widths. The transmittance

$$T = \frac{P_s}{P_i} \tag{9.43}$$

is usually expressed as a percentage. In practice a standard blank is used to establish the 100% transmittance level, and the unknown samples are compared to the standard sample. The standard blank is normally filled with a portion of the solution used to dissolve the material samples to be measured. This means that it is essential to specify the standard used to establish the 100% transmittance level with which the sample was compared.

In absorbance spectrometry the absorbance A is the negative logarithm to the base 10 of the transmittance. As an equation,

$$A = -\log T = \log (1/T) \tag{9.44}$$

where T is expressed as a decimal fraction and not as a percentage. A standard is used for absorbancy, just as for transmission. A zero absorbance measurement is equivalent to a 100% transmission measurement.

The spectral slit width of a spectrometer is defined as the total range of wavelengths emerging from the exit slit. This definition neglects stray light and spherical aberrations in the optical system. The spectral slit width indicates the band of wavelengths involved in a particular spectrometer transmittance measurement.

A schematic representation of a spectrometer is shown in Figure 9.34. A light source A is focused by means of mirrors B and C through a variable slit D to a mirror E. The mirror E focuses the beam on a prism F which can be positioned to produce a desired wavelength of the visible spectrum, which is returned to the exit slit by the mirror E. The selected wavelength of light is then passed through the sample compartment and collected by a photocell. The addition of the photocell to a spectrometer produces a spectrophotometer. The output of the photocell is measured on a meter calibrated to read transmittance or absorbance.

As mentioned earlier, a blank is used to establish the zero absorbance or 100% transmittance. A standard is prepared in three concentrations to calibrate the unit for quantitative measurements. A sample concentration-versus-absorbancy curve is shown in Figure 9.35.

In quantitative measurements the constituents of the sample are usually known, and it is the concentration of one or more of the constituents

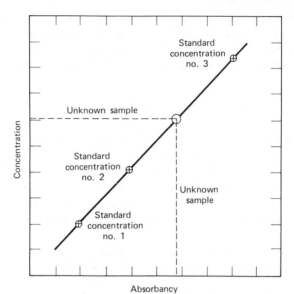

Figure 9.35 *Absorbancy versus concentration curve for a spectrophotometer.*

that one wants to determine. In spectrophotrometry the basis of quantitative measurement is the fact that the absorbance of an absorbing material depends on its concentration. The system will obey Beer's law if the
absorbance is directly proportional to the concentration. Beer's law states
that the intensity of a ray of light is inversely proportional to the thickness of the absorbing medium through which it passes. In the same manner the absorbance of a sample is proportional to the length of the path
of the light within the sample. This means the cells used to hold the
sample during the measurements must be constant if the comparisons
are to be valid. Since there are slight deviations from Beer's law in practical equipment in use, three concentrations are used, as graphed in Figure 9.35, and the unknown to be measured is plotted on the same graph
to establish its concentration for the same wavelength and slit width. A
new curve must be established for each wavelength and slit setting. The
highest accuracy is obtained by using standards of known concentration
of the sample element to be measured. Spectrophotometers find extensive
use in chemical analyses and in medical diagnostic laboratories.

Infrared Spectrometers.   Infrared spectrometers operate in the infrared
wavelength spectrum of 1 to 25 μm. An infrared spectrometer, as shown
in Figure 9.36, has an infrared source, an optical system, a sample cell,

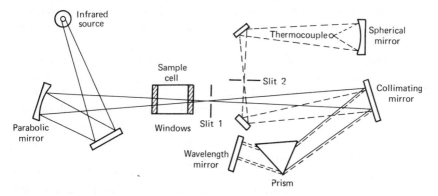

Figure 9.36 *Schematic of an infrared spectrometer system.*

a prism, a detector, and a recording system. The selected infrared source is operated at a temperature ranging from 900 to 1800°C, depending on the source material, to provide approximately blackbody radiation. The power output is controlled by a constant power input.

The sample cell material and the prism are chosen for the wavelength range most adaptable to the substance to be analyzed. Five common materials used as prisms and cell windows are listed in Table 9.1

Table 9.1   Cell Material and Usable Range

Material	Range (μm)
Quartz	1.0– 4.0
Lithium fluoride	1.0– 5.5
Calcium fluoride	2.0– 8.5
Sodium chloride	3.0–15.0
Potassium bromide	15.0–25.0

The infrared detector is often a bismuth-bismuth tin thermocouple encased in a highly evacuated housing with a potassium bromide window. Other types of detectors are bolometers, photoconductors and pneumatics. A bolometer is a device which changes its electrical resistance with temperature in response to the radiant energy incident to it. Infrared photoconductors are similar to bolometers in that they change their resistance as a function of radiant energy falling on them. Pneumatics operate on a gas-pressure principle with a diaphragm to indicate the pressure variation generated by the gas volume expansion caused by the radiant energy striking the gas-pressure cell.

The output of the infrared detector is usually recorded on a thermo-couple or resistance bridge self-balancing type of strip chart recorder, and the output recorded can be of two types. One type of recording is for various wavelengths within the capability of the prism and the cell window and a second type is for continuous recording of a flowing stream of liquid or gases at a specified wavelength. The second type is used for continuous analysis on process flows.

The angle at which the wavelength mirror is mounted determines the portion of the reflected beam which falls on the exit slit. By rotating the wavelength mirror slowly, the total wavelength spectrum of the prism can be scanned and recorded.

Beer's law is used as a basis for determination of the concentration of components in a sample at any particular wavelength. Beer's law can be expressed as an equation in the form:

$$C = \frac{1}{\alpha x} \log \frac{I_0}{I_x} \tag{9.45}$$

where α = absorption factor of the substance
   C = concentration of the substance
   x = sample thickness along the optical path
   I_0 = initial beam intensity without the sample
   I_x = intensity of beam after it passes through the sample

The terms α, x, and I_0 are empirical constants established by trial cali-bration of the spectrometer. I_0 is established by passing the light beam wavelength through the empty sample cell, x, is measured for the partic-ular sample cell, and α is determined by using a known composition of a known concentration. These values must be evaluated for each wave-length. For continuous analysis the wavelength is chosen to be the most sensitive for the particular analysis to be performed. The sample cell is also arranged to accommodate a gas or fluid sample flow for continuous process analysis. Just as in visible spectrophotometer analysis, the known concentrations are plotted for the calibration curve and the analytical values of the unknown are compared to the calibration curve to obtain values for the unknown.

Ultraviolet Spectrometers.   Visible spectrum spectrometers can be used as ultraviolet spectrometers by proper choice of light source and prism. Ultraviolet absorption spectrometers normally are operated in the 0.2 to

0.8 μm range. One ultraviolet light source is a hydrogen discharge lamp, useful in the 0.22 to 0.35 μm range. A quartz prism is normally employed to disperse the light for proper ultraviolet wavelength selection. The ultraviolet absorption spectrometer, like the infrared spectrometer, can be used to analyze static or continuous flow samples.

Emission Spectrometers. In emission spectrometery sample analysis, the sample is made to emit a spectrum by means of an electric arc, or by introducing atomized samples into a flame. An electric arc is used primarily for metallic samples such as those of aluminum, magnesium, and steel, and a flame for boron, calcium, copper, iron, and inorganic compounds containing potassium and sodium. Emission spectrometery finds its most extensive use in analyses of the elements, and measurements are made of the intensity of the spectral lines. An empirical calibration can be made so that the intensity of a single line is made proportional to the concentration of the element for rapid analysis applications when such measurements are adequate for product control. In cases in which unknown concentrations have to be measured, it is necessary to record the entire spectrum and then analyze the record for the wavelength and intensity of many lines to determine the unknown concentration of the sample.

Fluorescent Spectometers. When it is essential to avoid physical or photochemical deterioration in the analysis of the sample, fluorescence spectrometers can be used. In this type of analysis, the sample is irradiated with light of a given wavelength, and the absorbed energy causes the sample to emit radiation of a wavelength peculiar to the sample. This emitted radiation is then recorded after it has been refracted by the atomic planes of the detecting crystal on which it is focused by a collimating system. Appropriate standards must be chosen for adequate calibration.

Spectrometric techniques are generally used when less expensive techniques fail to produce the necessary analysis or control. The comparatively high cost of the instruments and their operation probably restricts their use in industrial applications far more than their technical capabilities.

9.7 pH MEASUREMENTS

In many industrial applications water is used in large quantities for cooling, flushing, mixing, diluting, dissolving material, washing, and such functions as humidity control. In many of these applications it is important to control both the acidity and alkalinity of the solution. The measure of the acidity and/or alkalinity is called pH measurement. The actual measurement is of the hydrogen and hydroxyl ion concentration.

In a neutral solution such as pure water the pH scale value is 7.00. Acid solutions increase in strength as the pH value decreases from 7.00 to 0.00, and alkaline solutions increase in strength as the pH value increases in value above 7.00. A change of 1 pH unit represents a 10-fold change in concentration. Thus the pH of a solution can be defined in equation form as

$$pH = \log \frac{1}{\text{hydrogen ion concentration in moles/liter}} \qquad (9.46)$$

The acidity of a solution is due to the presence of hydrogen ions, so if a solution of acid contains 1 g/l of ionized hydrogen, it has the same strength regardless of what acid is used to make the solution when a neutralization reaction is required. Strong acids have a high degree of ionization, and weak acids have a low degree of ionization. The same holds true for the hydroxyl ions in alkaline solutions. A neutral solution having a pH of 7.00 is a solution in which there is a balance between the number of hydrogen ions (H^+) and hydroxol ions (OH^-) present in the liquid.

There are two general ways to measure the acidity or alkalinity of a solution. These are chemical indicators and pH meters. Chemical indicators are discussed briefly, and the pH meter measuring technique in more detail.

Chemical Indicators. Chemical indicators have been found to change color as a function of hydrogen ion concentration. In general they are weak acids and bases and their salts. Under the most ideal conditions the best limit of accuracy is 0.1 pH unit. To obtain this limit it is necessary to prepare standards for each 0.2 pH unit over the range to be measured. Where this degree of accuracy is acceptable, the chemical indicator is an inexpensive method. However, it must be used with the knowledge that the indicating solutions are not stable and change color. This instability and color change can lead to gross error if not closely controlled.

For gross qualitative indications as to whether a solution is neutral, acid, or alkaline, use can be made of pink and blue litmus paper. For example, if a piece of blue litmus paper turns white in a solution, the solution is acidic. If it stays blue, the solution is neutral or alkaline. If the solution is strongly alkaline, the blue turns a darker shade of blue. Conversely, if a piece of pink litmus paper turns white, the solution is neutral or acidic. Strong acidic solutions are indicated by a deeper pink shade. It can be readily seen that it is difficult to establish pH values on such techniques.

pH Meters. The need for quantitative measurements of pH values led to the development of special hydrogen ion detectors and associated circuitry called pH meters. The majority of these meters are of a basic potentiometer type from which an electrical readout is obtained. One such basic system is shown in Figure 9.37. As shown, the basic system consists of a measuring cell or electrode, a reference cell or electrode, a potentiometer circuit, an amplifier, and a voltage measuring meter. This is a direct reading type of system. Industrial pH meters are usually of the direct reading type with pH scales for either visual indication or recording. Recording is usually employed on continuous process applications.

Reference Cells. Two types of reference cells or electrodes appear to have the widest usage. The calomel cell seems to be more widely used than the silver–silver chloride electrode. These electrodes must be stable, and are designed so that they do not change with the pH of the sample being measured. Both these reference electrodes are unlimited as pH references and can be used with any pH sensitive electrode. The useful life of the calomel cell can be shortened by high temperatures. Both these reference cells operate at atmospheric pressure or at reduced pressure for

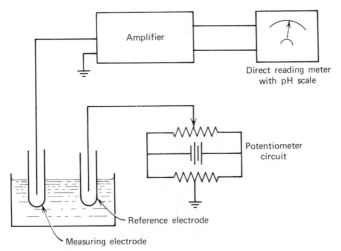

Figure 9.37 *Schematic of a basic potentiometric pH meter system.*

normal construction units. Special designs can be obtained for high temperature applications. High pressure units are subject to contamination from high pressure test solutions. The construction of the calomel cell is shown in Figure 9.38, and the silver–silver chloride cell is shown in Figure 9.39.

Salt Bridges. In each of the standard reference cells it should be noted that the electrical connection between the solution in which the reference is placed and the saturated potassium chloride in the standard is by means of a salt bridge. Salt bridges are used so that there is a minimum of dilution of the salt bridge solution and a minimum of diffusion of the potassium chloride into the salt solution or the salt solution into the potassium chloride or the solution being measured.

Salt bridges should be designed to provide a positive pressure between the salt solution in the bridge and the solution outside the bridge, so that the direction of flow opposes diffusion into the bridge.

Salt bridges should have a barrier at the termination point with the outside solution, such as the asbestos fiber in Figure 9.38 and the capil-

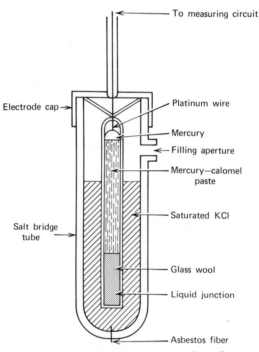

Figure 9.38 *Cross section of a calomel pH reference electrode.*

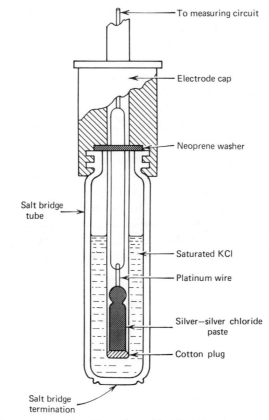

To measuring circuit

Electrode cap

Neoprene washer

Salt bridge tube

Saturated KCl

Platinum wire

Silver–silver chloride paste

Cotton plug

Salt bridge termination

Figure 9.39 *Cross section of a silver–silver chloride pH reference electrode.*

lary in Figure 9.39. Other barriers are agar-agar plugs, ground glass stoppers, porous ceramic, and glass frit. These barriers are used to provide high flow velocities at small volumes to prevent excessive dilution and contamination of solutions being measured.

Measuring Cells. Choice of a pH measuring electrode depends on the application. Four types are available that cover most industrial requirements. The four types are the glass electrode, the antimony electrode, the quinhydrone electrode, and the hydrogen electrode.

Glass Electrodes. Glass electrodes have a wide pH and temperature range. They cover the 0 to 13 pH range and can operate at temperatures from 0 to 100°C and at pressures from 0 to 100 psi. These electrodes are not affected by oxidizing or reducing acids, but like all other types of glass they are attacked by fluoride solutions. They are not affected by dissolved gases or suspended solids in the solutions, and work well in fluid flows except at high velocity. The glass electrode construction is shown in Figure 9.40.

Glass is an excellent electrical insulator, and for this reason the pH sensitive glass bulb is made porous enough to obtain electrical conductivity between the buffer solution inside the glass envelope of the electrode and the solution into which the electrode is placed for the purpose of making a pH measurement.

Glass electrodes have other limitations besides being attacked by fluorides and restricted use in high velocity flows. Glass has high internal resistance; shielding is required because of the high impedance, but the electrical insulation resistance is excellent. These electrodes are subject to error in highly concentrated alkaline solutions.

Glass electrodes are sealed to reduce leakage currents and to eliminate refilling operations. Early types required refilling in the same manner as the calomel reference electrode.

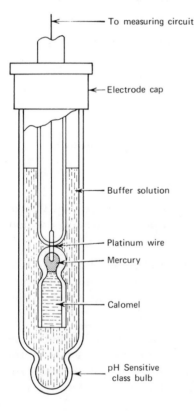

To measuring circuit

Electrode cap

Buffer solution

Platinum wire
Mercury

Calomel

pH Sensitive
class bulb

Figure 9.40 *Cross section of a pH sensitive glass electrode.*

Antimony electrodes have a limited pH range from 4 to 11.5 pH and should not be operated at temperatures over 60°C. These electrodes are rugged and durable where the pH of abrasive slurries must be measured. The electrode cell has low resistance, shielding is not required, and special measuring circuits are not necessary.

Some of the disadvantages in using these electrodes are that the active surface must be cleaned and scraped regularly, dissolved oxygen must be present to maintain the pH sensitive oxide coating, certain materials such as silver, mercury, lead, and copper poison the electrode, and some oxidizing and reducing solutions cause errors in measurements.

Antimony electrodes fall into a special usage class, and care must be exercised in their application. When used with calomel reference cells, the voltage values are not accurate. Accurate measurements require standardization against solutions of known pH.

Quinhydrone electrodes release an equivalent number of hydrogen ions as they are oxidized. The change in voltage of the system is responsive to the hydrogen ion concentration. This type of electrode was used extensively in early pH measurements by electrical voltage methods. These electrodes are simple and have low resistance, but they also have a limited pH range and must be used at temperatures below 37°C. These electrodes cannot be used in alkaline solutions or in the presence of strongly oxidizing or reducing agents. They are generally not acceptable for industrial applications, because they may change the pH of buffered solutions being measured.

Hydrogen electrodes are used only at atmospheric pressure, have an unlimited pH range, and can be used at all reasonable temperatures up to the boiling point of the solution being measured for pH. These electrodes are not useful in the presence of oxidizing and reducing agents. The hydrogen electrode is slow in reaching equilibrium, and is not useful in solutions of elements lower in the electromotive series.

A hydrogen electrode is constructed much like an inverted test tube with a central glass stem through which a platinum wire is passed. A platinum tab is attached to the end of the platinum wire, and is coated with platinum black to increase the surface area. For measuring electrodes the platinum must be kept free of any materials that would cause the electrode to act as a catalyst in the solution. Hydrogen electrodes can be maintained as reference electrodes if they are submerged in a solution of 1 mol/liter hydrogen ion concentration and a flow of hydrogen gas is maintained at 1 atm. Contact with the test solution is made through a salt bridge.

pH Measuring System.   A complete industrial pH measuring system includes an electrochemical cell, an electronic circuit, and a readout device.

The electrochemical cell consists of the reference electrode, a salt bridge connecting the reference electrode to the solution being measured, the measured sample solution, the glass electrode membrane, the glass electrode filling solution, and the internal cell of the glass electrode. The voltage measurement across this total cell is used as the pH indication, although it is the algebraic sum of all the voltages in the system. This voltage is actually generated by the flow of electrons in the system. Actually, the voltage is a good measure of the pH, because conditions are maintained so that the only variable is the pH difference across the glass electrode membrane. The other voltages are negligible or are steady values for a given set of conditions. The primary function of a pH sensitive electrode is to produce a voltage change of 59.1 mV = 1.00 pH at 25°C. It is also covenient to adjust the composition of the pH sensitive inner cell solution so that the voltage difference between the pH sensitive electrode and the reference cell is zero for a pH value of 7.00. In the majority of commercial applications, a pH of 7.00 has been chosen to represent a perfectly neutral solution such as distilled water.

Older pH instruments, which use electrometer grid detecting tubes and tube amplifiers, are subject to a voltage drift which is read on the indicating meter as a pH change. The use of solid state detectors and amplifiers with zener diode controlled voltages has eliminated dc amplifier drift problems. Calibrations are normally good for extended periods of time, and high accuracy can be obtained and maintained.

There are two general types of electronic pH meter readout systems. One is the *null type* pH meter, and the other is the *direct reading* type.

Null readings can be obtained quite readily using a potentiometer circuit. The slidewire on the potentiometer is moved to obtain a bucking voltage to the voltage developed across the electrochemical scale. The slidewire scale is marked in pH units, and the pH value is read directly from the potentiometer slidewire sliding indicator when the indicating meter indicates a null condition or zero voltage reading. A typical slidewire null balance system is shown in Figure 9.41. pH meters of this type can be made to give the highest precision available in pH measurements. Both temperature and asymmetry in the electrode tubes are sources of error in pH meter measurements, so most system circuits provide a means of compensation or correction. In the potentiometer circuit a portion of the slidewire is made variable to correct for temperature changes, and a correcting feature can likewise be made to correct for

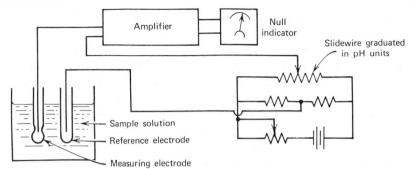

Figure 9.41 *Schematic of a typical null balance slidewire pH meter.*

asymmetry voltage, where asymmetry voltage is defined as the voltage required to obtain a zero reading when identical electrodes and solutions are placed on both sides of the glass membrane. The asymmetry is caused by small variations in available commercial electrodes. Also, in older type instruments using a battery power supply, a standard cell is normally used to standardize the instrument. The use of zener diode controlled power supplies, as stated earlier, eliminates this requirement.

Calibration of pH Instruments. To obtain accurate pH measurements the pH measuring system has to be calibrated. As discussed earlier, the pH value of 7 is normally chosen as the neutral solution value on the pH scale. A pH of 7.000 is neither acidic nor alkaline. Values lower than 7.000 are an indication of an acidic condition, and values higher than 7.000 are an indication of alkalinity. Calibration is carried out by placing the electrochemical cell in a buffer solution with the proper pH value for either an acid condition or an alkaline condition. During calibration in which a wide range is to be measured, it is recommended that the electrochemical cell be thoroughly cleaned in distilled water and the pH readout adjusted to a pH value of 7.000 within the accuracy of the system being calibrated. Only the most precise instruments are within ±0.002 pH unit, and the majority are not more accurate than ±0.02 pH unit. The value of 7.000 is a theoretical value, just as the 59.1 MV potential being equivalent to 1.00 pH at 25°C represents a theoretical value. Buffer solutions are commercially available or can be prepared for each pH value to be used in the calibration. The electrochemical cell should be rinsed several times in distilled water before being placed in a new solution and allowed to stabilize in each new solution during calibration. A drop or two of one buffer solution in another buffer solution causes a pH value change that results in a calibration error. Good electrodes become very stable after a few days of operation, so that with good quality electrodes a calibration is good for an extended period of days or weeks, depending on the application. In continuous operations the reference cells contain adequate salt bridge solution and a positive flow of the solution to maintain reliable liquid junctions. Usually, the electrodes are designed for a continuous process application, and the output is recorded. In these cases the recorder is normally designed to operate control functions to maintain the established pH value.

pH measurements are used in such industrial applications as control of the pH in sewage treatment, water purification, and chemical processes. New uses are being developed in oxidation–reduction processes, and in the determination of chloride ion concentrations. With the development of more stable solid state circuits, the reliability and accuracy of measurement will be increased and will generate the need for new types of electrodes and better techniques for electrode maintenance.

9.8 VISCOSITY

Viscosity is the property of a liquid that presents a resistance to flow. When a liquid flows over a surface, a portion of the liquid wets the surface and the wetting layer of particles adheres to the surface and remains stationary. As the liquid flows over the surface, the layer above the wetting layer moves slowly and, as the layers of liquid increase in depth, they move faster than the layer they cover, the top layer moving the most rapidly. If the top surface moves at a faster rate than the lower level next to the surface over which the flow occurs and the flow occurs over a given depth, an equation can be developed to express the condition. Assume the upper surface has a speed v_f faster than the lower surface and the depth is a height h. Then there is a shear flow rate of v_f/h, and this rate remains constant as long as none of the parameters change.

If some cubical portion of the flowing liquid is chosen as being representative of the flowing stream, the force causing it to flow against the resistance becomes a function of the area of the chosen cube. This can be expressed as a shearing stress F/A, where F is the total force on the cube, and A is the area on which the force is acting.

The shearing stress is proportional to the rate of shear in a liquid under similar conditions, but is not the same value for different liquids under the same conditions. A proportionality constant called the *coefficient of viscosity* is required to have an equality.

Thus

$$\frac{F}{A} = \frac{\eta v}{h} \qquad (9.47)$$

where η = the required constant.

Thus

$$\eta = \frac{Fh}{vA} \qquad (9.48)$$

In the metric system the dimensional value of an unit coefficient of viscosity is 1 dyn·S/cm² and is called a *poise*. A coefficient of viscosity for a given liquid is always stated for a given temperature, because the viscosity of a liquid decreases with increases in temperature.

The theoretical explanation of determining viscosity just discussed is difficult to prove directly with existing instrumentation. The practical approach to such a measurement is to flow the liquid through a tube and measure the rate of flow. Under such conditions the liquid flows faster in the center of the tube, since there is an adhering layer on the interior surface surrounding the center. A vertical tube of small bore can be used to make the measurement, but a rigorous analysis involves the use of integral calculus applied to each annular layer of radius r from r^1 to r^n extending from the surface to the center of the tube. The rigorous analysis yields a relation known as Poiseuille's law, in which a rate of flow q in centimeters per second passes through a tube l, centimeters long and r centimeters in radius at a pressure p in dynes per square centimeter at the bottom of the tube:

$$\eta = \frac{\pi p r^4}{8ql} = \frac{Ar^2}{8qL} \qquad (9.49)$$

where $A = \pi r^2$ for a circular orifice. To be a valid equation the flow velocities must not set up turbulence in the liquid.

Some rough-and-ready methods of measuring viscosity, not normally considered the best of industrial methods, can be readily demonstrated by dropping a ball bearing through a column of viscous material in a clear glass or plastic tube and measuring the time it takes the ball to reach the bottom. Another familiar method seen on television is the dropping of a pearl through a bottle of hair shampoo. It is also possible to estimate the viscosity of a fluid by sending air bubbles up through a column and timing the rate of rise of the bubbles through the fluid. These techniques are an offspring of methods of viscosity measurement involving Stoke's law and are not valid except for flows at very low Reynolds numbers; the accuracy is questionable unless elaborate calibration comparisons are made.

Viscosity can be measured practically on an intermittent basis on discrete samples or continuously in a system. Usually, intermittent measurements are made by laboratory instruments such as the one shown in Figure 9.42. They are usually relatively inexpensive and easy to operate, and under controlled temperature conditions are quite accurate. The instrument shown in Figure 9.42 is a Kepes balance rheometer. This type of rheometer subjects the substance to be measured to a sine wave shear strain, although no part of the instrument describes nonuniform motions.

Figure 9.42 *Kepes balance rheometer. Courtesy Olkon Corp. Contraves Viscometer Division.*

As shown in Figure 9.43, the measured substance is contained between a hemispherical cup and a spherical bob, both of which are concentric to each other. They rotate in the same direction at the same uniform angular speed. The driving shaft of the cup can be swiveled around the center of the measuring system by a small angle a (up to 6°). The bob is held in a measuring arc B which can swivel around the x axis. The shear stress acting on the bob causes a torque on the bob shaft of which only the component Md about the x axis is measured on the measuring arc B.

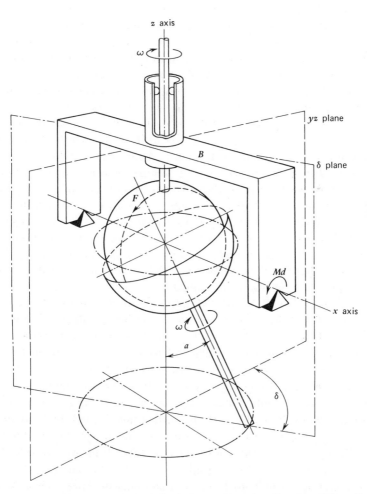

Figure 9.43 *Operating principal of the balance rheometer. Courtesy Olkon Corporation, Contraves Viscometer Division.*

The plane on which the two shafts forming the angle a lie (δ plane) can be swiveled about the z axis. At a given frequency the viscosity measurement can be determined by measurement of the torque Md at $\delta = 90°$.

Where viscosity must be controlled under dynamic conditions, a continuous flow viscometer should be selected. The selection is based on the range of viscosity and whether or not the flowing medium is a newtonian fluid. A newtonian substance is one that, when subjected to shear stress, undergoes deformation such that the ratio of the shear rate or flow to the shear stress or force exerted is constant. In comparison a non-newtonian substance does not have a constant ratio of flow to force.

Viscosity measuring systems frequently used for continuous and/or remote recording and control use pressure drop techniques which are simple and so-called foolproof. The liquid being measured is pumped through a friction tube or orifice plate at a constant rate and at a constant temperature. The viscosity is measured in terms of absolute viscosity by a pneumatic force balance type differential transmitter which gives a direct solution to Poiseuille's equation, or by a viscosity sensitive rotameter calibrated directly in viscosity units. Other methods that have found some industrial application are ultrasonically vibrated probes, torque elements, and constant volume pumps for constant flow rates.

In the application of an orifice plate to obtain the pressure drop to measure viscosity, the Reynolds number introduced and demonstrated in Chapter 7 is defined with respect to viscosity. A Reynolds number is a dimensionless index of the ratio of inertia to viscous forces. Inertia forces are predominant when Reynolds numbers are high, and the coefficient of the venturi tube, the flow nozzle, and the orifice plate becomes a constant for practical measurement purposes.

If the viscosity is variable, the variation in the restriction device coefficient seriously complicates flow measurements for low Reynolds numbers. Low Reynolds numbers are encountered when flow measurements are made of viscous oils or gases of high hydrogen content. This means that as the viscosity effects increase the Reynolds number decreases.

A torque element industrial viscometer is shown in Figure 9.44. It is a flowthrough system with the flow upward or vertically into the measuring system which is housed in a stainless steel flowthrough body. A breakdown view in Figure 9.45 shows the measuring system components. Essentially, the total system consists of a housing, a measuring cell made up of a measuring cup (the liquid being measured and a measuring bob), a magnetic coupler made up of a diaphragm separating a permanent magnet and a magnetic follower, and a torque meter made up of a

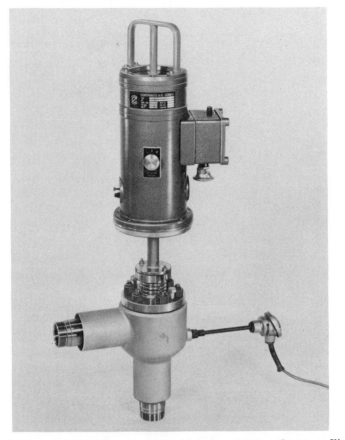

Figure 9.44 *DC-49 process viscometer. Assembled view. Courtesy Contraves Viscometer Division, Olkon Corp.*

three-speed gearbox, a synchronous motor, support bearings, a potentiometer, a torque spring, and a suspended pointer with an indicating scale. The indicating meter with an electrical output for control and/or recording is shown in Figure 9.46.

Since viscosity depends on the temperature of the fluid, the measuring system and the stainless steel housing can be temperature controlled with the resistance thermometer shown being used either to measure the temperature at which the viscosity is measured or to control the temperature in the measured fluid to maintain a constant viscosity.

In this system the arm of the potentiometer is attached to a shaft extending from the synchronous motor housing and the gearbox arrange-

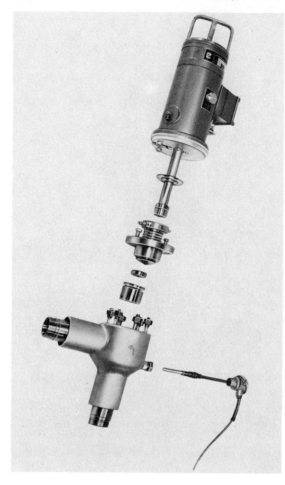

Figure 9.45 *DC-49 process viscometer. Breakdown assembly view. Courtesy Contraves Viscometer Division, Olkon Corp.*

ment. This shaft is restricted to swing in an arc of 350° as the torque measuring arm. The position of the arm of the potentiometer determines the magnitude of the signal for the control and/or measurement desired.

This type of measuring system is being favorably accepted for in-line industrial viscosity measurement of polymers, food, photographic products, pharmaceuticals, petroleum products, and chemicals having a viscosity ranging from 1 to 4,200,000 cP.

Figure 9.46 *Viscosity readout and control unit for a process viscometer. Courtesy Contraves Viscometer Division, Olkon Corp.*

9.9 MOISTURE MEASUREMENTS

Moisture measurements have been made since the fifteenth century with respect to the atmosphere. It is not difficult to make moisture measurements unless a very accurate measurement or close control is a requirement. There are three general methods of making relative humidity or air moisture measurements. These are the psychometer, the electronic sensor, and the dew point methods.

In the psychrometer method a human hair or animal membrane that changes dimensions with respect to moisture is used as the humidity sensor. These *hygromechanical* elements have been used as operating indicators and as control switches for many years.

The electric sensor meets the industrial need for speed, versatility, accuracy, and high sensitivity using low mass, nonmechanical components.

Where the actual water content of the air is important, or where actual condensation of moisture is to be avoided, dew point control can be applied most effectively.

An electronic humidity sensor is a precision device capable of detecting a fraction of 1% change in relative humidity.

One type of electronic sensor is constructed of two intermeshing gold grids stamped on plastic, and is coated with a complex film containing hygroscopic salts. As the relative humidity (RH) rises, the film becomes more conductive and the electrical resistance between grids is lowered. The variation in resistance is calibrated in RH units, and the associated controller interprets the changes to activate the proper humidity control equipment.

One type of dew point sensor consists of bifilar wire electrodes wound on a cloth sleeve which covers a hollow tube or bobbin. (Bifilar refers to a winding that consists of two separate wires wound side by side at a uniform distance apart.) The cloth sleeve is impregnated with a lithium chloride solution and allowed to dry. The bifilar wires are connected to the secondary of an integral transformer. The bifilar electrodes are not interconnected. They depend on the conductivity of the atmospherically moistened lithium chloride for current flow.

Lithium chloride has two unique characteristics that make it suitable for dew point measurements. It is highly hygroscopic, meaning it has a strong affinity for water vapor, and it has an inherent ability to maintain itself at a constant value just above 11% RH when it is in a moist atmosphere and is heated by an electric current passing through it. At values of 11% RH or below, the lithium chloride in the sleeve material dries out into a crystalline solid and becomes nonconductive.

A second type of dew point detector uses an observation chamber into which a gas sample, containing moisture vapor, is introduced. A pressure ratio gage indicates directly the relation between the gas sample and the atmospheric pressure. The gas sample is maintained at a pressure somewhat above atmospheric. When an operating valve is released, the gas is allowed to escape into the observation chamber and expand at atmospheric pressure. A lamp is illuminated at the time of gas release, so that if the gas has cooled below its dew point a distinctive fog is formed in the chamber. The procedure is repeated to establish the end point, or vanishing point of the fog. This end point can be accurately determined as measured by the pressure ratio of the vanishing point.

Other dew point techniques involve observation of the formation of dew on a polished surface, and lowering the temperature by refrigeration techniques to obtain a dew deposit from confined gases.

The measurement and/or control of moisture is desirable in establishing an environmental comfort zone for humans (such as temperature and humidity controlled areas for special work), in controlled storage areas, in compressed gases used for instrumentation and analytical work, in controlled atmosphere furnaces, and in drying ovens. Moisture control is also essential in the paper industry, so that paper can be rolled to correct thicknesses and stored without mildewing and molding. Without proper moisture control the paper could be stretched as it is being rolled on reels and later shrink and break on the roll.

Moisture measurement by means of radioactive materials and detecting devices is discussed in Chapter 10.

9.10 CHROMATOGRAPHY

Chromatography is a term that describes several physical methods used in chemistry and biology to separate and identify mixtures of chemical compounds. The name implies that color is involved in the detection procedure, but color is not mandatory. The leading methods use photometric readout and are based on either flame photometry or chemiluminescence.

In industrial measurement and control chromatography is a word used to describe an analysis method or technique. Chromatographic analysis in its simplest form consists of first separating the unknown sample into its constituents and then performing a quantitative analysis on each of the separated components. The same techniques have been applied for volatile liquids and gases. Analysis of the gas phase is more rapid, and there has been a continuous effort to develop systems for a reliable, continuous, on-line method of measurement rather than a sampling or batch method.

Techniques used for the gas phase have been developed where the separation is on a time basis. In general the gas sample to be analyzed is introduced into a column. This column may be packed with a porous solid or other materials in such a manner that the different components in the sample pass through the column at different rates. This gives a time separation. Some constituents enter a stationary phase as others move along, which permits the detector to handle only binary mixtures of the constituent and a carrier. This in turn permits the use of simple, nonspecific detectors in many cases, as discussed in Chapter 12.

The types of packing include those that operate by adsorption of the gas directly onto an activated surface, thin films of liquid in which different components have different solubilities, molecular sieves which separate the constituents on the basis of molecular size, and ion exchange, which operates on an ion-in-solution basis. In each case the packing is designed to space the different constituents on a time base principle so that they can be detected separately.

Normally, a carrier is used to keep the sample column moving at a predetermined rate. It is true that the gas or vapor would diffuse through the column by natural diffusion, but this is a very slow process and cannot be tolerated for most types of process measurements. The carrier is chosen so that it will push the component of interest through the column by displacement, or drag the component of interest through the column. The dragging method, termed *elution,* is most commonly used because it leaves the column clean and ready for the next sample. The displacement method leaves the column filled with the carrier, which must be purged from the system before a new sample and carrier can be introduced into the column. This wastes time and requires more complicated equipment, such as permeation tubes to provide controlled concentrations.

Chromatography is a simple technique conceptually, and it can be combined readily with computer control. The chromatograph also appears to be a more powerful tool than other instruments for performing quantitative analyses of gases and volatile liquids. As a result, more methods of improving its use as an on-line instrument are being developed.

The basic system is presented in Figure 9.47, showing a sampling valve with the carrier and sample input, a separation column, a detector, and a recorder. A more complicated system is shown in Figure 9.48, which includes a programmer, a computer to extract the samples from the process, and an analysis recorder. Actually, it is still a sampling method programmed for the selection of samples on a continuous basis. Each sample is analyzed separately.

Chromatography, as defined for industrial applications, may be located in reference handbooks under gas chromatography for gas or vapor mixtures having a stationary phase of liquid held by a solid with the gas moving; partition chromatography for mixtures of liquids or dissolved

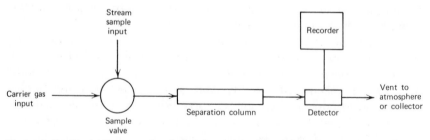

Figure 9.47 *Block diagram of a basic chromatography system.*

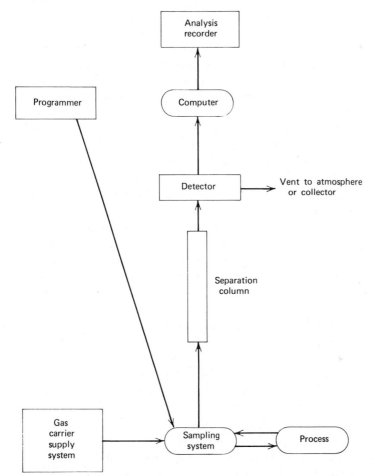

Figure 9.48 *Block diagram of a programmed and computer controlled chromatography system.*

solids having a stationary phase of liquid held by a solid with the liquid moving; paper chromatography for mixtures of liquids or dissolved solids having a stationary phase of liquid held by a solid with the liquid moving; adsorption chromatography for mixtures of gases or vapors or liquids or dissolved solids having a solid stationary phase with either a gas or liquid moving; and ion exchange chromatography for ions in solution having a solid stationary phase and a liquid moving. The mixture is moved physically by the carrier past the stationary phase, which may be either a solid or a liquid immobilized in the pores of a solid. The various ingredients of the mixture migrate from the moving phase into the stationary phase, and then back into the moving phase. Since the process is repeated many times during the course of travel, a complete separation can be accomplished because the more strongly adsorbed or more soluble constituents gradually lag behind the less strongly adsorbed or less soluble constituents.

When high speed response with minimum flow sensitivity is required, and the sampling system can be well temperature controlled, a thermal conductivity cell makes a suitable detector. Thermal conductivity detectors depend on the cooling effect of a gas flowing over a heated element. There is a different cooling effect when a constituent is present in the carrier gas than when the carrier gas has no constituent present. By passing pure carrier gas through one cell and the carrier plus the constituent through another cell, a differential signal is generated in proportion to the amount of constituent present. Other types of detectors, such as ultrasonic crystals, which operate on the attenuation principle caused by the amount of constituent present in the carrier, flame ionization detectors using electric conductivity changes in a flame which is produced by the carrier gas and the amount of constituent present, and the beta ray argon detector using a strontium 90 or krypton 85 source to produce metastable atoms in an argon carrier, are also in use.

Chromatographs are now available with the capacity to identify 100 or more constituents in a mixture. There appear to be few applications in which full capacity is required, but with such capabilities available it seems quite practical to set up small systems in which only a few constituents of a mixture require separation and quantitative measurement. These are applicable where detection and analysis are wanted for specific elements and their concentrations, such as for sulfur products, carbon monoxide, nitrogen oxides, and total hydrocarbons polluting the air.

9.11 SUMMARY

In summary, analytical instruments are used for industrial evaluation and control to measure stress, strain, vibration, acceleration, shock, motion, spectral quantities, pH, viscosity, relative humidity, and dew point, and for chromatographic separation of gaseous or liquid mixtures. Many more types of analytical measurements may be necessary for special cases, and new techniques will be developed to cover these measurements. Such areas may be intergranular corrosion in metals and degradation in other materials that lead to operational failure as aging and fatiguing occur.

Review Questions

9.1   What characteristic makes a metallic strain gage a useful analytical tool?

9.2   What is a gage factor? On what principle is its value based?

9.3   What is the technical difference between the gage factor and the K factor?

9.4   On what factors does the magnitude and sign of a semiconductor resistivity change depend?

9.5   How does the gage factor of a semiconductor and that of a metal foil gage differ?

9.6   How can a semiconductor strain gage be made independent of temperature and strain?

9.7   What desirable feature is provided by the enlarged tabs at the end of the grid on a foil gage?

9.8   Explain how offset error in a strain gage can be minimized.

9.9   Name the features that make a foil gage a first choice in strain gage applications. Why is a foil gage not always a good first choice?

9.10 What type of measuring circuit is used for static strain applications? Can the same circuit be used on dynamic measurements? If so why and if not why not?

9.11 What equation best describes an equivalent output for a given strain input when a constant voltage power supply is used?

9.12 What type of strain gage is recommended when the axis of strain is not known?

9.13 Explain how a strain gage can be calibrated.

9.14 What errors exist that must be recognized when calibration is considered?

9.15 Explain the difference between a uniaxial, biaxial, and triaxial strain gage.

9.16 What type of light is essential for a holographic image? How can the viewed image be changed?

9.17 What is the basic principle in the use of photoplastics? How can strain be determined by this technique?

9.18 What features make transducers attractive as sensors?

9.19 What characteristics are desirable in an ideal transducer?

9.20   What is the difference between a piezoresistive and a piezoelectric transducer? Explain why both types are considered necessary for physical parameter measurements.

9.21   In a linear seismic piezoelectric accelerometer, at what frequency is the displacement directly proportional to the mass and independent of frequency?

9.22   What construction feature is a must in a compression accelerometer?

9.23   What is a cantilever accelerometer? Where are such accelerometers used?

9.24   Explain the operating principle of a dual-output bender accelerometer.

9.25   Why is a source follower needed in some types of accelerometers? What function does it perform?

9.26   What is the maximum temperature at which a piezoelectric material can be used?

9.27   Explain how accelerometers can be mounted to a test sample for the best results.

9.28   What mounting techniques should not be used for accelerometers exposed to shock environments?

9.29   When do acceleration measurements affect the systems being measured?

9.30   Explain the difference between uniaxial, biaxial, and triaxial accelerometers.

9.31   What is an omnidirectional accelerometer? Where is such a unit used?

9.32   What is the advantage of a constant voltage accelerometer system? What are its disadvantages?

9.33   Give the advantagest and disadvantages of a charge amplifier system.

9.34   Explain how an accelerometer is calibrated.

9.35   Explain the operating principle of an LVDT.

9.36   What value does a central core position in an LVDT represent? Is it always an absolute value? Why?

9.37   What ratio of carrier frequency is considered a minimum when sinusoidal or oscillatory motion is measured with an LVDT?

9.38   Why is a well regulated input power system required for an LVDT?

9.39 What precautions must be observed if an LVDT is mounted in metal parts?

9.40 How is the sensitivity of an LVDT rated? How is it usually described?

9.41 How can the voltage or power output of an LVDT be increased beyond its normal rating?

9.42 What is meant by the linear range of an LVDT? Is this its only useful range? Explain.

9.43 How is an LVDT checked out for applications when stray magnetic fields are present?

9.44 Explain how an LVDT is calibrated.

9.45 Explain the difference between absorbance and transmittance in spectrometry.

9.46 What conditions must be maintained when an unknown is being identified by spectrographic techniques?

9.47 What determines the choice of material for sample cells and prisms used in infrared spectrometers?

9.48 If Beer's law is valid, which values must be evaluated for each wavelength to determine a sample concentration?

9.49 What is the difference between an infrared and an ultraviolet spectrometer?

9.50 What is the difference between a spectrometer and a spectrophotometer?

9.51 When is a fluorescent spectrometer used in analytical measurements?

9.52 Explain how a spectrometer is calibrated.

9.53 What is a pH measurement? Where are such measurements used?

9.54 How can pH be determined? How accurate is each measurement technique?

9.55 Why are reference electrodes required when a pH meter is used to determine hydroxyl ion concentration?

9.56 When is it unadvisable to use a glass electrode for pH measurements? Why?

9.57 Give the advantages and disadvantages of a quinhydrone electrode.

9.58 Give the advantages and disadvantages of an antimony electrode.

9.59 What is assymmetry and how is it compensated for in a pH measuring system?

9.60 Explain how a pH meter system is calibrated.

9.61   How is Poiseuille's law related to viscous flow?

9.62   What is the relationship between viscosity and Reynolds numbers?

9.63   What is the difference between a newtonian and a nonnewtonian fluid?

9.64   What are the units of a poise?

9.65   How can you measure relative humidity?

9.66   What is a dew point measurement?

9.67   What is the difference between a hygrometer and an electronic humidity sensor?

9.68   For what type of moisture measurements is a dew point measurement most effective for control?

9.69   Would you use a dew point control for environmental comfort control? Explain your answer.

Problems

9.1   Assuming a gage resistance of 350 Ω, a lead resistance of 10 Ω, a gage factor of 2, and a strain of 2000 μin./in., find the R_{cal} value.

9.2   What is the R_{cal} value for a 120 Ω gage with a 10 Ω lead resistance and a gage factor of 2 for a 4000 μin./in. strain application?

9.3   If a circuit using a 350 Ω strain gage, having a 2.0 gage factor, is to be measured by means of a 51 Ω galvanometer type strain measuring instrument in a bridge circuit having matched gage resistors, what current will be measured for a 1500 μin./in. strain when 28.0 V is applied to the circuit?

9.4   If a solution shows a pH change from 7.0 to 3.5, what change in concentration of the solution has occurred?

Bibliography

Baldwin-Lima-Hamilton, *Foil Strain Gage Specifications*, BLH Electronics Division, September 1964; *Strain Gage Handbook*, BLH Electronics Division, Waltham, Mass., Bulletin 4311A.

Considine, Douglas M. (ed.), *Process Instruments and Controls Handbook*, McGraw-Hill, New York, 1967; *Scientific Research*, March 1966, pp. 20–23.

Consolidated Electrodynamics Corp., *CEC Piezoelectric Accelerometer Users Handbook*, Bulletin 4200-96, Pasadena, Calif.

Dorsey, James, *Semiconductor Strain Gage Handbook,* Section I, Theory, Baldwin-Lima-Hamilton Electronics Division, May 1963; *Semiconductor Strain Gage Handbook,* Section II, Data Reduction, Baldwin-Lima-Hamilton Electronic Division, August 1964.

Frank, Eugene, *Strain Indicator for Semiconductor Gages,* B & F Instruments, Cornwall Heights, Pa., September 1961.

Fuller, David H., Gas Chromatography in Plant Streams, *ISA Journal,* Vol. 3, No. 11, November 1956, pp. 440–444.

Kinnard, Isaac F., *Applied Electrical Measurements,* John Wiley, New York, 1956.

McCoy, R. D. and Ayers, B. O., On-Stream Analysis with a Chromatograph, *Control Engineering,* Vol. 17, No. 7, July 1970, pp. 44–49.

Noble, Frank W., An Ultrasonic Detector for Gas Chromatography, *ISA Journal,* Vol. 8, No. 6, June 1961, pp. 54–57.

Pennington, Dale, *Current Piezoelectric Technology, Endevco Corp.,* Pasadena, Calif., 1965.

Soisson, Harold E., *Electronic Measuring Instruments,* McGraw-Hill, New York, 1961.

Stein, Marshall L., *Typical Strain Gage Data Acquisition System Requirements,* B & F Instruments, Cornwall Heights, Pa., November 1963.

CHAPTER 10

Radiation Measurement and Instrumentation

The detection and measurement of radioactivity is entirely dependent on instrumentation. Radioactivity from alpha and beta particles, gamma rays or neutrons cannot be detected with the human senses. This means that an individual does not know that radioactivity is present, or the level of radiation if it is present, unless he has adequate means of detecting and measuring it. This is equally true whether or not an industrial application is involved. Therefore we discuss the different types of radiation emitted by radioactive substances, the various means of detection used, and the instrumentation associated with the detector used to measure the particular type of radiation.

Alpha radiation has relatively low energy and has a very short range of penetration. It can be shielded out or stopped by a sheet of paper such as a page of this book. This means that alpha emitting material has to be introduced into the sensitive area or volume of a detector, or that very thin walls or windows must be used to permit the radiaton to pass through them into the sensitive area or volume of a detector. An alpha particle has a maximum range in air of less than 4 in., a maximum velocity of less than 2×10^9 cm/s, and a positive charge equivalent to that of two electrons. Alpha particles have been identified as the nuclei of helium atoms, and have a mass of 6.644×10^{-24} g. As they pass through air or gas, they produce ionization by knocking electrons out of atoms with which they collide. Each collision reduces their speed, and after a sufficient number of collisions they cannot cause ionization. They also produce tracks on photographic emulsions.

Beta particles carry negative charges and have a range of several feet in air. Beta particles are effectively shielded out by a $\frac{1}{4}$ in. thickness of aluminum, in contrast to the book page required for alpha particles.

Beta particles are really high speed electrons with velocities up to approximately that of light. They cause ionization when they collide with atoms of air, but they are light and are bounced around. This means that they do not travel in straight line paths. Detectors are more rugged and have heavier windows than for alpha particles. However, the entrance to the sensitive area or volume of the detector should not be heavy enough to shield out low energy beta particles when a complete spectrum of beta radiation is to be detected and measured. The mass of an electron at rest is 9.11×10^{-28} g, and it is higher at speeds comparable to that of light. Beta particles are useful as tools in industrial applications such as thickness measurements.

Gamma rays are electromagnetic radiations having the same characteristics as x rays, but they have shorter wavelengths. X rays are in the 10 to 150 Å (10^{-8} cm) range, and gamma rays are in the 0.01 to 1.40 Å range. Gamma rays are much more penetrating than beta particles. They can pass through several inches of lead. Gamma radiation is emitted when an atom nucleus passes from a higher to a lower energy state. Gamma radiation is also useful in industrial applications such as nondestructive testing operations for quality control.

Neutrons do not have a charge, so they cannot be detected in the same manner as alpha, beta, and gamma radiation. They also have more penetration power than gamma radiation, and in most cases several feet of concrete or the equivalent is required to provide adequate shielding. Neutrons are generated during atomic reactions, and methods of detection and measurement are necessary to check the adequacy of shielding, and for proper atomic reactor operation and control.

10.1 RADIOACTIVITY DETECTORS

Radioactivity may be detected by means of photographic film, ionization chambers, proportional counters, Geiger-Mueller counters, scintillators, semiconductors, and combinations or modifications of these basic detectors for special applications. The type and intensity of the radioactivity determine the basic type of detector to be used. The application dictates the calibration method of the readout device.

Photographic Film. Photographic film is sensitive to electromagnetic radiation emitted by alpha, beta, and gamma emitting sources. A microscope can be used to measure the actual length of the track on the processed film. Beta and gamma radiation also cause darkening of the film. The relative exposure of the film to the two types of radiation can be measured on a densitometer. The density of the exposed film is compared to a known density produced by a known standard of radiation, so that the total exposure of the film can be determined. This technique is used in personnel badges and in film packs placed on the outside of shielding for protection against overexposure, and to determine the amount of radiation to which personnel may be exposed. A typical film pack used for personnel exposure is shown in Figure 10.1. A densitometer for reading the density of an exposed film is shown in Figure 10.2. A film pack is an *after the fact* radiation exposure record. It does not tell an individual how much radiation he is receiving at the time he is being exposed. It tells him how much radiation his body has received after he has been exposed.

Ionization Chambers. An ionization chamber consists of a cylindrical housing and a central axial wire; the housing acts as a cathode, and the axial wire acts an an anode. The housing distributes the potential around the anode and forms the walls of the sensitive volume in which

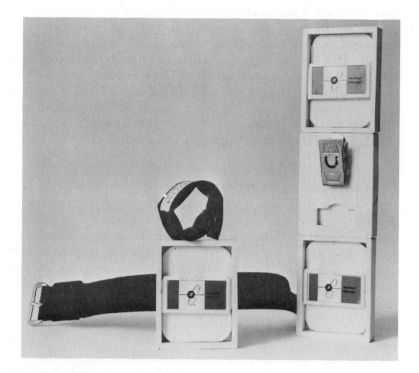

Figure 10.1 *Film pack used to collect radiation exposure for personnel. Courtesy Nuclear Chicago Corp.*

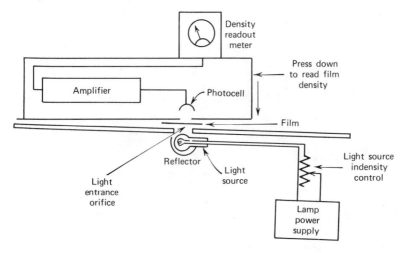

Figure 10.2 *Densitometer for reading film densities.*

an electrical field is defined by the geometry of the electrodes. Ionization chambers, such as the one shown in Figure 10.3, operate in what is termed a *low voltage region*. A simple ionization chamber circuit is shown schematically in Figure 10.4. As shown in the figure, a potential V is applied across the ionization chamber electrodes. A resistor R is placed in the output circuit to cause a voltage drop, so that a voltage pulse is formed when radiation entering the sensitive volume causes ionization of the gas contained in the volume. The gas can either flow through the chamber at approximately atmospheric pressure, or be sealed in the chamber at pressures either greater or less than atmospheric pressure. The capacitance C includes all the distributed capacity in the circuit. The recovery time cannot be shorter than the RC time constant (resistance R in ohms times capacitance C in farads equal to time t in seconds). For example, let $R = 100,000 \ \Omega$ and $C = 100 \times 10^{-12}$ F. Time $t = 1 \times 10^5 \times 1 \times 10^{-10} = 1 \times 10^{-5}$ s for a recovery to 0.368 of its full recovery value. This type of instrument detects the pulse ΔV as an individual pulse. The ionization chamber requires a voltage between the anode and cathode ranging from a few volts to several hundred volts, depending on the chamber design.

If the circuit is arranged to make the anode (wire) positive with respect to the cathode (housing), and a flux of radiation is held constant, the voltage can be varied and the negative ions formed in the gas by ionization, as well as the electrons in the volume, will be collected on the anode. With no potential on the central wire, there will be no detect-

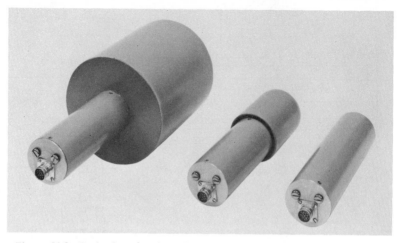

Figure 10.3 *Ionization chamber. Courtesy Tracerlab/West.*

able current flow and no detectable voltage fluctuations on the anode, although a few random ions will be collected. As soon as voltage is applied, a field is created between the anode and the cathode. Any negative ions formed by the radiation rays striking the gas atoms tend to be collected by the anode. Any positive ions formed during the collisions tend to drift toward the cathode. Any electrons formed in the chamber are collected by the anode.

The number of charged paritcles that arrive at the wire, and the total distributed capacity of the central wire, determine the size of the voltage pulse ΔV. Electrons are much more mobile than positive ions. As a result, the slow moving positive ions can form a space charge around the

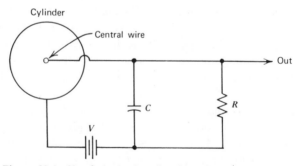

Figure 10.4 *Simple ionization chamber schematic circuit.*

wire and reduce the electric field. The arrival of a charge ΔQ at an anode of capacity C causes a change in potential ΔV. In terms of the unit electronic charge, and assuming no space charge, the rise in potential on the anode can be expressed as an equation:

$$V = \frac{Q}{C} = 1.60 \times 10^{-7} \frac{N}{C} \qquad (10.1)$$

where
N = number of electrons arriving at the anode
C = capacitance in picofarads
V = potential rise in volts
1.60×10^{-19} = unit electronic charge in coulombs
$1.60 \times 10^{-7} = 1.60 \times 10^{-19}/1.0 \times 10^{-12}$

In terms of the instantaneous parameters, the voltage pulse

$$\Delta V = \frac{\Delta Q}{C} = 1.60 \times 10^{-7} \frac{N}{C} \qquad (10.2)$$

At the low voltages applied to ionization chambers electrons do not create additional ions by collision during the collection process. This means that the number of electrons or negative particles collected on the anode is equal to or less than the number produced by the initial ionization event. As long as the voltage is maintained at a value below the point of secondary electron production by collision of the electrons with the filling gas, but is adequate to give the ions enough velocity to cause ionization events, the voltage pulse generated on the anode will be independent of the collecting voltage.

Ionization chambers can also be used as pulse totalizers and integrate the total ionization over a specific time interval. Either function, counting or integrating, can in principle be performed by the same chamber, depending on the RC of the circuit. The final potential of a counting ionization chamber can be expressed as

$$V_F = V_0\, \epsilon^{-t/RC} \qquad (10.3)$$

where V_F = final potential
V_0 = initial potential
t = time
R = resistance in ohms
C = capacitance in farads
ϵ = natural exponential = 2.718

This equation shows that V_F depends on the initial voltage, the time, and the RC value. For a counting chamber to return to $1/\epsilon$ of its original value, that is, $t = RC$, a total reading is made of the radiation detected for that time period.

Ionization chambers can be used for the detection of alpha particles when the alpha emitting material is placed inside the chamber. In area monitoring, an ionization chamber can be easily made with a very flat energy response to gamma radiation, and a single detector can cover a wide range of flux (radiation) levels, for example, a million to one. Ionization chambers are extremely simple and likewise extremely reliable. However, the output of an ionization chamber is a very small current, so it requires very special and sensitive readout circuitry. It has one advantage; since it operates on a very long, flat voltage plateau, voltage regulation need only be nominal for proper operation. A typical environmental system using an ionization chamber is shown in Figure 10.5.

Proportional Counters. A proportional counter is a tube filled with a special gas which produces a small output pulse, when properly biased, for each burst of ions created by a radiation ray or particle passing through the sensitive volume. A typical proportional counter is shown in cross section in Figure 10.6. The proportional counter operates at a higher voltage than the ionization chamber. The voltage is raised to a value where secondary electrons are formed in the gas. This is also termed *gas amplification*. Proportional counters operate in the voltage region whose lower limit is the lowest voltage capable of producing gas amplification, and their upper limit is the Geiger threshold. The Geiger region is discussed in the next section. Gas amplification A takes place as soon as the instantaneous field $\Delta v / \Delta r$ is strong enough to give the electron a high enough velocity to create ionization on collision with more than one gas particle. The instantaneous voltage pulse for gas amplification can be expressed as

$$v = 1.60 \times 10^{-13} \frac{AN}{C} \qquad (10.4)$$

or the voltage of the pulse on the central wire anode in terms of the electron charge e is

$$V = \frac{ANe}{C} \qquad (10.5)$$

where A = gas amplification factor = $\epsilon \int x \, dx$
 N = number of electrons
 e = electron charge
 C = capacitance in microfarads or picofarads

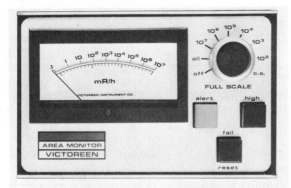

(a)

(b)

Figure 10.5 *Environmental monitoring system using an ionization chamber detector. Courtesy Victoreen Instrument Co.*

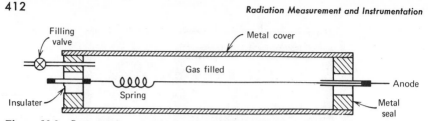

Figure 10.6 *Cross section of a typical proportional counter.*

In a proportional counter the pulse size is developed through progressive ionization in an avalanche formation known as the Townsend avalanche. In a cylindrical proportional counter, if we define the radius at which the ionization starts as r_0 and assume a linear potential dependence, r_0 can be related to the applied potential V, the threshold voltage V_T, and the radius of the central wire r. For the point where the avalanche starts,

$$r_0 = r_1 \frac{V}{V_T} \tag{10.6}$$

The gas amplification factor as defined for equation 10.5 is read "*A* equals ϵ to the integral power of *X dX*." The term *X* is the first Townsend coefficient and depends on the field strength and the pressure of the gas. The term *dX* is the distance traveled by the electron. The number of ionizing events that occur for each initial start depends on the mean free path of the electron, and this mean free path is a function of energy. The mean free path is not necessarily a straight line path, because the electrons are bounced around when they collide with heavier gas particles.

Proportional counters have the advantage that the output pulse is proportional to the energy of the incoming radiation ray or particles dissipated in the detector. This permits the proportional counter to give a differential measurement of alpha radiation versus beta-gamma radiation. Even with the added gas amplification the output pulses are quite small, so a proportional counter system requires an amplifier with considerable gain. The proportional counter system also requires a fairly well regulated power supply for proper operation.

To use a proportional counter for detection of alpha radiation, the counter has to be built with a very thin window which is usually not entirely gas-tight. This means that a special gas has to be continuously provided to the detector at a continuous rate. The thin windows are fragile and pose a maintenance problem, but are useful where a dual detection system is needed for detection of alpha and beta-gamma radiation. One such system is shown in Figure 10.7. This system employs a

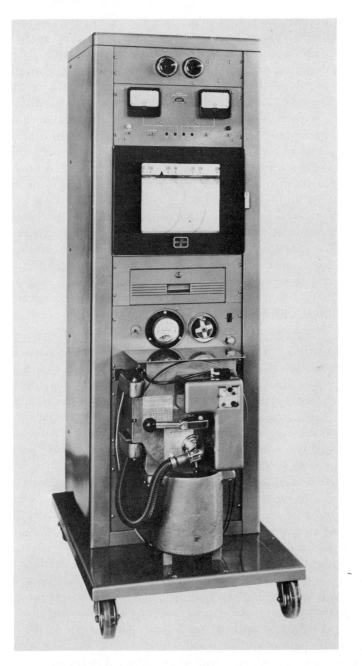

Figure 10.7 *Dual detection and readout system for alpha and beta-gamma radiation. Courtesy Nuclear Measurements Corp.*

P-10 gas mixture and is used to monitor the radioactivity of fission product particulates collected on a filter. A portable survey meter using an unsealel air proportional detector is shown in Figure 10.8.

Proportional counters are also useful for measuring neutrons. These counters are normally sealed and are lined with boron or filled with boron trifluoride gas. The neutron interacts with the boron to produce an alpha particle which is readily detected inside the proportional counter. This is one of the most successful ways of measuring low level neutron fluxes and at the same time discriminate against beta and gamma radiation. These detectors are used for monitoring the neutron flux in nuclear reactor operations, and can be used for both slow neutrons and fast neutrons. In high neutron fluxes, compensation is needed to remove the effect of gamma radiation. By properly selecting the pulse height threshold, meaning that any pulses below a certain magnitude are not detected, it is possible to block the smaller gamma pulses and count only neutron pulses with gamma backgrounds as high as 10^4R/h. R is the symbol used to represent the roentgen which is a unit of measure for radiation. This and other terms used in radiation measurements are discussed later in the chapter. Two different models of boron trifluoride chambers are shown in Figure 10.9.

Geiger Counters. A voltage-versus-counts-per-minute curve is shown in Figure 10.10. The units are arbitrary to show that the Geiger threshold is at the upper limit of the proportional region and that a voltage plateau is reached where there is little or no change in counting rate for a given voltage increase. If the voltage is raised to too high a value, a continuous discharge condition is reached and no useful information can be obtained. It is possible to damage a Geiger Mueller tube permanently by applying too high a voltage.

Counters operating in the Geiger region are usually called GM counters or, more precisely, Geiger-Mueller counters. Geiger-Mueller counters used in industry are of the self-quenching type and have either a side window or an end window. Both types are shown in Figure 10.11, and in cross section in Figure 10.12.

The Geiger-Mueller tube is a gas filled tube which produces a rather large, easily counted pulse for each ionizing event that enters the sensitive volume and triggers the avalanche action. Depending on the design and construction, the tube requires 500 to 2000 V for operation, but fortunately it generally operates on a long, flat plateau so that it works satisfactorily with reasonably good voltage regulation. The windows can be made thin enough to allow beta radiation from most radioactive materials or isotopes to enter with only a small attenuation or loss of energy. This makes the Geiger-Mueller tube a good detector for beta or

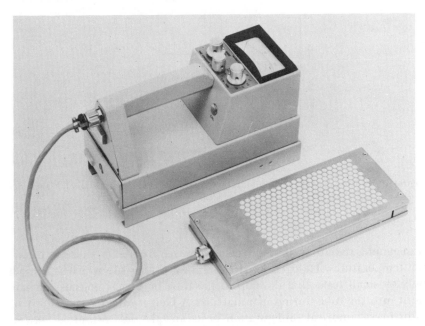

Figure 10.8 *A survey meter using an unsealed air proportional detector for measurement of surface alpha radiation. Courtesy Nuclear Chicago Corp.*

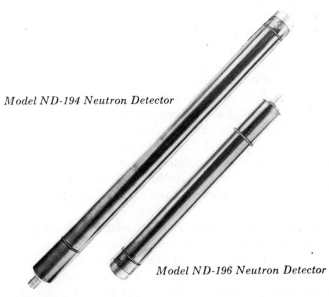

Model ND-194 Neutron Detector

Model ND-196 Neutron Detector

Figure 10.9 *Two boron trifluoride proportional counters used for neutron detection. Courtesy Baird-Atomic, Inc.*

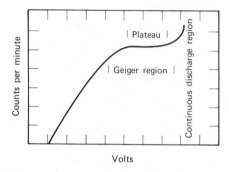

Figure 10.10 *Graph showing the ionization chamber region, proportional region and Geiger region, with the plateau in the Geiger region.*

beta-gamma radiation. It is also relatively free from effects due to changing temperatures. Halogen filled tubes have a long life which is limited only by small leaks that develop in the tube or by gas poisons that are built into the tube during manufacture. A Geiger-Mueller tube must be built in a cylindrical configuration so it can take only a limited view of a planar sample. It can never see more than 50% of the radiation emitted by the sample it views.

An area monitoring system using halogen filled tubes is shown schematically in Figure 10.13. This system consists of 10 area monitors whose power requirements are furnished by a central power supply. This system uses noncontact type meters such as those discussed in Chapter 5 for on-off temperature control, except that no physical stops are used so that radiation values up to full scale can be read. As shown in the diagram, the outputs of the 10 units are recorded on a sequential basis every $2\frac{1}{2}$ s. The main power supply has a memory light which must be reset manually. However, when a memory event occurs, it does not indicate which monitor caused the event and, since the monitors automatically reset, it is very difficult or impossible to determine where the event occurred unless it lasted long enough to be recorded. A Tracerlab system using halogen filled detectors and the nonconact meter approach is shown in Figure 10.14.

Geiger-Mueller counters such as the portable one shown in Figure 10.15 are also used in a variety of radiation survey meters. This type of instrument is used to detect beta or beta-gamma radiation in work areas where radioactive materials are produced, or is employed in industrial applications. Other versions of the Geiger-Mueller tube are used in locating brain tumors, for tracing radioactive materials fed to plants in

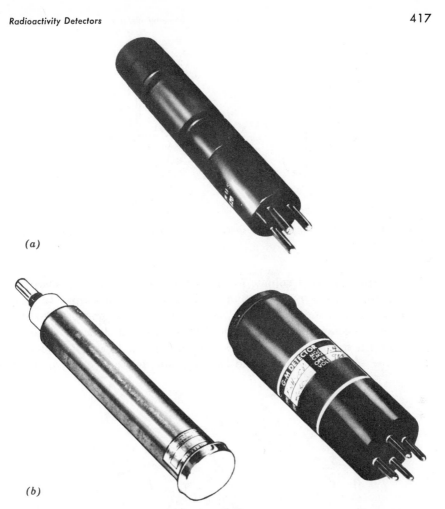

(a)

(b)

Figure 10.11 *Geiger-Mueller tubes. (a) Side window type. (b) End window type. Left, survey end window; right, large end window.*

fertilizers, and for gaging the thickness of materials in industrial processes such as the rolling of steel.

Scintillation Counters.   A scintillation counter consists of a very dense crystal such as sodium iodide which converts radiation into flashes of light. This light is detected by a photomultiplier tube. The crystal has two important advantages. First, its density gives it a gamma ray sensitivity approximately 100 times that of the Geiger-Mueller tube, because

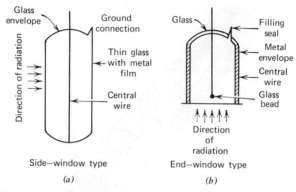

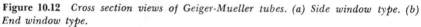

Figure 10.12 *Cross section views of Geiger-Mueller tubes. (a) Side window type. (b) End window type.*

of the density per unit volume in the crystal compared to the relatively thin gas fill in the sensitive volume of the Geiger-Mueller tube. Second, the scintillation counter produces an output pulse proportional to the energy of the gamma ray photon that causes the pulse. When a scintillating crystal is struck by a radioactive ray, photons of light are produced. Each type of crystal continues to emit photons for a certain length of

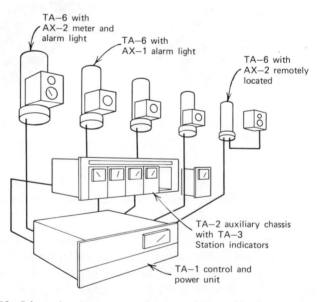

Figure 10.13 *Schematic of an area gamma monitoring system using halogen filled Geiger-Mueller tube detectors. Courtesy Trapelo Division, LFE.*

Figure 10.14 *Tracerlab system using a central power supply, noncontact meters, and halogen filled Geiger-Mueller tube detectors. Courtesy Tracerlab/West.*

time after it has been activated by the incoming ray or particle. The faster the decay time of a scintillator, the more incoming events it can detect and resolve. Since the output pulse is proportional to the energy of the gamma ray photon, a single or multichannel pulse height analyzer can be used to analyze for, or discriminate against, various gamma ray energies. The photomultiplier tube used with the scintillator requires between 1000 and 2000 V for operation. It is extremely voltage sensitive, and a well regulated high voltage supply is required for proper oper-

Figure 10.15 *Portable survey meter using a Geiger-Mueller tube detector. Courtesy Nuclear Chicago Corp.*

ation. The photomultiplier is also very light sensitive and must be used in a light-tight detector system. In addition, the photomultiplier tube is quite temperature sensitive, so to obtain constant results in spectrum analysis the scintillation counter must be used in an environment where its temperature is controlled within a few degrees between standardizations.

Scintillation detection and counting systems are used in area monitoring applications where gamma radiation is to be measured. An area system is shown schematically in Figure 10.16. This system consists of 10 area monitors connected to a central control point and to a radiation control center. At the central control point each unit reports area radiation activity with a series of lights indicating above-normal conditions. At the radiation control center the area radiation reading is recorded every 2.5 s. In the event of power failure, instrument failure, or radioactivity above normal, an alarm is sounded and a memory system indicates the condition until it is reset. This particular system is all solid state and operates on an electronically actuated alarm system for relay operations. Well designed units of this type can employ remote meter systems up to 1000 ft from the actual monitor, using a minimum of two wires. Each monitor contains its own power supply and can operate on rechargeable batteries for a period of 14 h during a main power failure. This gives it an advantage over the system shown in Figure 10.13, because if the monitors are connected to different power sources only those monitors on the affected power source will be disabled or shift to their self-contained rechargeable power supply without loss of area coverage.

Scintillation counters are also available with alpha sensitive phosphors having an overall alpha efficiency of approximately 50% with a collection efficiency to approximately 32%. One such detector is shown in Figure 10.17.

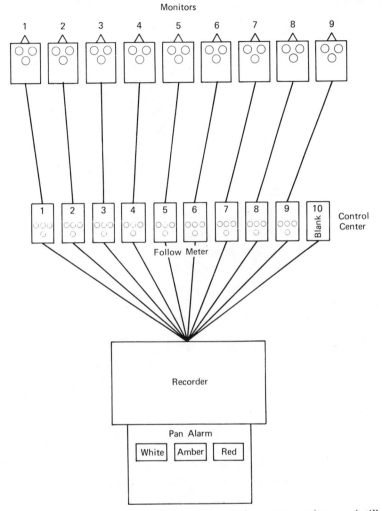

Figure 10.16 *Schematic of an area gamma monitoring system using a scintillation detector.*

Scintillation detectors are likewise available with anthracene crystals and plastic scintillators for the detection of beta radiation. A thin anthracene crystal is insensitive to alpha radiation and has low gamma radiation efficiency because of its thinness. The beta sensitive plastic scintillator gives increased beta sensitivity and significantly lower gamma efficiency than the anthracene crystal, which makes it ideal for beta spectroscopy work. The short decay time of the scintillator permits counting at high

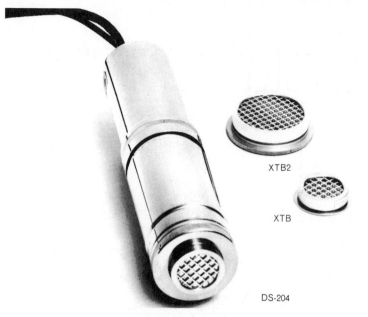

Figure 10.17 *Alpha sensitive scintillation detector. Courtesy Nuclear-Chicago Corp.*

rates with low coincident losses. This means that very few events occur simultaneously at these high counting rates, so that few losses are caused by two or more events being counted as a single event. It should be noted, however, that there are times when it is desirable to record or use only those events that occur simultaneously. This is discussed in the section dealing with reactor control. The spherical neutron dosimeter shown in Figure 10.18 contains a 4×8 mm lithium (or Europium) crystal surrounded by a 10 in. spherical polyethylene moderator for slowing down the neutrons. The scintillation crystal is coupled to a magnetically shielded photomultiplier through a Lucite light pipe. The polyethylene density is closely controlled to ensure accurate response over the 10^{-8} to 10 meV range which covers the intermediate energy range where some personnel dosimeter units are insensitive. This unit uses conventional auxiliary equipment such as a preamplifier, an amplifier, a rate meter, and other accessories for direct reading in rems (roentgen equivalent man).

Semiconductors. The semiconductor diode has been used successfully as a detector for alpha particles and has a proportional energy response. The semiconductor diode detector has been used in physics research to measure fission fragments, high energy particles, neutrons, and gamma radiation. This is a relatively new field and application for semicon-

Figure 10.18 *Nemo spherical neutron dosimeter using a scintillation detector. Courtesy Texas Nuclear Corp.*

ductors, but as the technology widens and the manufactured product becomes more uniform in performance, they are being more widely used in industry. In most industrial applications of radiation detectors there is a health and safety consideration, so better known and more widely used systems are generally favored.

The semiconductor diode requires a good amplifier system to keep the electronic noise smaller than the detected signal, and the present semiconductor diode is limited in the physical area it can cover. This type of detector has the advantage that it may allow the selective monitoring of specific alpha particle admitters in the presence of other type radiation. For example, it has a possible use in the measurement of plutonium in the presence of natural radioactivity. In the present state of the art, these detectors offer no great advantage over other detectors in environmental monitors, with the possible exception of alpha detection and discrimination.

10.2 READOUT INSTRUMENTS AND ACCESSORIES

The choice of readout instrument and/or accessories depends on the application and the output signal of the detector. Very small output signals require preamplifiers and amplifiers. Medium output signals may require an amplifier before they can be fed into conventional readout equipment. Readout instruments may range from a simple indicating meter to a complete spectrum analyzer or computer operation if the ultimate in control is needed for the application. Each unit also requires some type of power supply.

Preamplifiers. A preamplifier is located as close to the detector as possible. There are at least three types of preamplifiers: trigger preamplifiers, linear preamplifiers, and proportional preamplifiers.

A trigger preamplifier is a device that produces a large low impedance output pulse of constant amplitude and pulse width for each impulse fed into it by the detector. This type of preamplifier is used in portable transistorized monitoring equipment and other monitoring equipment where total radiation or dosage values are the only requirements. In some cases, in addition to amplifying the detector pulse, the preamplifier is used to match the impedance of the detector to a coaxial line to prevent the degradation of pulses in the line and to prevent loss in long lines.

A linear preamplifier is used when the output is fed into an amplifier or when a single-channel analyzer is used. A linear preamplifier is used to preserve the linear relationships of pulse heights produced by various energy radiations.

Proportional preamplifiers are used for proportional counting. These units can now be completely transistorized with low noise and nonoverloading characteristics. They are available with a choice of gain control, so that their output may be fed into amplifiers, amplifier-analyzers, or directly into scalers or count rate meters.

Preamplifiers can have gains ranging from 1 to 1000, a pulse rise time from less than 0.1 μs to a few microseconds, a pulse decay time of less than 0.3 μs to a few microseconds, random noise as low as 5 mV for a maximum output of 4 V negative, input impedances in the order of 50,000 Ω, and output impedances from 30 to 50 Ω. The choice of characteristics for the preamplifier ultimately depends on the application requirements. In general, the more rigid the specifications on accuracy and time response, the higher the price.

Amplifiers. The type of amplifier needed for radiation readout systems depends on the output pulse of the detector or preamplifier in the system. A pulse amplifier is required where pulses are produced and, if the application requires spectrometry (measurement of a spectrum of energy), the amplifier must be linear in its operation and should be nonoverloading. In applications where a wide range of radiation must be measured with the same instrument, it may be advantageous to use a logarithmic amplifier. If a continuous output detector is used rather than a pulsed output, a dc amplifier is required.

Pulse amplifiers are fast response ac amplifiers. Although radiation detectors are dc devices, the output pulses are amplified as ac quantities. Pulse amplifiers can be designed to accept either positive or negative pulses. In some cases the output pulse of the detector must be reshaped to be properly amplified. If the pulse cannot be reshaped to give the proper rise time and base width, an amplifier must be used that accepts the pulses as they are generated by the detector. Pulse amplifiers designed to accept positive pulses are normally nonconducting when no pulses are present, and they do not conduct with a negative pulse input.

A typical tube negative pulse amplifier is shown schematically in Figure 10.19. This amplifier is essentially a one-shot multivibrator for amplifying and shaping pulses. This circuit accepts negative pulses with amplitudes as low as 200 mV, and has an optimum performance with pulses of 500 mV. It produces an output pulse of 20 to 50 V amplitude with a maximum decay time of 13.5 μs. As mentioned before, the pulses should have short decay times so that random generated pulses can be resolved and amplified. It is also important to keep the time cycle of the circuit short enough to prevent noise pileups. This means that small noise signal pulses can charge up a capacitor and produce a pulse equivalent in

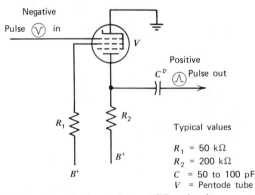

Typical values

R_1 = 50 kΩ
R_2 = 200 kΩ
C = 50 to 100 pF
V = Pentode tube

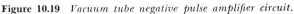

Figure 10.19 *Vacuum tube negative pulse amplifier circuit.*

amplitude to the amplitude of an accepted signal from the detector if the time cycle is of sufficient length.

Pulse amplifiers can be designed with feedback to add stability to the circuit. When a wide range of pulse frequencies as well as pulse amplitude must be amplified, compensation is necessary to overcome both the low frequency and high frequency drop-off.

A typical logarithmic positive pulse amplifier circuit is shown in Figure 10.20. The input to the first transistor is the output pulse of a scintillation detector photomultiplier. The pulses are measured as the output of a transistorized electronic voltmeter for a full scale value of 1 mA. The circuit shown can be adjusted to give either a three- or five-decade response as integrated counts per minute. This maximum signal can be converted to a 10 mV output by replacing the 1 mA meter movement with a precision 10 Ω resistor and reading the voltage developed across the resistor.

Linear amplifiers are designed so that the amplified output signal is directly proportional to the input pulse. The output developed should also have the same wave shape as the applied exciting voltage. Linear amplification can be achieved by altering a class C vacuum tube amplifier to meet linear conditions. A class C vacuum tube amplifier is an amplifier in which the grid bias is appreciably greater than the tube cut-

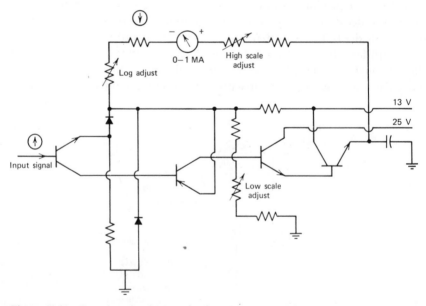

Figure 10.20 *Transistor positive pulse logarithmic amplifier circuit.*

off value. Under these conditions no plate current flows when there is no grid excitation signal and, when an alternating grid voltage is applied, the plate current flows for somewhat less than the positive half cycle. Linear response is accomplished by altering the class C operation so that the tube bias correpsonds to the projected cutoff, as shown in Figure 10.21. By applying a bias voltage, which corresponds to the bias that would be obtained if the plate-current-versus-grid-bias-voltage curve of the amplifier tube, for the the operating potential, were projected as a straight line from the linear portion of the characteristic curve, one has projected cutoff. In other words, the straight line, if projected to the bias voltage line, gives the bias voltage that would cause the tube to stop conducting. The linear amplifier is designed to eliminate harmonics that could alter the output wave shape with respect to the input wave shape. The relationship between output and input voltages for a typical linear amplifier under different load conditions is shown in Figure 10.22. The relationship is quite linear up to a critical input value. At the critical input value, the output levels off and is no longer proportional to the input. This occurs when the peak voltage developed in the tube plate circuit across a load impedance is approximately the same as the plate supply potential. As shown, a greater exciting voltage is required for a given output voltage of low impedance.

Linear amplifiers are required when detector pulses are analyzed for the various amplitudes to determine the energy output relationships of the emitted radiation. A digital linear amplifier is shown in Figure 10.23.

DC Amplifiers. The vacuum tube in a dc amplifier conducts in proportion to the amplitude of the voltage of the input signal. The vacuum tube is normally in the conducting condition and, as the grid potential is raised or lowered, the plate current increases or decreases in proportion to the applied signal. These amplifiers must be direct coupled to pass the plate current to the grid of the next stage of amplification or to to the input of the indicating, recording or controlling device.

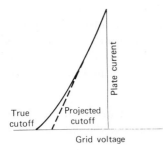

Figure 10.21 *Projected cutoff.*

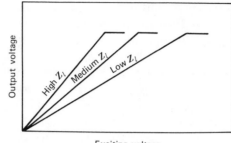

Exciting voltage

Figure 10.22   *Linear amplifier curves.*

DC amplifiers are required when the dc output signal of a detector is too small to drive the actuating mechanism of an indicator, recorder, or controller. They are also used in regulated high voltage power supplies required for radiation instrument systems. Vacuum tube dc amplifiers are subject to drift, which is caused by aging of the tubes, variations in the plate voltage, and changes in temperature. This is a serious limitation where accuracy is needed. These amplifiers must be warmed up for reasonably long periods of time, ranging from 20 min to several hours, to give reliable and reproducible data. Compensation methods used to minimize drift are balanced circuits, negative feedback, chopper stabilization, and aging of the vacuum tube before installation.

Transistors can be and are used in dc amplifiers and in power supplies. These are current sensitive devices and cannot be used when the output of the detector is so small that it is difficult to separate it from the current generated by the noise present in the detector or the amplifier itself. When the output current is large enough, these amplifiers offer the advantage of very short warmup, reasonably good stability, small size, and less weight in portable equipment.

Electronic Counters.   Counters may be simple binary circuits using a flip-flop principle, or they may be more complex decimal systems. In either case they are used to read out the number of events that occur in a predetermined time. The output of the electronic counting circuit may be integrated in counts per second, counts per minute, milliroentgens per hour, roentgens per hour, or any convenient unit of measure most suitable for the application. These values may be indicated, recorded, or printed out.

Binary Counters.   Binary counters operate on the principle of multiples of two. They are usually referred to as scalers and use the common flip-

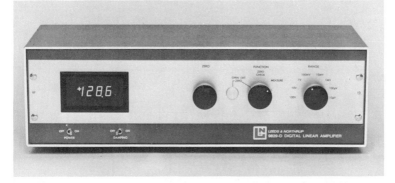

Figure 10.23 *An industrial linear amplifier. Courtesy Leeds and Northrup Co.*

flop circuit. The flip-flop circuit is a scale of two, where two tubes or transistors are used to detect the incoming signal as shown in Figure 10.24. When the first input event is received, it activates the first vacuum tube T_1, or transistor Q_1. When the second event arrives, it activates T_2 or Q_2 and deactivates T_1 or Q_1. The third incoming event activates T_1 or Q_1 and deactivates T_2 or Q_2. If lights are used to indicate the counts, an array such as that shown in Figure 10.25, which is a duplicate of the groups shown on the binary scaler in Figure 10.26, is used. The sequence of lights appearing, as events generate signals to actuate the flip-flop circuit, are as follows:

Number 1 lights for one event.
Number 2 lights for two events, and number 1 goes out.
Number 1 and 2 light for three events.
Number 4 lights for four events, and numbers 1 and 3 go out.
Numbers 1 and 4 light for five events.
Numbers 2 and 4 light for six events, and number 1 goes out.
Numbers 1, 2, and 4 light for seven events.
Number 8 lights for eight events, and numbers 1, 2, and 4 go out.

This sequence continues up to 255 events, and then the cycle is repeated. When the cycle repeats, a register is activated by event 256. The registered number represents 256 events, and all the lights are out. Event 257 lights number 1 and the cycle continues. Counters of this type can be designed to record a given number of counts and stop, or to record the events occurring in a given length of time and stop. This selection is shown on the instrument in Figure 10.26. When the counting starts, the

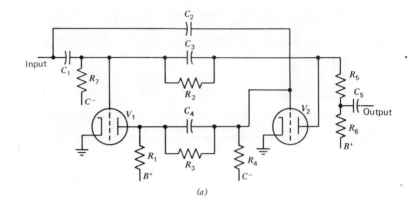

(a)

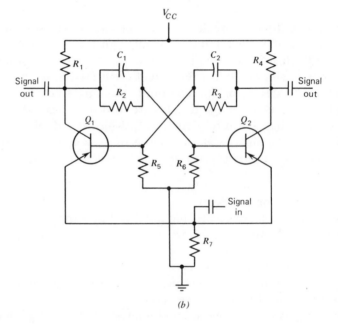

(b)

Figure 10.24 *Schematic circuits of scaling pairs. (a) Vacuum tubes. (b) Transistors.*

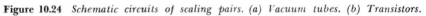

Figure 10.25 *Binary neon light array.*

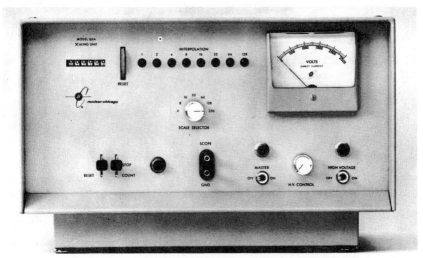

Figure 10.26 *Basic binary scaler. Courtesy Nuclear-Chicago Corp.*

total registration must be counted and recorded. Both methods are time consuming and subject to human error. The register value has to be read and multiplied by 256, and the glowing lights have to be counted and added to the register reading. Then the total has to be manually recorded. There are chances for mistakes in reading, multiplying, and adding the numbers.

The binary counter was the first system designed to count radiation events, but it is seldom used for industrial applications because of the possible errors and the high operation cost to read, multiply, add, and record every reading. Decimal and direct readout systems were developed, and are favored in industrial applications.

Decimal Counters. A typical decimal system is shown in Figure 10.27. A neon array is arranged in a vertical column with numbers from 0 through 9 starting at the bottom. On the tenth event all the lights go out and a second array is actuated. In the unit shown a total of 10^8 events can be recorded for a full cycle. For a full count before recycling, 999 counts can be read on the lights. Recycling occurs in the 1000th event, and all the lights go out. The count register can record 9999 × 1000 events and recycles when the count is 10,000,000. These counters are designed to count a given number of events or to count for a specified length of time. Decimal units can also be designed for automatic print-

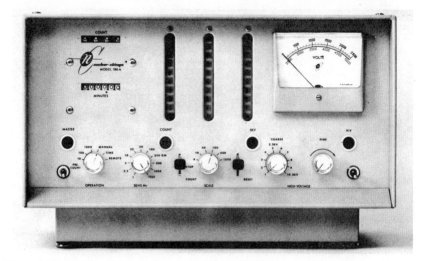

Figure 10.27 *Decimal counting unit using vertical neon arrays. Courtesy Nuclear-Chicago Corp.*

out of the total count and for resetting at the of the counting period. When large volumes of samples have to be counted, a sample changer can be used and the operation sequenced for automatic reading of samples, recording of data, and automatic stopping of the system when the last sample has been counted. A block diagram of such a system is shown in Figure 10.28.

A modular design decimal system is shown in Figure 10.29. The actual number of events detected is shown as a number using readout tubes displaying numerals instead of neon light arrays on the scaler. A separate timer module, a high voltage supply module, and an amplifier module are shown. These features were incorporated into the neon array unit of Figure 10.27. To demonstrate the programming and analyzer applications of such a system, the programmer, analyzer, and time base modules have been included to make up a logic system with automatic printout and digital recorder modules. The preamplifier is included to indicate that this system can be used with low output detectors. Other modules that can be substituted in the system are graphic recorders, linear rate meters, and logarithmic rate meters. Fewer modules can be used as well as more. A basic event counting system can be made up of the scaler, high voltage supply, and timer modules, if the detector output is large enough to drive the scaler. The modular concept permits the use of only those modules necessary for the application. The latest counters are

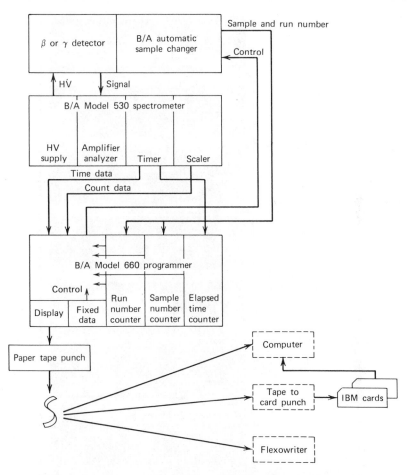

Figure 10.28 *Block diagram of an automatic counting and recording system. Courtesy Baird-Atomic, Inc.*

digitized and have in-line readouts. The most recent innovations include liquid crystal and led readouts and the use of integrated circuits instead of discrete circuit construction. This type of construction produces the same or a better degree of accuracy, requires less power to operate, and in general reduces the physical size of the modules needed in a total system and may reduce the number required.

Count Rate Meters.   In applications in which it is not necessary to know the exact count, but it is necessary to know the rate at which the radio-

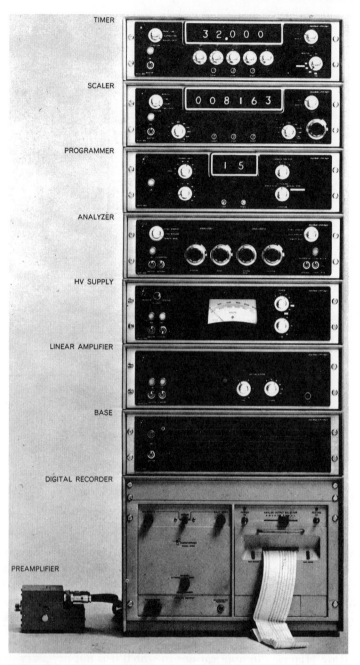

Figure 10.29 *Modular system for automatic counting and recording. Courtesy Nuclear Chicago Corp.*

434

activity occurs, a count rate meter is used. The count rate meter is a unit designed to take the incoming pulse and analyze the repetition rate and then display the result in terms of counts per unit time on an indicating meter. The readings can also be displayed on a graphic recorder to give a graphic record of the changes in the counting rate. Count rate meters may also be equipped with alarm devices to activate an alarm when a preset count rate is exceeded or when there is too rapid a change in the counting rate.

Count rate meters may be either linear or logarithmic. The linear count rate meter gives a linear readout which starts at zero and has an upper limit depending on the range of the instrument. In many cases the instruments are designed for multiple ranges, and a range switch is used to select the range needed. In most instances the ranges are changed manually, but automatic range change circuts are available. The accuracy of a linear count rate meter is generally good in the upper third of the scale, but rather poor in the lower portion of the scale. This is one reason for having a multiple-range unit. The reading can be kept in the accurate portion of the scale by switching to the appropriate range. A linear count rate meter with a ± 0.25 V input with 10 ranges from 100 to 3,000,000 counts/min is shown in Figure 10.30. This unit has 10 time constants from 0.1 to 100 s and an output for 1 MA, 10 MV, and 100 MV chart recorders. As shown, it also incorporates a pulse height discriminator and zero suppression for one full scale of suppression. A built-in test signal is also provided in the unit for self-calibration.

The time constant of a count rate meter determines its ability to measure accurately the statistical data it receives and integrates. Longer time constants are needed at low counting rates so that more data can be measured by the count rate meter to avoid the uncertainty of statistics. This means that the time constant is an inverse function of probable error. The linear count rate meter shown in Figure 10.30 has such an adjustment for the time constant.

A logarithmic count rate meter has a log scale output which may cover two, three, five or more decades. As an example, the scale can cover 10^5 counts where the first decade is 10 to 100 counts/min, the second 100 to 1000 counts/min, the third 1000 to 10,000 counts/min, the fourth 10,000 to 100,000 counts/min, and the fifth 100,000 to 1,000,000 counts/min. Such a circuit does not have a zero left-hand reading, and the meter does not have a zero reference point. This means that a zero reading is an infinite distance to the left of the finite value shown on the left-hand side of the scale. This type of unit requires one or more calibration signals to be provided for checking operational performance. In a well designed logarithmic count rate meter, the percentage of reading accuracy

Figure 10.30 *Linear count rate meter. Cambridge series Model CS400. Courtesy Baird-Atomic, Inc.*

is constant across the entire scale. The time constant of a logarithmic count rate varies depending on the counting rate. Under such conditions it is difficult to express a time constant or probable error for a logarithmic count rate meter. The slow response of a logarithmic count rate meter in the lower part of its span is often assumed to be sluggishness. The sluggishness occurs at the low counting rates where it is desirable to maintain a low probable error and where the accuracy of measurement for monitoring purposes is of less importance. At high counting rates, where monitoring information is most important, the log count rate meter responds much more quickly, which is the reason it is used almost exclusively in dynamic monitoring applications. The count rate meter shown in Figure 10.31 combines the features of both a linear and logarithmic count rate meter. It has 13 linear count ranges and a 10^6 counts/min log range with a 1 μs resolving time.

Power Supplies. Power for radiation detectors and associated equipment can be supplied by batteries or ac power packs.

Batteries are used in portable instruments and may range in voltage and size from 1.5 V penlite cells to small 1.3 and 1.4 V mercury batteries for the older tube type portables. These batteries have a definite usable ampere-hour life and have to be replaced at regular intervals. In some

Figure 10.31 *Combination linear/logarithmic count rate meter. Model 435. Courtesy Baird-Atomic, Inc.*

transistorized monitoring equipment, rechargeable batteries are used to maintain the proper power for the units when regular ac power source failures occur. The batteries are always on charge, and the nickel-cadmium type have a long life of several years compared to a 3 to 6 month life for dry cells in field use. Care must be taken to remove batteries from service before they are likely to leak or corrode, to prevent damage to battery holders and other circuit components.

Ac power packs normally supply both the low voltages and high voltages needed by the instrument or system. In many cases the power supply is built right into the total circuit. It may also be a separate unit. The usual method is to rectify the ac power to produce a pulsating direct current which is then filtered and regulated as required for the specific application. Dc power supplies are subject to the same limitations as dc amplifiers, so they have to be stabilized and regulated to give good stable outputs under varying load conditions. Otherwise, the reproducibility or a system is not reliable and the data produced are questionable.

Rectification can be either half wave or full wave. The half wave rectifier only rectifies either the positive or negative half of the ac signal. In a 60 cycle system a half wave rectifier rectifies 60 half waves per second. A typical vacuum tube half wave schematic circuit is shown in Figure 10.32a, and a solid state version in Figure 10.32b.

Typical full wave schematic circuits are shown in Figure 10.33a and Figure 10.33b. These circuits rectify both the positive and negative half cycles. Thus in a 60 cycle system each of the 60 cycles produces 120 rectified pulses per second or twice as many as the half wave circuit. This not only gives a higher averaged dc output, but has less ripple than the half wave circuit.

Power supplies are often classed as nonregulated, regulated, and well regulated. The characteristics that indicate the degree of regulation are ac ripple, noise, line regulation, short term stability, long term stability, load regulation, and temperature coefficient. These values are usually expressed as a percentage in terms of a known reference: ripple is expressed as percentage of the output voltage or millivolts rms at the output; line regulation as percentage of a change in power input voltage, say from 95 to 135 V for a nominal 120 V 60 hertz source; short term stability in percentage change per hour; long term stability in percentage change per day or week; temperature coefficient as percentage per degree

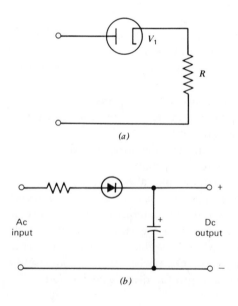

(a)

(b)

Figure 10.32 *Typical half wave rectifier schematic circuits. (a) Tube. (b) Solid state.*

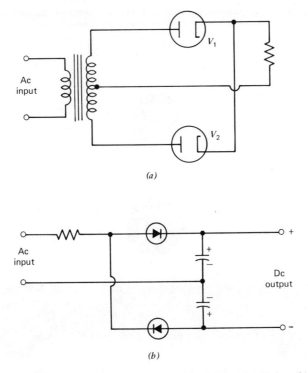

(a)

(b)

Figure 10.33 *Typical full wave rectifier schematic circuits. (a) Tube. (b) Solid state.*

Celsius or Fahrenheit; and load regulation as percentage of the no load to full load condition. The lower the percentage in each case the better the regulation and almost invariably the higher the cost.

Power supplies are also characterized by the voltage range and current they can supply, and by the load impedance they can match. Power supplies are available with either positive or negative outputs, and with both positive and negative outputs. The design of the radiation detection and measuring unit or system determines both the type and capacity of the power supply required. A high voltage supply for use with scintillation detectors is shown in Figure 10.34. This power supply has both positive and negative polarity outputs in 11 selected steps from 595 to 1445 V with a 1 mA current output over the entire range. A continuously variable high voltage supply for use with gas counters is shown in Figure 10.35. This power supply is all solid state and is built on the modular concept for a 600 to 6000 V range with approximately 1 mA current output. These two power supplies illustrate that there is a large variety of power supplies to choose from for a specific industrial application.

Figure 10.34 *High voltage power supply for use with scintillation counters. Model 345. Courtesy Baird-Atomic, Inc.*

10.3 REACTOR RADIATION INSTRUMENTATION

Nuclear radiation instrumentation performs two major functions. It detects and measures radiation, and it is used for controlling, measuring, and indicating the operation of a reactor's control rods.

Circuit reliability and stability considerations require at least two, and quite often three, safety channels of measurement and control on each reactor. Two of the channels must have coincident signals to shut down the reactor. In other words, it is not considered safe to have only one channel of safety control, because false alarms due to instrument malfunction or failure are expensive and serve no useful purpose. When two channels of control receive the same signal simultaneously, action takes place rapidly to shut the system down safely.

Reactor control instrumentation is usually divided into several channels, and often two channels are used as in safety monitoring. Different instrument ranges may be required for different types of nuclear reactors,

Figure 10.35 *Solid state high voltage power supply for use with gas counters. Model 365. Courtesy Baird-Atomic, Inc.*

but generally the same types of instruments are used. The usual channels for operation are the counting rate channel used during start-up, the period channel, the linear channel, and the level safety channel.

The counting rate channel normally has a fission chamber detector which supplies pulses to a linear amplifier, a pulse height discriminator, a scaler, and a log count rate meter. This channel is a low level radiation system. The log count rate meter is used to obtain a wide range of operation. A block diagram of this system is shown in Figure 10.36.

A typical period channel uses a compensated ionization chamber detector which feeds into a preamplifier and a log N amplifier. This amplifier feeds into a pulse height discriminator and scaler or logarithmic count rate meter. The input section of the log N amplifier consists of a diode operated in the logarithmic portion of its plate characteristic to provide a wide range of operation. The derivative of the voltage developed across the diode is proportional to the reactor period. The period channel is normally operated so that, when a reactor period occurs in less than 1 s, the reactor is scrammed (shut down). A period channel block diagram is shown in Figure 10.37.

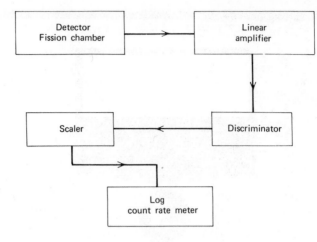

Figure 10.36 *Reactor counting rate channel block diagram.*

A linear channel is used to measure the reactor level during operation. It normally uses a compensated ion chamber and a micromicroammeter readout.

The level safety channel shown in the block diagram of Figure 10.38 is a primary nuclear safety device. In a level safety system, as stated earlier, two or more channels are used and, when coincident signals are

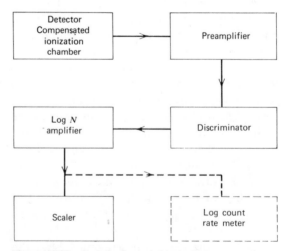

Figure 10.37 *Period channel block diagram.*

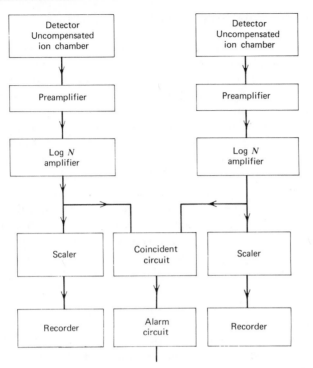

Figure 10.38 *Level safety channel block diagram.*

received by at least two channels, the reactor is scrammed by dropping the control rods into the reactor core to stop the nuclear chain reaction. This type of channel consists of an uncompensated ion chamber, a preamplifier, a log N amplifier, a scaler, and a recorder. In some applications a compensated ion chamber and a magnetic amplifier are used.

For survey purposes one or more proportional counters with associated preamplier, pulse height discriminator, count rate meter, scaler, and recorder are part of the control room equipment and the health physics analytical area. Optional equipment includes scintillation and Geiger-Mueller counter detectors with pulse height analyzers, count rate meters, and recorders.

10.4 PROCESS RADIATION INSTRUMENTATION

Radiation instrumentation is used in industrial processes for density, moisture, thickness, and liquid level measurements and control. This type of instrumentation is required when the detector cannot contact the product to be measured, either because the product is too bulky or because the detector might contaminate it.

When radioactive sources are used in industrial applications, care must be exercised to shield the source so that operators and other personnel in the area are not exposed to excessive radiation. This exposure is usually minimized by storing the source in a shielded container which is completely closed when measurements are not being made. The majority of source holders and storage containers contain lead shielding in sufficient thickness to limit the radiation level to the limit value established for industrial exposure. All source containers and storage vessels are labeled with a warning that a radioactive source is enclosed. Radiation values are not necessarily shown on the label, because most sources decay with time.

The shield and mechanism for exposing the source during measurements are often complicated and ingenious in manipulation. The complexity depends on the source strength and the application. Some of the more elaborate systems are discussed in Chapter 11.

Density Measurements.   Continuous and accurate density measurements of liquid or granular material in pipes, bins, tanks, hoppers, or other types of containers can be made. In such a system the density gage consists of two units which can be clamped on a process pipe at any desired location. A source containing the isotope cesium 137 emits gamma rays which are detected at a rate inversely proportional to the density of the material being measured. The gamma rays are detected by means of an ionization chamber type of detector which produces a proportional output signal. The electrical output signal is linked to a controller to provide a complete control system. Two typical process applications are shown in Figure 10.39. These are gamma signal attenuation techniques.

Solid material density can also be measured, and one unique application is measuring the density of asphalt, as it is placed on roads, from the surface. A surface density probe is shown in Figure 10.40. The density of soils, aggregates, and concrete, as well as asphalt, can be measured either from the surface or with a depth gage. A depth density gage is shown in Figure 10.41. This method of measuring density is more accurate than older methods, and in some cases provides measurements that cannot be obtained by other methods. Such measurements are made by using backscattered gamma energy. Actually, this is a nondestructive

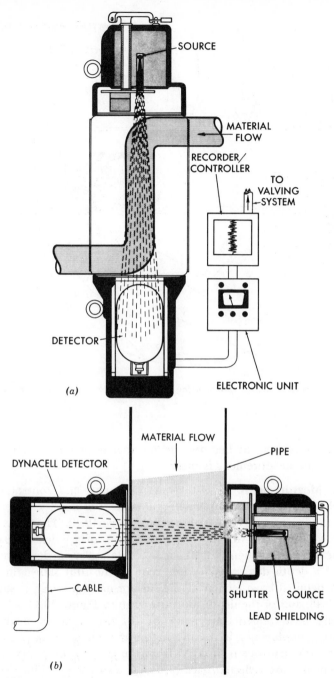

Figure 10.39 *Liquid density source and detector. (a) Measuring system for vertical measurement. (b) Measuring system for horizontal mounting. Note that the collimated source covers only the detector area. Courtesy Nuclear-Chicago Corp.*

445

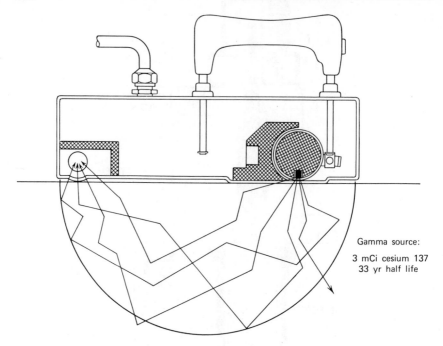

Figure 10.40 *Surface density probe using a gamma source. Courtesy Nuclear-Chicago Corp.*

method of measurement because the product is present in its original form after the measurement is made.

Moisture Measurements. Moisture measurements and control use the same techniques as the measurement and control of density. A combined moisture and density system is shown in Figure 10.42. The units shown from left to right are a surface moisture gage, a surface density gage (the same one shown in Figure 10.40), a portable scaler, a depth moisture gage, and a depth density gage (the same as the unit shown in Figure 10.41). A detailed cross section of the surface moisture gage is shown in Figure 10.43, and of the depth moisture gage in Figure 10.44.

Moisture measurements are made by the use of either neutrons or gamma rays, employing a backscattering measurement technique. A beam of high energy neutrons is directed from the gage to the material. Some of the neutrons are reflected backward or backscattered. These reflected neutrons lose part of their energy and become "moderated" by the material being measured. These moderated or low energy neutrons are meas-

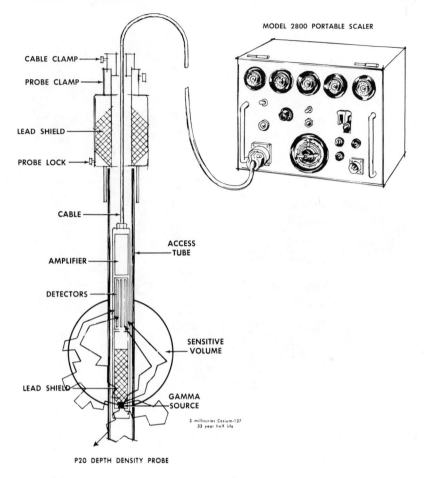

MODEL 2800 PORTABLE SCALER

CABLE CLAMP

PROBE CLAMP

LEAD SHIELD

PROBE LOCK

CABLE

AMPLIFIER

DETECTORS

LEAD SHIELD

ACCESS TUBE

SENSITIVE VOLUME

GAMMA SOURCE

3 millicuries Cesium-137
33 year half life

P20 DEPTH DENSITY PROBE

Figure 10.41 *Depth density probe. Courtesy Nuclear-Chicago Corp.*

ured by the neutron detector shown in the cross section system of Figure 10.45. All materials serve as neutron moderators, but some materials are more efficient than others. Hydrogen is the most efficient. When most of the hydrogen in a material is associated with water, determination of the number of moderated neutrons reflected from the material provides a measurement of the amount of moisture in the material per unit volume.

Most materials vary in density as well as in moisture content, so that moisture cannot be measured as an independent variable unless the density of the material is known. One method of determining the density is to employ a gamma source and direct the gamma ray beam into the

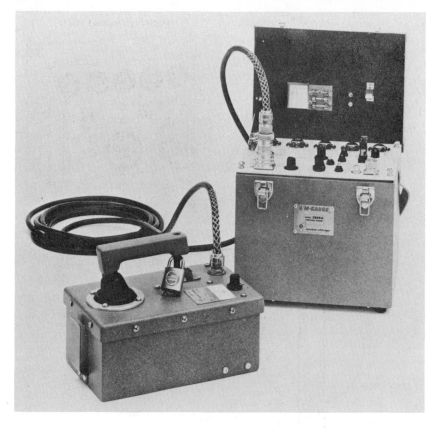

Figure 10.42 *Combination system for both surface and depth measurements of density and moisture. Courtesy Nuclear-Chicago Corp.*

material simultaneously with the neutron beam. The number of gamma rays reflected by the material or transmitted through the material is an inverse function of its density.

In the system shown in Figure 10.45, the measuring head contains a neutron source detector assembly in one section and a gamma source detector in another compartment. The neutron reflection measurement provides a signal which is directly proportional to a volumetric moisture percentage. The gamma reflection measurement provides a signal which is inversely proportional to density. These independent signals are sent to an electronic unit where they are scaled to pounds per cubic foot and presented simultaneously to a simple ratio computer. The ratio computer output signal is presented continuously in terms of percent water by

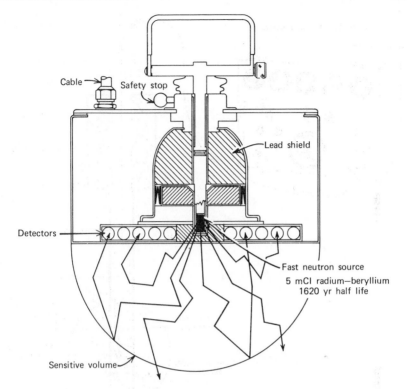

Figure 10.43 *Surface moisture gage. Courtesy Nuclear-Chicago Corp.*

weight, and the electronic unit also provides a proportional voltage signal for driving a recorder and a moisture control system.

The plutonium-beryllium neutron source with an output of approximately 3×10^6 neutrons per second and the cesium 137 gamma ray source with a radioactivity of 50 to 200 mCi are housed in dustproof and shielded containers to provide maximum safety in industrial environments for the complete protection of operating personnel.

Such a system can generally be expected to have an accuracy of better than 0.2% over a moisture range of 2 to 80% for materials that contain a relatively fixed percentage of bound hydrogen. The measurement becomes less precise with materials having larger variable amounts of bound hydrogen. For a moisture content of less than 2%, a larger time constant is required to maintain high accuracy.

The neutron method of moisture measurement offers the advantage of not contacting the material being measured. There are no probes to clean and maintain, and there is no danger of contaminating the product. The material does not have to be sampled for analysis. The measure-

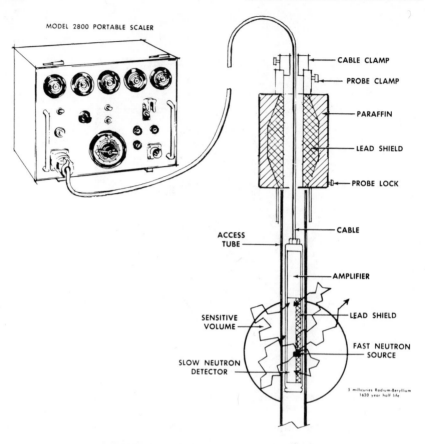

MODEL 2800 PORTABLE SCALER

CABLE CLAMP

PROBE CLAMP

PARAFFIN

LEAD SHIELD

PROBE LOCK

CABLE

ACCESS
TUBE

AMPLIFIER

SENSITIVE
VOLUME

LEAD SHIELD

FAST NEUTRON
SOURCE

SLOW NEUTRON
DETECTOR

5 millicuries Radium-Beryllium
1620 year half-life

Figure 10.44 *Depth moisture gage. Courtesy Nuclear-Chicago Corp.*

ment is continuous, and there is a continuous density correction for variations in product. This method is insensitive to process changes in chemical composition, temperature, pressure, or flow. Large volumes can be measured easily, giving more accuracy for total volume. The response time of the unit can be changed to meet the exact process demand, and the unit is extremely stable. The output signal can be used to control the integrated dry weight of the product or other variables.

Thickness Measurements. Measurement of the thickness of a moving sheet of material can be made using a beta or gamma source. A beta gage has the source on one side of the moving sheet and the detector on the opposite side. A beta source can be used on material whose density is

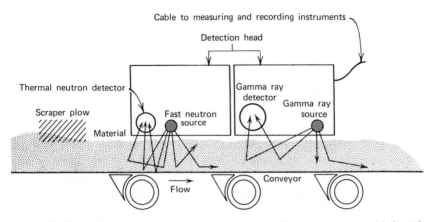

Figure 10.45 *Bulk moisture measuring gage mounted over a conveyor belt. The scraper plow maintains a uniform thickness of material. Courtesy Nuclear-Chicago Corp.*

low enough that it does not completely shield the beta source from the detector. Any thickness of material with the equivalent stopping power of $\frac{1}{4}$ in. of aluminum cannot be measured with a beta thickness gage. Thin sheet steel, paper, and other materials can be measured by means of a beta gage. Thickness measurements are made on the basis of their shielding power. Variations in thickness can be recorded or indicated in counts per minute which can be calibrated to read in mils of thickness for the particular material being measured.

Materials that are too thick or dense for measurement with a beta gage require the use of a gamma source and detector. Thicknesses of several inches can be made with a gamma source by passing the gamma rays through the material in the same manner as the beta rays for thinner materials, or by the reflection method discussed for density measurements in the determination of moisture or depth.

Radiation thickness gages are used where there is a large area to be covered without contacting the material or where physical contact is not desirable on moving materials. The output signal can be used to control the pressure on rolls or the speed of the rolls to roll a thinner sheet or to stretch the material to obtain the proper thickness.

Liquid Level. Radiation sources and detectors can be used to control precisely the level between predetermined fixed points, or as continuous level monitors and controls. Accuracy can be held consistently to $\pm\frac{1}{8}$ in. for wide band or narrow band control. Continuous level gages provide electrical signals which vary as the related levels change within the vessel. The electrical signal is calibrated for application to a particular process vessel and can be used, as part of a control link in the system, to maintain required levels.

Nuclear level sensing systems have a fast response and are useful in such applications as maintaining acid levels at critical points in high pressure, high temperature reactor vessels in a chemical plant, to maintain the level of molten steel, in the mold, in a continuous casting process, to sense the interruption of the flow of coal to a boiler in a power plant, and to prevent an overload condition in the primary crushing of boulders in a mining operation. All these measurements and control signals are completely external to the process material or environment, and are suitable for vessels ranging from 3 in. to 60 ft in diameter. They are reliable, rugged, sensitive to level changes measured through the walls of the containing vessel, have no moving parts, and use solid state electronics.

10.5 CALIBRATION OF RADIATION INSTRUMENTATION

Radiation detectors are calibrated by using a radioactive source of known strength placed at a given distance from the sensitive volume or area of the detector. With the detector connected to an appropriate readout indicator or recorder, the former can be calibrated in counts per unit time, in radiation units per unit time, or in values of current. This type of calibration is based on evidence that 1 Ci of activity produces 37.1×10^9 disintegrations per second. Careful preparation of sources of known radioactive strength for the type of radiation to be detected can result in accurate calibrations. Commercial radiation sources are available for calibration purposes.

A calibration source of known strength at a fixed distance from the detector produces a given output value. This value can be checked by moving the source a fixed distance and noting the change. The counting rate should decrease in proportion to the square of the distance the source is moved from the detector being calibrated. The calibration units usually used are counts per second, counts per minute, roentgens per minute, milliroentgens per hours or rems. The calibration units used depend on the application of the detecting unit and the strength of the radiation being measured or monitored.

In the selection of a detector with a calibrated value for a specific application, the readout equipment can also be specified. For example, if a monitor is needed to measure the activity collected in a fixed filter through which the air passes, the activity increase in counts per hour can be expressed in counts per minute as S, where

$$S = YRA(3.74 \times 10^{10}) \qquad (10.7)$$

where Y = yield of the detector arrangement in percent

R = airflow rate in cubic feet per minute

A = activity in microcuries per cubic centimeter

For a detector with a yield of 11.3% with an airflow of 5 ft³/min having a beta activity of 10^{-9} μCi/cm³, a counting rate of

$$S = 11.3 \times 5 \times 10^{-9} \times 3.74 \times 10^{10} = 2100 \text{ counts/min}$$

increase per hour would be expected. Equipment capable of handling this increase can be selected, and the fixed period for filter changes can be specified.

If a moving filter is used, and the filter change rate is 2.5 in./h, a different equation will be needed, but if the filter is not moved after the first hour the fixed filter equation will apply. By using the same airflow and same activity used for the fixed filter, the continuous counting rate C can be calculated in counts per minute:

$$C = YRAT(1.87 \times 10^{10})$$
$$= 5.3 \times 5 \times 10^{-9} \times 2.5 \times 1.87 \times 10^{-10} = 1240 \text{ counts/min}$$

where the yield Y is 5.3% and T is the transit time of the filter and is equivalent to the transit distance divided by the transit time V. In this case it was 2.5/1, or the 2.5 figure used. The transit time V is defined as the advance rate of the filter paper in inches per hour, and the transit distance is measured in inches.

The constants and the yield differ in the two equations, because in one case we were looking at an increase per hour for a fixed filter which has a different geometry than the geometry on the moving filter system, where the actual count is made after one transit of the sample.

Count rate meters can be calibrated using either the pulses from a 60 cycle ac source as 3600 counts/min, or from a calibrated pulse generator.

Calibration can be made either for individual sensing units or for the entire system. When an integrated system is to be used, it is preferable to calibrate the whole system. When more than one type of detector is used on a system, the system should be calibrated with each detector.

10.6 SUMMARY

The types of radiation encountered in industrial applications, namely, alpha, beta, gamma, and neutron radiation, and the sensing units necessary for their detection have been discussed. The auxilliary equipment needed to evaluate, indicate, and record the detected signals has been described. Some applications and advantages of using radiation in industry, and the means of calibration to ensure the accuracy needed for the application, have been covered. Further applications are discussed in Chapter 11 which deals with nondestructive testing for quality assurance.

Review Questions

10.1   How can one detect the presence of radioactivity?

10.2   Explain how alpha particles can be measured.

10.3   What is the radioactive measurement range of beta particles?

10.4   How can beta particles be measured effectively?

10.5   What is a gamma ray? How can it be measured effectively?

10.6   What features make a neutron behave differently for measurement than alpha, beta, or gamma radiation?

10.7   How is radioactivity detected by photographic film?

10.8   What is ionization and how does an ionization chamber work?

10.9   How can an ionization chamber be used as a pulse totalizer and integrator?

10.10   What is the difference between an ionization chamber and a proportional counter?

10.11   How is pulse size developed in a proportional counter?

10.12   What advantage does a proportional counter have over an ionization chamber?

10.13   What features are required in a proportional counter to measure neutron radiation?

10.14   What is a roentgen? What type of radiation does it represent?

10.15   How does a Geiger-Mueller counter work? What kind of radiation can be measured with it?

10.16   Explain what an area monitoring system is and how it works.

10.17   How is a scintillation counter constructed?

10.18   How does a scintillation counter system function to measure radiation?

10.19   What does the thickness of a crystal control in a scintillation counter?

10.20   Does a semiconductor diode have an advantage as a radiation detector? If so, what is it?

10.21   When are the following accessories required? (1) Preamplifier, (2) amplifier, (3) counter.

10.22 Explain how a binary counter works.

10.23 How does a decimal counter work in comparison to a binary counter?

10.24 What is the advantage of modular design? Are there any disadvantages?

10.25 Describe the types and discuss the applications of count rate meters.

10.26 Name and describe the types of power supplies used with radiation detectors.

10.27 Give the advantages and disadvantages of half wave and full wave rectification.

10.28 Define regulation in terms of line, load, and ripple.

10.29 What are the best choices of power supplies today?

10.30 Why is more than one channel of measurement and control required for nuclear reactor operation?

10.31 What functional areas are measured and/or controlled in a nuclear reactor operating system?

10.32 What feature makes radiation measurement and control desirable in a process industry?

10.33 Show how density can be measured using a radioactive source and detector.

10.34 How can moisture be measured using radioactivity techniques?

10.35 What determines how accurately a moisture measurement can be made using radioactivity techniques?

10.36 Explain how the thickness of a moving sheet of material can be measured using radioactivity techniques.

10.37 How would you measure and control liquid level with a radiation source and detector?

10.38 Explain how one would calibrate radiation detectors and control instrumentation.

Problems

10.1 What is the RC time constant for a circuit having a 10 MΩ resistance in conjunction with a 1.5 pF capacitance?

10.2 What is the value of a voltage pulse involving the following circuit parameters? $R = 1.5$ MΩ; $c = $ pF; $N = 1000$; applied potential $= 300$ V dc.

10.3 If the gas amplification factor $A = 3.5$, what is the instantaneous voltage pulse value for the parameters in problem 10.2?

10.4 What is the light array on a binary counter for a total count of 296 counts/min?

10.5 What value in counts per minute is indicated by a display in which lights 1, 3, 4, 5, and 6 are on?

10.6 How many counts would be expected for a detector yield of 89% in an airflow of 7 ft³/min having an average beta activity of 10^{-8} $\mu Ci/cm^3$?

10.7 What count would be expected for a gamma detector yield of 7.8% in a 5 ft³/min airflow with an activity of 10^{-6} $\mu Ci/cm^3$?

Bibliography

Baird-Atomic, Inc., *Atomic and Laboratory Instruments, Systems, and Accessories,* Catalog A-6, 3-65, Cambridge, Mass.

Isotopes, Inc., *Solid State Detectors,* SSD-1-1065; *Proportional and Geiger Counters* GC-1-965, Westwood, N.J.

Kohl, J., Monitors for Nuclear Reactor Fission Products, *ISA Journal,* Vol. 5, No. 6, June 1958, pp. 43–46.

Mruk, W. F., Instrumentation and Control of the Oak Ridge Research Reactor, *ISA Journal,* Vol. No. 6, June 1957, pp. 206-211.

Nuclear Chicago Corporation, General Catalog, Des Plains, Ill., 1965; Technical Specifications: *Nuclibadge A5C1 =1; Asphalt Density Gauge D1/2; Bulk Moisture Gauge D2/3; Level Switches D2/5; Density Moisture d/M-Gauge D1/3; Portable Alpha Survey Meter A5n4; Portable Neutron Meter; Survey Meters A5a3; Portable Alpha, Beta, Gamma Survey Meters A5a2-1; Cutie Pie Survey Meters A5a1-1.*

Nuclear Measurements Corporation, *NMC Air Monitors,* 6P150, Indianapolis, June 1963.

Soisson, H., Electronic Instruments in Nuclear Power Reactors, *Electronics,* June 12, 1959, pp 62–63.

Soisson, H. E., *Electronic Measuring Instruments,* McGraw-Hill, New York, 1961.

Texas Nuclear Corporation, *9140 Nemo Spherical Neutron Dosimeter Systems,* Houston.

Tracerlab/West, Radiation Telemonitoring Systems General Discussions and Specifications, Richmond, Calif., July 1965.

Victoreen Instrument Company, *Environmental Radiation Monitoring,* Cleveland, 1963.

Victoreen Instrument Company, *Nuclear Instrumentation,* Catalog 62, Cleveland.

CHAPTER 11

Nondestructive
Testing Equipment

In any industrial application it is highly desirable to test a product and still have the product to sell to the customer. Thus nondestructive testing (NDT) generally meets the requirement. In the words of many NDT advocates, "NDT doesn't cost. It pays." As we progress with this chapter, it will have to be admitted that NDT equipment is not exactly inexpensive, but in many instances it is the only way to obtain test information on product quality and still have the product. NDT techniques and measurements are valuable as in-process as well as final product testing for warranty guarantees. However, there are still areas of testing that need new or improved NDT methods.

Standard types of industrial DNT are magnetic particle, penetrant, x ray, gamma ray, ultrasonic, and eddy current. Some other types that are being developed and are in use for special industrial applications include infrared, microwave, signature analysis, and ultrasonic holography.

11.1 MAGNETIC PARTICLE

Magnetic particle testing requires that the parts to be tested have magnetic qualities. The parts should be demagnetized after testing. This implies that magnetic particle testing can be used on all types of ferromagnetic materials such as castings, forgings, tubing, and pipe. This method can be used on parts of all sizes, shapes, and composition or heat treatment. This type of testing is also useful for in-process checkouts, and for inservice testing for fatigue cracks on surfaces.

Magnetic particle testing is useful in the detection of surface or subsurface flaws, cracks, porosity, nonmetallic inclusions, seals, laminations, hot tears, laps, flakes, heat treat and grinding cracks, machining tears, and quenching and straightening cracks in new parts.

The principle of operation is quite simple, as shown in Figure 11.1. A dc current is passed through the part and generates a magnetic field within the part. This is circular magnetization. If there is a defect in the part, such as the crack shown in Figure 11.2, the crack causes a discontinuity in the magnetic field which creates a magnetic field outside the part. This magnetic field holds an iron powder and builds up an indication of the crack, as shown in Figure 11.2a. Figure 11.2 illustrates longitudinal magnetization. Longitudinal magnetization is created by passing the current through a coil, and the part is passed through the center of the coil. As the current is passed through the part, or as the part is passed through the coil, iron powder is sprayed over the part either as a dry dust or in a fluid suspension. The part must be clean, so that the iron powder will build up only where defects exist. A greasy spot could give a false indication, since iron powder would be trapped and held by the grease.

The test is simple, easy to perform, and is positive in disclosing defects on clean material. Apparatus is available for relatively fast production testing, and portable equipment is available for field testing. Inspection is important, and if the technique is used it should be applied as early as possible in the manufacturing cycle to avoid wasting machining and processing costs on defective materials.

The parts to be inspected by this technique must be able to withstand a slurry wash or a powder application and a final cleaning after testing and demagnetization. If these conditions cannot be met, a different type of testing has to be used. The technique and its application are relatively inexpensive compared to other methods and techniques to be discussed. The Russians have developed a traveling magnetic tape technique wherein the defect is registered on a magnetic tape indexed to the part being inspected. This eliminates the iron particle wash or slurry flow requirement, but adds a readout and analysis cycle to locate the defect.

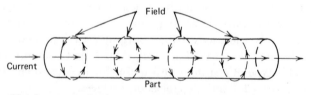

Figure 11.1 *Circular magnetization of a part for magnetic particle NDT. Courtesy Magnaflux Corp.*

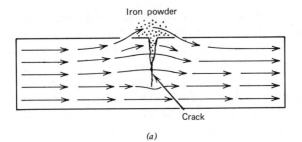

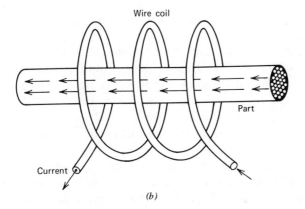

Figure 11.2 *Longitudinal magnetization of a part for crack detection by magnetic particle NDT. Courtesy Magnaflux Corp.*

The magnetic particle technique does not have the capability for detecting internal voids and other defects, nor for establishing the depth of the surface cracks or other defects it discloses by the iron powder buildup. Such defects can be removed by grinding, and additional magnetic particle tests made until the defect is removed. Care must be exercised, because grinding operations generate heat in the part and may cause the crack to propagate beyond its original depth or length.

11.2 DYE PENETRANTS

Dye penetrants are limited to the detection of surface defects. Such defects may be surface cracks, porosity, laps, cold shuts, lack of weld bond, and fatigue and grinding cracks. The defect may be capable of trapping dye applied to the surface.

Dye penetrants can be used on all types of metal, glass, ceramics, castings, forgings, machined parts, and cutting tools. In general it can be used almost anywhere and on any materials that do not absorb the dye permanently.

Application of the technique requires that the surface be clean and that it can be wiped or washed after the dye has been applied. In practice a dye penetrant is applied to the part by brushing or spraying. The dye is then removed from the surface of the part, and a white absorbing material is sprayed on the surface being examined for defects. The principle involved is that if there is a defect into which the dye can penetrate it will be trapped in the defect when the excess dye is removed from the surface. When the white absorbent coat is applied, any trapped dye in the defect bleeds out and shows on the white surface as indicated in Figure 11.3. The size and shape of the spot on the white coating indicate the relative size and shape of the defect. A crack is indicated as a line with the same contour as the crack, and porosity is indicated as spots. Experienced dye penetrant inspectors can tell something about the depth of the defect by the size of the indication, the depth of the color, and the length of time it takes the dye indication to appear.

This technique does not require an elaborate setup, and the cost per test is fixed by the amount of dye required. It is simple to apply, fast in action, and accurate in indication. However, it is a qualitative test to detect a defect, but is not a quantitative test to give actual size or depth of the defect.

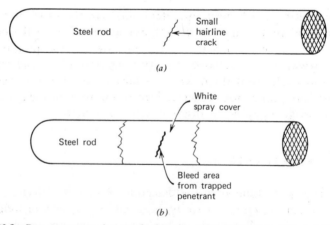

Figure 11.3 *Dye penetrant for crack detection. (a) Steel rod with hairline surface crack. (b) Dye bleed spot showing area of crack. Courtesy Magnaflux Corp.*

Dye penetrants are used extensively to inspect for welding defects such as porosity or bond cracks. The weld areas must be brushed clean, so that indications are not observed for scale spots. It is also used to inspect brazes, glass-to-metal bond edges, ceramic-to-metal bond edges, and for small sand holes in castings and cold shuts on forging.

11.3 X RAYS

X rays can be used to detect internal flaws and defects in any material that can withstand the x-ray radiation and give a contrast on x-ray film, a fluoroscope, or an imaging electronic device such as an oscilloscope.

X-ray techniques are useful for the detection of welding flaws such as internal voids, slag inclusion, and cracks; voids and cracks in castings, forgings, and tubing; defects in formed parts and assemblies; lack of bonds and thickness variations. X-ray techniques are also used extensively for the examination of parts in such operations as blind hole deep drilling where wall thickness and straightness of drilling are of prime importance.

In x-ray NDT x-ray film provides a permanent record of the part undergoing inspection. Film can be used only for static sample inspection. To make dynamic inspections use must be made of fluoroscopic techniques or imaging on electronic devices, where motion of the part can be observed. To obtain the best image, an image intensifier can be used on the fluoroscope.

X-ray NDT requires a relatively high initial expense for the x-ray equipment. It is also relatively expensive to operate and maintain the equipment. The equipment requires a good power source, a trained technician to operate the equipment, the maintenance of a photographic dark room, and the continued cost of film and maintenance of a special facility to contain the scattered radiation generated during the operation of the x-ray equipment.

Reading of the x-ray film or the fluoroscopic and electronic images requires a trained inspector to interpret the condition of the object or objects. It is also necessary to establish standards to obtain quantitative values. Usually, these standards are included as a part of the record and are a ready reference for the individual performing the image analysis.

A basic x-ray setup can be as simple as the schematic representation shown in Figure 11.4. While all the essentials are shown, one does not obtain the best definition on the radiograph. Scattered radiation and variations in sample thickness cause blurring of edges, which adds to in-

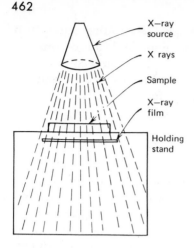

X—ray
source

X rays

Sample

X—ray
film

Holding
stand

Figure 11.4 *Basic x-ray equipment arrangement. Courtesy General Electric Co.*

accuracies in interpretation of the image parameters. The use of filters and screens can produce far more definitive radiographs, but there is less contrast because the soft x rays are absorbed by the filters and the scattered x rays are absorbed by the screens. A filter and screen setup for a hemispherical body is shown schematically in Figure 11.5. A setup for using a fluorescent screen is shown in Figure 11.6.

Definition of the depth and angle of a flow by x-ray radiographic techniques usually requires several radiographs. To obtain depth it is necessary to obtain the radiographs from different angles. One radiograph may be made head-on or at a 0° angle, a second at 45° with respect to the head-on, and another at 60° or 90°, depending on the type of defect or suspected defect area to be investigated. This type of technique is used for determining the number of voids or porosity spots and the spacing between them. This technique is applicable only to the radiography of static objects.

Under ideal conditions a resolution of 4 mils in a $\frac{1}{2}$ in. defect is possible. In thin sections in which shadowing is not a problem, resolutions of 1 mil in a 10 mil thick sample are possible under ideal conditions.

Magnification of the image size on film can be accomplished by placing the film below the specimen at some fixed distance, as shown in Figure 11.7. In this type of setup, scattered radiation can cause serious resolution problems unless very special precautions are exercised to minimize the scatter. There is always an optimum film-to-focus distance, and at least the minimum distance should be observed.

Another technique for obtaining more intensity on film traces is to employ a photocell dodging technique in which two photocells and light-

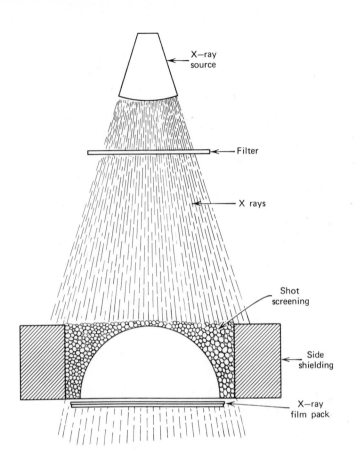

Figure 11.5 *X-ray setup showing the location of filters and screening for a hemispherical shape.*

ing are used to compensate for undesirable variations in the radiograph. One photocell is used to detect the variations during a scan of the object, and the second photocell is activated to provide the proper lighting to produce a second film with improved intensity contrasts for better product definition.

Calibration Standards. A calibration penetrameter, normally to represent 2% of the object film density, is exposed with the object and placed on the top surface as shown in Figure 11.8. A typical penetrameter is shown in Figure 11.9. The holes are used to evaluate the definition of the object, and the size of the holes represents some multiple of the

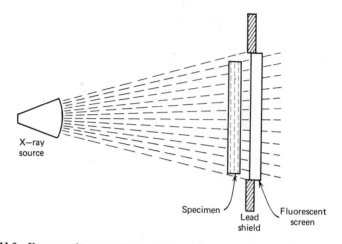

Figure 11.6 *X-ray equipment using a fluorescent screen for observing the object.*

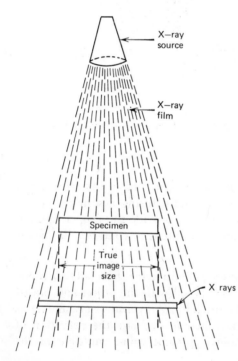

Figure 11.7 *X-ray arrangement to obtain an enlarged image of the object.*

object thickness for example, 1T, 2T, 3T, or 4T. These are defined in ASTM standards.

In x-ray radiographs the type and speed of the film are important. As a general rule, the faster the film the larger the grain size. Slow films have smaller grain sizes, but they also require longer exposure time. Film development also plays an important part in securing a good radiograph. Poor film processing can ruin a good radiograph. In most industrial processes two films are used, so that film defects are not analyzed as product defects. Different speed films are sometimes used to obtain simultaneous radiographs when the product under investigation has a large thickness variation. Separate penetrameters are used for the representative thicknesses in such applications.

Although standards have been established to cover a large number of materials, there is still a heavy dependence on the experience of the person who interprets the radiograph. There is no iron clad definition of what constitutes a good product or a bad one. The decision is usually

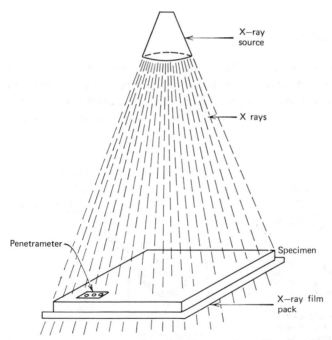

Figure 11.8 *X-ray arrangement to show location of the calibration penetrameter. Courtesy General Electric Co.*

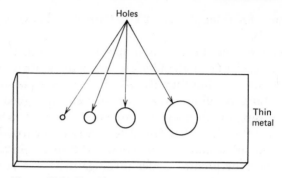

Figure 11.9 *Penetrameter for x-ray calibration.*

arbitrarily based on the application of the product. The more stringent the safety rule, the more definitive and stringent the product specification. Specifications have been rigidly established for work subject to boiler codes for unfired vessels and navships applications, especially in weldments for nuclear powered naval ships and submarines.

11.4 GAMMA RAYS

Gamma ray sources can be used to replace the x-ray tube. One advantage of the gamma ray source is its portability. These sources are usually handled as shown in Figure 11.10. There is a radiation hazard during the transfer of the source from the storage vessel to the position for the ex-

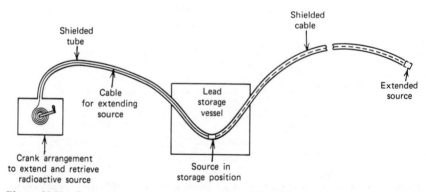

Figure 11.10 *Gamma ray source storage vessel and handling equipment.*

posure, during the exposure, and during the transfer back to the storage vessel. The length of the transfer tube is adequate for the size of the source, so that the operator is not exposed to more than the recommended radiation dosage established by law. These sources require a license, and safety rules are rigidly enforced. The source storage vessels are inspected regularly for radiation leakage.

Gamma ray sources emit a spectrum of energy. This means that when a radiograph is produced there is a blending of the contrasts. In other words, one cannot obtain sharp definition of the object because of the different gamma ray energies emitted. In the case of thin objects, the more energetic gamma rays go right through the material, and little or no energy absorption is indicated on the film.

A typical gamma source arrangement is shown in Figure 11.11. A penetrameter is shown on the source side of the object, just as for x-rays.

Gamma ray radiography interpretations for defects are the same as those for x-rays. The same rigid specifications for boiler code and navships applications apply.

It should be noted that, when the x-ray tube is shut off and the gamma ray source is returned to its storage vessel, there is no radiation hazard. No residual radiation remains in the objects that have been radiographed. This is not true of neutron radiography.

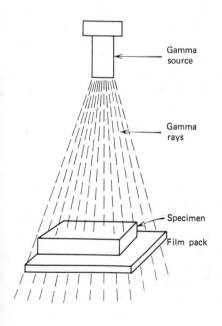

Gamma source

Gamma rays

Specimen

Film pack

Figure 11.11 *Gamma source arrangement for radiographs. Courtesy General Electric Co.*

11.5 NEUTRON RADIOGRAPHY

A source of neutrons replaces the x-ray tube or the gamma ray source in neutron radiographic techniques. Care must be exercised in regard to exposure times, because it is possible to induce radioactivity in the object receiving the neutron bombardment.

Normally, neutron beams are produced by nuclear reactors. This means that the objects to be neutron radiographed, or "neutrographed," must be transported to the site of the reactor. This could become an expensive exercise in itself. Maintenance and operation of reactors are also very high. Thus "neutrography" must be a by-product use of the reactor application. As a result, only a small amount of NDT is accomplished by this technique.

The neutron technique offers advantages in a radiographic presentation of an object, because all parts are present in the radiograph. As an example, if a telephone were "neutrographed," the entire unit would be present in the picture. This includes the plastic covering as well as the metallic parts. In a conventional x-ray or gamma ray radiograph, all the plastic parts would be missing. Only the metal parts would show on the film. This is a relatively new technique which has been demonstrated but is now too expensive to be a widely used nondestructive tool for industrial applications.

The possibility of developing sources of neutrons by gamma–neutron conversion could greatly expand the use of this technique as a nondestructive tool in industry. Conversion of x-rays to neutrons by bombardment of conversion materials should be adequate for thin materials if x-ray equipment can withstand the operating times required to obtain such an exposure and the converted beams can be adequately collominated to produce good definition.

11.6 ULTRASONICS

Ultrasonics are usually defined as frequencies higher than 20,000 Hz, and for nondestructive techniques the frequencies are in the 1 to 15 MHz range. Ultrasonics are sound waves, but are higher in tone than the frequencies detected by the human ear and, like radiation, must be detected and measured by instrumentation.

Ultrasonic sound waves, like audible sound waves, are mechanical vibrations involving movement of the medium in which they are traveling. Theoretically, any medium that behaves in an elastic manner can transmit sound. It has also been postulated that a wave is propagated through a medium by particle motion. In solids these particles do not move away from the exciting source, so they must vibrate about some fixed mean position. The excited particles then excite other particles and set up a transmission wave. These may be longitudinal waves in which the particle motion is in the same direction as the wave being propagated; transverse or shear waves in which the particle motion is perpendicular to the direction of propagation; surface or Rayleigh waves in which the particle motion is elliptical and confined to a depth of approximately one wavelength below the surface; or plate or Lamb waves which are produced in thin plates whose thickness is comparable to one wavelength.

The velocity of the propagated sound wave depends on the type of wave being transmitted and the density and elastic constants of the medium in which it is traveling.

Ultrasonic Transducers. For applications to NDT ultrasonic waves have to be generated for transmission and detected. Most ultrasonic flaw detection systems depend on piezoelectric materials for the production of energy in a transducer system, just as an accelerometer uses a piezoelectric transducer as a driver or sensor. A piezoelectric transducer for ultrasonic frequencies can be used both as a generator and a receiver, or one can be used as a generator and one as a receiver. Piezoelectric transducers for ultrasonics are commonly referred to simply as ultrasonic transducers. Mounted ultrasonic transducers are usually referred to as search units, crystals, or probes. Two examples are shown in Figure 11.12.

Since ultrasonic waves are a form of energy that is transmitted, it is essential that good transmission coupling be used between the generator and the medium through which the energy is transmitted. In NDT appli-

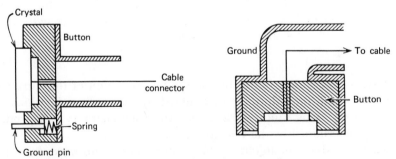

Figure 11.12 *Typical ultrasonic transducers in cross section.*

cations this is accomplished by using very flat surfaces with a thin oil or grease, or by means of a couplant such as water or glycerine. Air is an extremely poor couplant.

Attenuation. The intensity of ultrasonic sound, like audible sound, decreases as the distance from the source increases. In ultrasonics one exception exists in the near field of a transducer where destructive and constructive interferences cause intensity fluctuations. The normal decrease in intensity is due to attenuation caused by geometrical factors and absorption or scattering mechanisms.

In ultrasonic testing, the absorption in solids is primarily caused by scattering. Coarse grained structures with porosity absorb more energy than fine grained structures without porosity. Attenuation in a given solid can be reduced by decreasing the frequency of the sound waves within the ultrasonic test range. However, transducers are normally designed for a given frequency, and one has to change a transducer to make a frequency change and still maintain a reasonable energy transfer for test purposes. Often the higher frequency may be required to make the desired sample penetration, so the attenuation must be tolerated.

Energy Transfer Across a Boundary. As stated earlier, a couplant is normally required to transmit generated energy efficiently. When an ultrasonic wave impinges on a boundary between two media, part of the energy is reflected and part is transmitted. The intensities of the reflected and transmitted energy at normal incidence depend on the acoustic impedance z of the two media. The characteristic acoustic impedance z of a material is defined as the product of the density ρ and the velocity v:

$$z = \rho v \qquad (11.1)$$

The velocity v is the rate at which the ultrasonic wave travels in a medium, usually expressed as inches per second, feet per second, or meters per second. The velocity is also associated with the frequency f and the wavelength λ of the energy transmitted. The wavelength is the distance occupied by one complete cycle. Thus

$$v = \lambda_f \qquad (11.2)$$

The frequency is determined by the rate of oscillation of the ultrasonic transducer, and the velocity is determined by the medium in which the wave travels. This simply means that the wavelength is a dependent variable determined by the velocity and the frequency.

When it is desirable to obtain shear, surface, or Lamb waves for a particular test application such as shear waves in weldment testing, longitudinal waves are directed into the material at some angle from the normal to the material surface. Part of the incident energy is reflected, and part is refracted. It is possible in some cases to produce four separate waves, two reflected and two refracted. When a coupling such as water is used, only one reflected wave is produced. The relationship between the various angles of reflection and refraction can be determined by means of Snell's law:

$$N \sin i = N' \sin r \qquad (11.3)$$

where $i =$ angle of incidence measured from the normal to the interface of the two media

$r =$ angle of refraction into the second medium also measured from the interface normal

Ultrasonic Measurement Systems. Ultrasonic measurement systems require the basic equipment shown in Figure 11.13. The transducer for emitting the ultrasonic wave requires a driver oscillator, and the receiving transducer output must be displayed on an oscilloscope or an oscillograph. The basic system shown uses only one transducer which acts as both a transmitter and a receiver. In many applications two separate transducers are used, as shown in Figure 11.14.

Ultrasonic measurements can be made by using a through transmission technique as shown schematically in Figure 11.15, by a reflection technique using a single probe as shown in Figure 11.16, by a reflection technique using two probe units as shown in Figure 11.17, by an angle beam technique as shown in Figure 11.18, by a delta techniques as shown in Figure 11.19, and by a bubbler technique as shown in Figure 11.20.

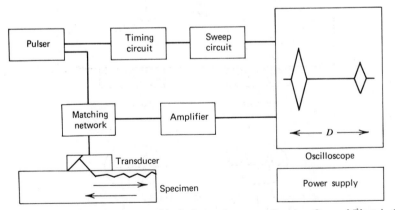

Figure 11.13 *Block diagram of a typical ultrasonic system. Courtesy General Electric Co.*

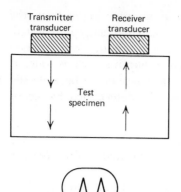

Figure 11.14 *Separate transducers for transmitting and receiving.*

The through transmission technique uses a couplant between the transmitter transducer and the test piece, and between the test piece and the receiver transducer. If there are no defects, the receiver transducer detects a pulse of approximately the same amplitude as the transmitted pulse at some Δt time later, depending on the thickness of the test piece. If there is a defect, the receiver transducer does not detect the transmitted pulse. In this case the size of the defect and its position can be

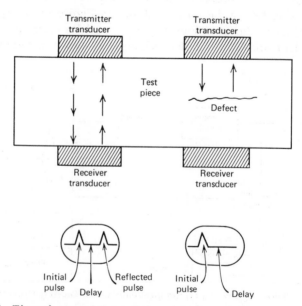

Figure 11.15 *Through transmission technique for ultrasonic measurements.*

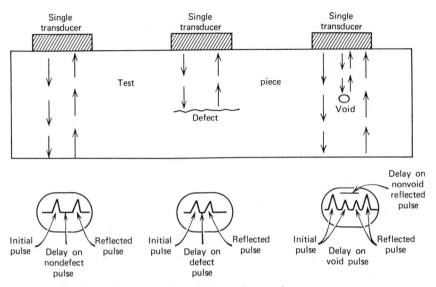

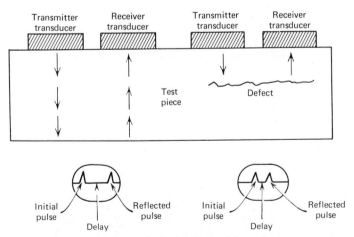

Figure 11.16 *Reflection technique using a single transducer.*

determined, but its depth from the transmitter transducer cannot be determined. Other techniques are required to obtain this information.

The reflection or pulse echo technique, using either a single transducer as a transmitter and receiver or two separate transducers, also requires a couplant between the test piece and the transducer. When no flaws are present, the delay time between the transmitted signal and the

Figure 11.17 *Reflection technique using two transducers.*

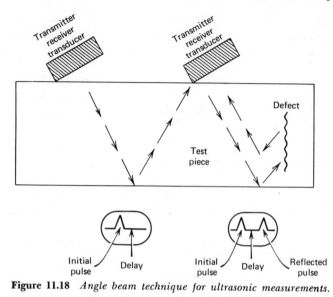

Figure 11.18 *Angle beam technique for ultrasonic measurements.*

received signal is twice as long as for the through transmission technique, because the signal has to pass through the test piece in the forward direction and is then reflected from the back face and passes back through the test piece to the receiver. When a defect is present, the transmitted signal is reflected back to the receiver from the defect. This delay time is shorter than that of a full test piece thickness. By proper

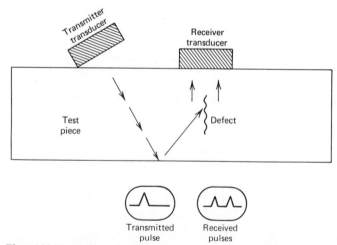

Figure 11.19 *Delta technique for ultrasonic measurements.*

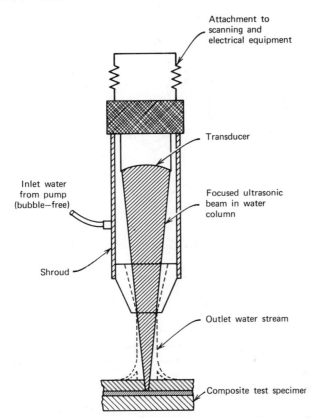

Figure 11.20 *Bubbler technique for ultrasonic measurements. Courtesy General Electric Co.*

calibration both the defect area and depth in the test piece can be accurately determined.

The angle beam technique is used on test samples that have cross grain orientation horizontal to the angle at which the probe is placed. To ensure good coupling, the probe can be shaped to match the part being tested. This can be expensive for a single part, so such parts are usually tested in a tank with water as the couplant. The angle technique usually employs one transducer as the transmitter and receiver, but in the delta technique a separate transmitter and receiver can be used.

The bubbler technique is used to provide a water couplant for good signal transmission in applications in which the part cannot be submerged in a tank for testing because of size or other considerations. A continuous flow of water around the transducer provides adequate couplant conditions.

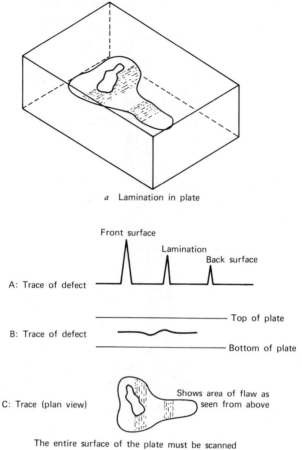

a   Lamination in plate

A: Trace of defect

B: Trace of defect

C: Trace (plan view)

The entire surface of the plate must be scanned
to produce the plan view

Figure 11.21 *Types of product scan by ultrasonic. (1) A scan; (2) B scan; (3) C scan.
Courtesy General Electric Co.*

Three types of scans can be made for the detection of defects by ultra-sonic techniques. An A scan produces an electrical signal deviation as an indication of the defect. A B scan produces a vertical trace of a defect showing its size and position by proper mapping. A C scan shows a plan view of the defect printed out as a halftone map. A comparison of the three types of scans is shown in Figure 11.21.

Ultrasonic scanning can be done manually or automatically. Manual scans are made by moving the search unit over the test pieces by hand, using the reflection or pulse echo technique and observing the reflected signal on a cathode ray screen or other type of projection device. Manual scans are not recommended when comparative scans of a test piece are to be made over an extended period of time. This is because of the extreme difficulty in correlating the results, even with the use of well defined grid systems. In such cases the search unit output must be recorded on a well defined ratio basis of scan distance to recorded signal.

Automatic scanning can be set up to perform repetitive scans, using an indexing system for recording the search unit output. These scans can be readily correlated to exact positions on the test piece. Through transmission units mounted for tandem operation are effective in automatic scanning systems for test piece submergence in the couplant. Such systems are definitely not portable, and in many cases are quite elaborate so that precision results can be obtained. Search unit outputs can be adapted to produce halftone maps showing defect shape, size, and position.

Ultrasonic testing methods, like all other measurement methods, require calibration standards and, like radiography, require interpretation of the signals and the maps. Standards such as the one shown in Figure 11.22 are used to establish the sensitivity of the measurements for interpretation of the signals for area and depth.

11.7 EDDY CURRENT TESTING

The basis of all eddy current testing is that changes in test part characteristics cause associated changes in the test coil *impedance* of the eddy current test equipment. Impedance is an electrical property which relates the driving voltage to the resulting current in an electrical circuit. We previously discussed matching impedance in the acceleratometer applications in Chapter 9 and for the ultrasonic transducer, using a couplant, to obtain a maximum power transfer with the test part. In the case of eddy current testing, impedance is a function of test signal frequency, coupling with the test part and the nature of the test part brought into proximity with the test coil. Impedance in eddy current testing is expressed in ohms, just as in other electrical circuit applications.

There are four material properties that affect eddy current distribution in a test part: (1) conductivity, (2) permeability, (3) size as related to diameter, mass, or thickness, and (4) homogeneity.

Conductivity in eddy current testing is the electrical property of the material related to its ability to conduct an electrical current. Silver and copper are regarded as highly conductive materials. Copper is widely used as an electrical power conductor, and silver plated switches are used

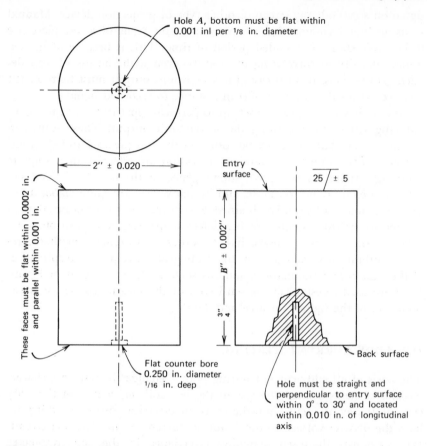

Figure 11.22 *Standard for calibration of ultrasonic equipment. Courtesy General Electric Co.*

to reduce resistance to current flow. These materials also conduct eddy currents more easily than stainless steel and other metals. Conductivity can be expressed mathematically as the reciprocal of the electrical resistivity of a metal:

$$\sigma = \frac{l}{RA} = \frac{1}{\rho}$$ (11.4)

where σ = conductivity in mhos
ρ = resistivity in ohm-centimeters
l = unit length
R = resistance in ohms
A = cross sectional area

Permeability μ of a material, as associated with eddy current testing, is actually magnetic permeability. This is a property of the material which modifies the action of magnetic poles placed in the material and modifies the magnetic induction resulting from an applied magnetic field or magnetizing force. The permeability of a material may be defined as the ratio of the magnetic flux density in the material to the applied magnetizing field.

For the purpose of establishing relative permeability values for eddy current NDT measurements, it is assumed that a vacuum or free space has unit permeability, diamagnetic materials have a permeability of less than unity, and paramagnetic materials have a permeability of slightly more than unity and are essentially independent of a magnetizing force. Ferromagnetic materials have a permeability which is considerably greater than unity and varies with the magnetizing force.

Homogeneity of a material involves like elements, grain size, and structure lattices which make it more uniform. Heterogeneous materials, however, have a mixture of elements, grain sizes, and structures that are not uniform. Each of these characteristics can affect eddy current testing, because they can change the permeability or the resistivity of a material being tested, as well as the eddy current distribution in the test part.

Test Setup Arrangements. An eddy current tester and setup is shown schematically in Figure 11.23. The test coil or probe is excited with an alternating current of a suitable frequency, ranging from 1 Hz to 1 MHz. The test coil generates an alternating magnetic field of the test coil frequency in the vicinity of the coil, and this generated alternating magnetic field induces eddy currents in the metallic part is shown in Figure 11.24. An eddy current test can be described as a simple transformer action. The excitation coil is the primary of such a transformer and produces an eddy current flow in any nearby conductive material (test piece) by means of the excitation field H_0 it radiates. The test piece is the secondary of the transformer, and may be considered to possess an infinite number of conductors through which induced eddy currents flow. Each of these infinite conductors is crossed by the time varying electromagnetic flux lines and sets up its own magnetic field. These eddy currents flow in paths shaped like the excitation coil, unless they are restricted by the geometry of the test piece. The intensity of the eddy current flow in the test piece has been shown to decrease with increasing depth. Thus the eddy current magnetic reaction H_r is a function of test material thickness. Also, if the spacing between the excitation coil and the test piece, usually referred to as lift-off, is too great, it may prohibit the use of eddy current testing in many production applications. Lift-off effect is most severe in ferromagnetic materials.

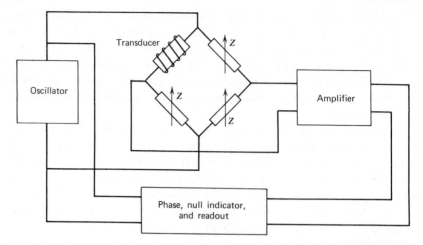

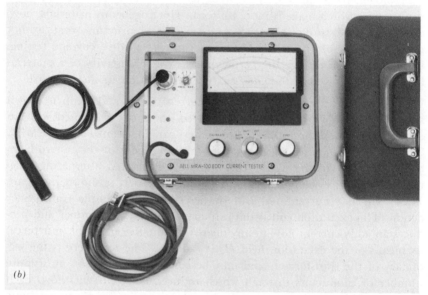

Figure 11.23 *Eddy current testing. (a) Schematic of an eddy current test setup. (b) Bell MRA-100 eddy current tester. Courtesy F. W. Bell, Inc.*

The electromagnetic fields generated by the eddy currents in the test piece partially cancel the original field generated by the probe coil to produce a *resultant* field. This resultant field then interacts with the coil current and associated flux linkages to create a specific coil impedance for the particular part and position being tested. This specific coil impedance is then indicated or recorded for the particular testing operation

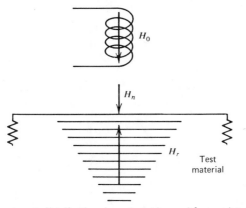

Figure 11.24 *Eddy current distribution in a test piece with associated magnetic field components. Courtesy F. W. Bell, Inc.*

as a function of the test coil impedance in the bridge arrangement shown.

Magnetic Reaction Analyzer Eddy Current Testing. Another technique for eddy current testing measurements uses the same signal transmission methods but employs a separate detector. The instrument involved can be used as an absolute or differential readout and is called a *magnetic reaction analyzer*. The basic detector is a Hall element, as shown diagrammatically in Figure 11.25. These detectors are only a few mils thick and rectangular in shape. A common size is 0.030 × 0.060 in. The Hall voltage V_H, measured across the element width, is proportional to the magnetic vector field normal to the element surface when the control current I_c is held constant:

$$V_H = KI_cB \qquad (11.5)$$

where K = a constant
  B = magnetic vector field in gauss

The Hall element responds to the instantaneous magnitude of magnetic flux density. This is different from the coil pickup which responds to the rate of change of magnetic flux. Thus the Hall detector performs with equal sensitivity even at dc (zero frequency) levels, while the sensitivity of a pickup coil reduces to zero at dc levels. The Hall element is small enough that the excitation coil unit can be assembled so that it encircles the Hall detector unit. Both the excitation field of the coil and the magnetic reaction field of the eddy currents in the test piece are detected by the Hall detector unit. Coupling efficiency approaches 100% when the

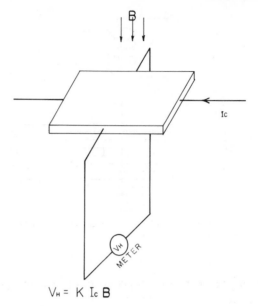

$$V_H = K \, I_c \, B$$

Figure 11.25 *Hall element. Courtesy F. W. Bell, Inc.*

Hall detector is located close to the test object. When a probe containing a Hall detector surrounded by the excitation coil unit is held remote from a test object, the Hall detector measures only the flux field H_0 established by the excitation coil. Although the frequency of the excitation field H_0 may be varied from 20 Hz to 100 Hz, the amplitude of this field remains constant, since the excitation coil is supplied by a constant current ac amplified. Hall element sensitivity is not influenced by changing frequency, so H_0 is indicated with a constant value at all test frequencies. By establishing this value of H_0 as the full scale reading of an absolute magnetic reaction analyzer, the unit is internally standardized.

Placing of the test probe on or near a conductive test specimen causes the induction of circular paths of eddy currents in the object. These eddy currents produce a reaction magnetic vector field H_r. H_r tends to oppose H_0 and produce a resultant net field component H_n. In terms of vectors, the relationship at the Hall detector is

$$\mathbf{H}_n = \mathbf{H}_0 + \mathbf{H}_r \qquad\qquad (11.6)$$

The net vector $\mathbf{H}_n$ is the same *resultant field* typically indicated by most eddy current systems as previously discussed, and is measured by a magnetic reaction analyzer using the circuit shown in the block diagram of Figure 11.26a. The actual unit is shown in Figure 11.26b.

The reaction vector $\mathbf{H}_r$ can be measured by the circuit of Figure 11.26 by closing the $\mathbf{H}_0$ subtraction input switch. This introduces a subtraction signal into the amplifier, which is equal in magnitude to the signal produced by $\mathbf{H}_0$. This can be checked by removing the probe from the test piece and observing a zero output signal. Placing the probe on the test piece then gives only the indication for $\mathbf{H}_r$, where

$$\mathbf{H}_r = \mathbf{H}_n - \mathbf{H}_0 \qquad (11.7)$$

The block diagram of an absolute magnetic reaction analyzer, shown in Figure 11.26, includes a variable frequency, constant current oscillator, a test probe containing an excitation coil and a Hall element, and a stable electronic unit with an indicating meter and dc outputs for recording.

The vector relationships of the three vectors that can be measured by a magnetic reaction analyzer are shown in Figure 11.27. By measuring the amplitudes of the three vectors, the three sides of the triangles are known, so all the phase or angle information is included in this type of readout. To study material properties and compare them, vector locus curves can be plotted on circular coordinate graph paper with coordinate systems radiating from the origins of $\mathbf{H}_n$ and $\mathbf{H}_r$, since both $\mathbf{H}_n$ and $\mathbf{H}_r$ vary with properties of the test material, as discussed earlier.

As shown in Figure 11.28, when the test frequency, electrical conductivity, or thickness of the test specimen is increased, there must be a corresponding increase in eddy current density within the test material. This causes an increase in the $\mathbf{H}_r$ vector, so that the circle locus is traversed in a clockwise manner for any of these increases. If inspection is performed at a fixed testing frequency, any material discontinuities that alter the conductivity or the thickness of the part will create a change from the normal, predictable vector plot. These changes or abnormalities can be studied on an absolute scale without continuous referral to a standard, as required in normal relative testing.

Differential Magnetic Reaction Analyzer. The differential magnetic reaction analyzer, Figure 11.29a, has been designed to operate reliably in production and can reject the majority of extraneous noise signals that might be common to manufacturing facilities. A differential magnetic reaction analyzer provides an excitation field for two Hall detectors. These detectors may be located within a single probe assembly for differential testing, or in two individual probe assemblies for comparison

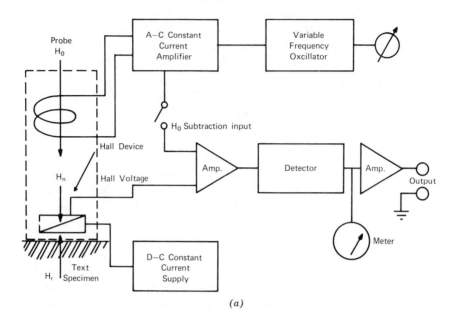

(a)

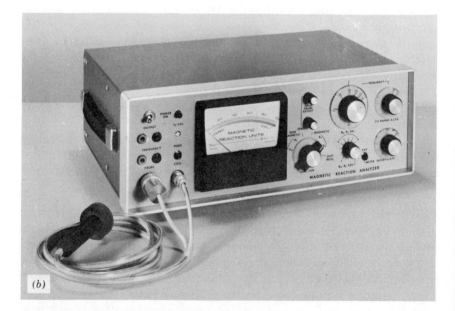

(b)

Figure 11.26 *Magnetic reaction analyzer. (a) Block diagram. (b) Actual unit. Courtesy F. W. Bell, Inc.*

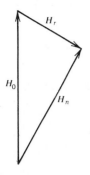

Figure 11.27 *Magnetic vector plane relationships. Courtesy F. W. Bell, Inc.*

testing. The outputs of the two Hall detectors are connected differentially as shown in Figure 11.29*b,* so that the indicated circuit output represents the difference between the two detector output voltages or the difference of the net magnetic vectors at two discrete test locations.

A high attenuation filter plug-in module is used to narrow-band, or tune, the differential analyzer selectively to the frequency that has been determined to be the optimum value for a particular application. A test frequency is chosen to provide the required sensitivity and to differ from the characteristic frequency or harmonics of any nearby equipment. Twenty-two frequency choices are available.

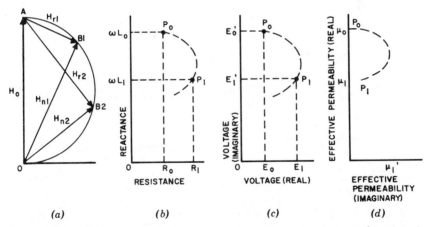

Figure 11.28 *Influence of the basic variables of an eddy current test on the magnetic vector locus curve. Complex planes for representing eddy current test coil characteristics (a) Magnetic vector plane. (b) Impedance plane. (c) Voltage plane (d) Effective permeability plane. Courtesy F. W. Bell, Inc.*

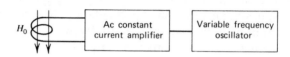

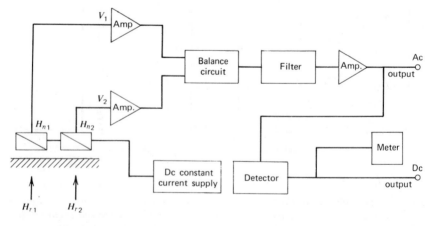

(b)

Figure 11.29 *Differential magnetic reaction analyzer. (a) Actual unit. (b) Block diagram of a differential magnetic reaction analyzer system. Courtesy F. W. Bell, Inc.*

Comparison Testing. Comparison testing with a differential magnetic reaction analyzer is usually performed with two separate probes, each having a separate Hall detector. A typical setup is shown in Figure 11.30a. Large area or long term property variations such as thickness, heat treatment, or hardness can be compared to a standard or a set of standards prepared for various materials or sizes used in a particular production process. Figure 11.30b shows some outputs for probe positions over a weld.

A differential magnetic reaction analyzer in a comparison test arrangement has flexibility and adequate sensitivity to expand a 1% change in an absolute analyzer reaction vector readout to a new full scale calibration. Comparison testing techinques enable one to select any point on an absolute analyzer data curve and treat it as a new system origin. Sensitivity from this point can be expanded 100 times for high resolution of the test variables.

Differential Testing. Differential testing with the differential magnetic reaction analyzer usually uses one probe containing two Hall elements, as shown in Figure 11.31. Local property variations within the sample under test, such as defects, voids, and metallurgical nonhomogeneities, can often be detected using this testing configuration to the exclusion of long term variations such as gradual changes in material thickness, physical properties, or heat treat conditions. The long term variables or variables that do not change abruptly affect the two closely spaced Hall detectors in the probe in a similar manner, so that no differential signal is generated. This type of differential probe is adaptable for formulating precise guidance systems for process control applications.

Eddy current testers require the use of standards to obtain quantitative information from a test specimen. Measurements on the standards are then compared to the results from a test specimen. These standards can become an expensive part of the testing, especially in small lot testing, because they have to be precise and made of the same material as the test specimens. When a defect is disclosed, there is always a question as to how bad it actually is with respect to acceptance or rejection of the piece part.

11.8 INFRARED AS A NDT TOOL

Infrared detectors were discussed in detail under pyrometry in Chapter 5. The same types of detectors are used as NDT tools. A test piece is heated by an infrared source of radiant energy, and the specimen surface is then scanned for hot and cold spots. See Figure 11.32.

In theory, if there is an inclusion or defect, the radiated body will not absorb the infrared energy uniformly and this nonuniformity will be identified by the infrared detector. Both line and area scan methods are applicable.

A bolometer detector has adequate sensitivity for NDT applications and can be used to monitor either the natural temperature of the sample or induced radiant energy to produce a temperature profile.

A common setup, shown in Figure 11.33, uses strip heaters for examination of the bond line of a test specimen as pictured in *A*. The thermal profile as detected is pictured in *B*. The thermal profile shows a higher temperature detected for the unbonded area, as would be expected because the diffusion of the applied heat is slower in that portion of the test specimen.

The infrared detectors used in NDT techniques follow the Stefan-Boltzman law in the same manner as in temperature measurement applications. In NDT applications the emissivity of the radiating test specimen is not a problem, since absolute temperature measurements are not required. Only comparison measurements are important to show that there is a change.

Infrared as an NDT tool has been in use for about 10 years and can be used to detect uniformity of heat flow in turbine blades, solder joints, and bonded and honeycomb laminates, among other things. It can be used for plastics, metals, glass, and combinations of these materials. It is more valuable for near-surface defects, preferably when their lateral area is two to three times their depth.

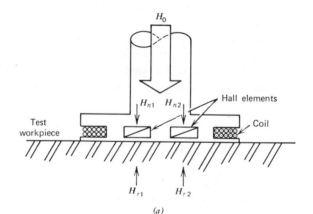

(a)

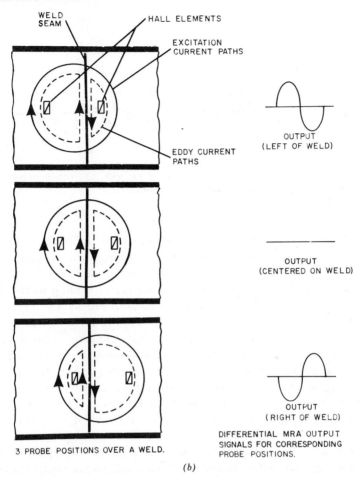

WELD
SEAM

HALL ELEMENTS

EXCITATION
CURRENT PATHS

OUTPUT
(LEFT OF WELD)

EDDY CURRENT
PATHS

OUTPUT
(CENTERED ON WELD)

OUTPUT
(RIGHT OF WELD)

DIFFERENTIAL MRA OUTPUT
SIGNALS FOR CORRESPONDING
PROBE POSITIONS.

3 PROBE POSITIONS OVER A WELD.

(b)

Figure 11.30 *Differential probes and output signals. (a) A typical dual probe for differential MRA output signals produced for various probe positions over a weld. Courtesy F. W. Bell, Inc.*

11.9 MICROWAVES AS A NDT TOOL

Microwaves have been used as a NDT tool for about 5 years. Microwaves can be considered analagous to radar, and depend on radio frequency detection. As such, they are totally reflected by metal. For testing purposes a well collimated and controlled microwave beam obeys all the laws of optics and can be treated accordingly with respect to the angles of incidence and reflectance.

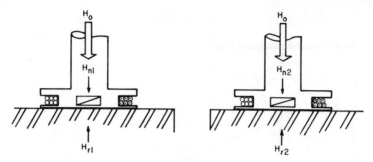

Figure 11.31 *A typical single probe arrangement for true differential measurements. Comparison test configuration. Courtesy F. W. Bell, Inc.*

There are two general microwave measurement techniques. One uses standing wave systems and the other uses traveling wave systems.

A standing wave system uses a transmitted wave which is reflected from a test surface and produces a standing wave corresponding to the signature of a detected flaw. Such a system is characterized by a voltage standing wave ratio (VSWR) which approaches infinity for a system with no losses. Standing wave systems are less susceptible to errors caused by mismatches than are traveling wave systems. A standing wave system is relatively simple to construct, but is extremely sensitive to both test specimen and system orientation. A typical standing wave system consists of a microwave signal generator for the gigahertz range (10 to 100 gHz), a broad band diode detector, wave guides, directional couplers, phase shifters, tuned shorts, receiver and exciter slots, tuned crystals, a signature diode detector, a dc amplifier, filters, and an oscillographic recorder.

A traveling wave system is one in which a microwave signal generator transmits a traveling wave which illuminates the test specimen, and the test specimen reradiates some of the energy as reflected waves. A microwave detector receives the reradiated energy in its different phase and

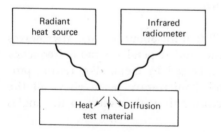
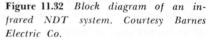

Figure 11.32 *Block diagram of an infrared NDT system. Courtesy Barnes Electric Co.*

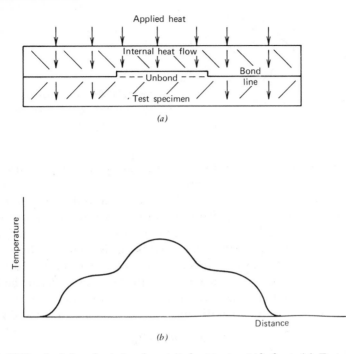

Figure 11.33 *An infrared setup using strip heaters to apply heat. (a) Test specimen. (b) Thermal profile. Courtesy General Electric Co.*

energy distributions. This detected pattern is the signature of the detected flaw in the test specimen, and must be analyzed to determine the nature of the flaw. The interaction physics of a traveling wave system can be accurately described as optical diffraction and reflection.

As in all NDT techniques, the problem of detection and recognition of the detected signals in terms of material characteristics or quality requires definition and interpretation. Standards with well defined flaws are required to establish the recognition of the detected signal in microwave systems.

A microwave test setup is shown in Figure 11.34. The incident beam is transmitted at some angle of incidence, and the reflected wave is detected at some corresponding angle of reflectance. If a flaw or defect is present, there will be a phase shift in the reflected wave from the normally detected signal. The phase shift in the detected signal must then be interpreted in terms of the type and magnitude of the defect causing it. A typical frequency representation of several cycles of the transmitted

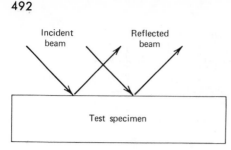

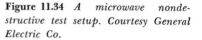

Figure 11.34 *A microwave nondestructive test setup. Courtesy General Electric Co.*

wave, the reflected wave, and the standing wave is shown in Figure 11.35. The standing wave is the signature of the flaw.

Microwaves can be used as a differential measurement to control thickness in a rolling or squeezing operation by using the setup shown diagrammatically in Figure 11.36. Two microwave units *A* and *B* are used. They are set up so that the distance d_A is equal to the distance from the transmitter to the upper surface of the metal at its proper thickness. The distance d_B is made equal to the distance from the lower surface of the metal at its proper thickness. The thickness t of the metal is then equivalent to the total distance L, less $d_A + d_B$. In equation form,

$$t = L - d_A + d_B \tag{11.8}$$

where

$$d_A = \frac{T}{2}v \tag{11.9}$$

$$d_B = \frac{T}{2}v \tag{11.10}$$

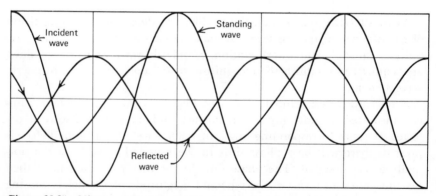

Figure 11.35 *Microwave frequency pattern. Courtesy General Electric Co.*

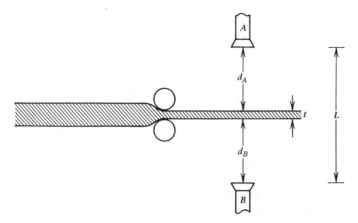

Figure 11.36 *Microwave thickness measurement and control. Courtesy General Electric Co.*

The variations in the received signals are used to control the pressure on the rolls to maintain the predetermined thickness of the metal. Responses are fast, and measured deviations for the corrections can be made extremely accurate at the small distances over which the measurements are made. Surface variations in the 10 to 100 μin. range can be detected by microwave techniques.

Microwave systems are considered complex. They are not widely used in industrial applications, because of the initial cost of equipment, difficulties in preparation of standards for the recognition of flaw signatures, and the time and cost involved in the training of personnel to read and analyze the signatures produced.

11.10 SIGNATURE ANALYSIS AS AN NDT TOOL

The microwave technique just described involves one type of signature that could be produced for consideration in signature analysis, but in general the term is used much more broadly to describe a system or component signature produced by a detector when the system or component is subjected to any planned stimulus such as a functional operation or exposure to a controlled vibration spectrum.

Detectors to produce a signature for analysis could be an accelerometer or a strain gage attached to the system or to the exciter used in the signature setup. Such a signature system is usually termed a mechanical signature. One type of mechanical signature setup is shown in Figure

11.37. This type of system can be used to determine whether or not any parts have been left loose in the specimen being tested. In such a test a specimen considered to be completely satisfactory is used to establish a standard acceptable signature. As other specimens are tested, the signatures are compared and, if there are deviations, the signature is analyzed to determine the type of loose particle causing it and this signature then becomes part of the analysis data for future tests.

Although the above test and system are considered a mechanical signature, the actual signature is an acoustic signal generated by mechanical motion or mechanical behavior in contrast to an electronic or electrical signal developed by the specimen being tested.

Mechanical signatures can be used to evaluate the performance of passive systems for loose particles, most rotating components such as gyros, fans, and motors, intermittent components such as stepping motors or switches, solenoid valves and camera shutters, and continuous motion devices such as bearings, whereas acoustic signal signature can be detected and recorded.

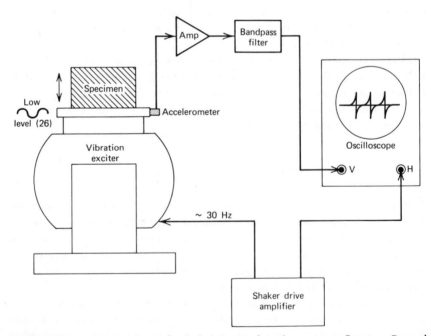

Figure 11.37 *Loose particle mechanical signature detection system. Courtesy General Electric Co.*

In each case care must be exercised in obtaining a "normal" signature from multiple unit measurements, or by analytical prediction from a mathematical model. For this reason mechanical signature analysis (MSA) is primarily an engineering laboratory tool but is being favorably considered in many industrial applications in which the utilization is high enough to make each test economically feasible. Ideally, the system could be set up so that the correlation of single test unit data could be performed by a computer in which the normal signature is stored and at the end of the test a conclusion would be made immediately available. Such a system would be ideal for testing the flight worthiness of every aircraft before it is placed on an airport flight line. In fact, such a system would be an ideal inspection tool for all types of motion devices, but unfortunately we do not have normal signatures and computers available to perform the comparison analyses so we use other methods and techniques.

11.11 ULTRASONIC HOLOGRAPHY

In an ultrasonic holography system the test specimen is "illuminated" by an ultrasonic wave instead of a coherent light beam as discussed in Chapter 9. At the present time liquid surface holography, using ultrasonic waves, gives excellent resolution in imaging in real time when an acoustic lens is used to image the object in the plane of the hologram. Other techniques that have been attempted or are under study are recording directly on a photographic plate, recording on thermoplastic film, and recording a modulated point light source, synchronized with the detector, directly on film.

Essentially, the problem is to convert the ultrasonic signal into a form that can be optically imaged either in real time or as a recorded signal which can be reconstructed later.

In ultrasonic holography the interference pattern is formed by acoustical waves. The trick is to record this pattern in such a manner that the resulting reconstruction is an optical representation of the device as it would be seen if the eye responded to the ultrasonic energy.

One type of setup for obtaining a hologram on a liquid surface is shown in Figure 11.38. In liquid surface holography the liquid surface performs as a square law detector. Variations in the elevation of the liquid surface are produced as variations in acoustical radiation pressure occur. In other words, the surface of the liquid in the small tank changes level and shape in response to the ultrasonic energy applied to the sub-

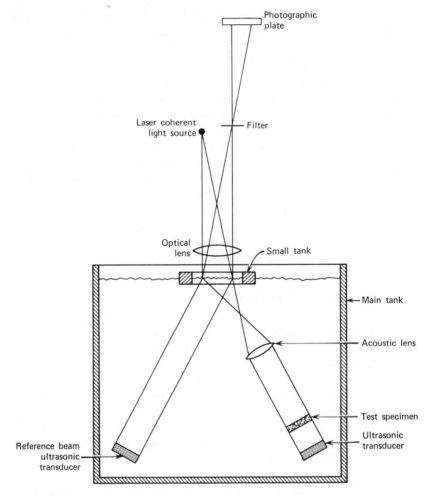

Figure 11.38 *Ultrasonic liquid surface hologram equipment Courtesy General Electric Co.*

merged test specimen. This surface wave shape is illuminated with coherent light, and the reflected light is phase modulated by the height variations in such a manner that an image of the ultrasonic field in or near the object is reconstructed. This image is then photographed for later reconstruction by optical techniques, or directly illuminated for real time imaging.

Good definition of the test specimen can be obtained and can be viewed from different angles in the same manner as the optical hologram.

11.12 ULTRASONICS IN BOND TESTING

A resonant frequency NDT system for the testing of structural adhesive bonding and composite material is shown in Figure 11.39. While this device is broadly classified as ultrasonic, it is ultrasonic only in the sense that the probing element is made of piezoelectric material and is caused to resonate at frequencies in the ultrasonic range. The resonating probe is basically mechanical. The unit operates on a sweep frequency ultrasonic energy pulse which drives a piezoelectric crystal at ultrasonic frequencies. Two types of resonant vibrations are employed; one is axial and one is radial. The axial is for testing sandwich structures, and the radial is used for testing all lap shear type joints.

Figure 11.39 *Fokker bond tester. Courtesy Shurtronics Corp.*

The sweep frequency is 60 cycles and acts as a carrier wave sweeping through resonant frequencies ranging from 80,000 Hz to approximately 2 MHz, depending on the specific probe being used. This unit is capable of evaluating the cohesive strength of the glue line and will produce actual joint strength output values in pounds per square inch for both lap shear and for honeycomb structures. Figure 11.40 is a typical strength diagram for a lap shear joint, and Figure 11.41 is a face-to-core bonding diagram.

The basic unit shown in Figure 11.39 when used alone requires a skilled operator to interpret the results. This always leaves some room for doubt about the final product quality. However, the unit can be supplemented by means of auxiliary equipment to set automatically limits predetermined by an engineering evaluation for the particular product. It is possible to set up grid patterns on the product surfaces to be evaluated and to record the values for permanent records. The system can be automated for large production operations, and multiple probes can be used on a time sharing basis to speed up inspection cycles. The amount of equipment used is usually determined by the economics involved, the availability of trained operators, and the quality of the product being controlled.

At present the largest industrial users of this type of bond tester are aircraft and missile structure manufacturers, although other uses in honeycomb structures and bonded layer products are feasible.

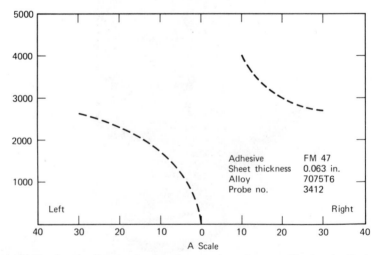

Figure 11.40 *A scale diagram for lap shear testing. Courtesy Shurtronics Corp.*

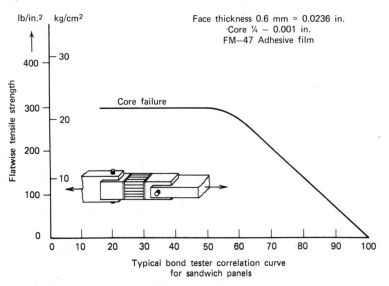

Figure 11.41 *B scale diagram for face-to-core bond testing. Courtesy Shurtronics Corp.*

11.13 SUMMARY

In summary the techniques used in obtaining NDT data depend on the information needed, the economics of the application, and the accuracy of the measurements required. The method and the techniques to be used also depend on the type of material to be tested and the type of defect for which the test was designed. As yet there is no universal test or equipment that will supply all the information needed, and there are existing requirements for which no good, reliable NDT methods, equipment, or techniques are available. There will continue to be a need for research and development to provide industry with new NDT equipment and techniques.

Review Questions

11.1   When is it practical to use magnetic particle testing on industrial products?

11.2   What are the limitations of dye penetrant testing? What materials can be exposed to such tests with positive results?

11.3   What advantages do x rays have over magnetic particles and dye penetrants for NDT?

11.4   Give at least three distinct disadvantages of the use of x rays.

11.5   How does one obtain a definition of depth and angle by x-ray techniques?

11.6   What is a penetrameter and how is it used?

11.7   What part does film play in x-ray NDT?

11.8   What advantages does neutron radiography have over gamma and x-ray radiography? Why is it not more widely used in industrial applications?

11.9   What is meant by ultrasonic NDT? What equipment is required?

11.10   How is attenuation defined in ultrasonic NDT and how can it be minimized in a given material?

11.11   What determines the frequency and velocity of a particular ultrasonic NDT setup?

11.12   What technique is used to obtain shear wave measurement responses in ultrasonic NDT weldment testing?

11.13   What purpose does a couplant serve in ultrasonic NDT applications?

11.14   Where is an angle beam technique of measurement advantageous?

11.15   Where is the delta technique applicable?

11.16   Name the three types of ultrasonic scans and give the advantages and disadvantages of each.

11.17   What is eddy current NDT? Where is it applicable?

11.18   What four material properties affect the eddy current distribution in a test piece?

11.19   What is lift-off and where it is most severe in NDT applications?

11.20   How is the Hall voltage determined as applied to magnetic reaction analyzer eddy current testing?

11.21   What is one of the main advantages of differential eddy current testing techniques as compared to absolute and comparison eddy current measurements?

11.22   What principles are involved in using infrared as an NDT tool?

11.23   What laws does an infrared detector follow to produce useful NDT information?

11.24   How are microwave techniques applied to NDT for metals?

11.25   What is the difference between a standing wave and a traveling wave NDT system?

11.26   What is a mechanical or an electrical signature with respect to NDT measurements? How are these signatures actually obtained?

11.27 How does ultrasonic holography resemble optical holography? How does it differ?

11.28 What advantages does ultrasonic holography offer as an NDT tool?

Bibliography

Brenden, Byron B., *Acoustical Holography as a Tool for Nondestructive Testing*, Battelle Memorial Institute, BNWL-SA-1912, August 1968.

Clotfelter, W. N., Brankston, B. F., and Zackary, E. E., *The Nondestructive Evaluation of Stress-Corrosion Induced Property Changes in Aluminum*, NASA TMX-53772, George C. Marshall Space Flight Center, Huntsville, Alabama, August 1968.

Feinstein, L., and Hruby, R., *Surface-Crack Detection by Microwave Methods*, NASA Technical Brief, Brief 67-10482.

Hochschild, Richard, *Microwaves Nondestructive Testing in One (Not So Easy) Lesson*, Bulletin, 1100, Microwave Instruments Co., Corona Del Mar, Calif.

Magnaflux Corporation, *Operator-Inspector Guidance Charts for Inspection with Magnaflux*, Magnaflux Corp., Chicago, 1952.

Magnaflux Corporation, *Electromagnetic/Eddy Current Testing*, ED-300, Magnaflux Corp. Chicago, Ill., July 1966.

Phelan, C S., *Resonant Frequency NDT System for Structural Adhesive Bonding and Composites*, Shurtronics Corp., October 1968.

Smith, George H., *Eddy Current Test Applications*, F. W. Bell, Inc., Columbus.

Smith, George H., *Eddy Current Test Systems*, F. W. Bell, Inc., Columbus.

Zinke, Willard L., *Nondestructive System for the Inspection of Solder Joints*, Materials Evaluation, May 1964, pp. 219–224.

CHAPTER 12

Environmental and Pollution Measurements

Environmental parameters and pollutant levels are measured to predict weather conditions, to protect health and safety, and to improve human welfare.

Many ecologists are alarmed at the existing conditions, and are demanding that something tangible be accomplished to clean up our environment to make it safer for all kinds of life. The United States government and many state governments have established an Environmental Protection Agency (EPA) to measure the concentration of gases, vapors, and particulates in the air and water, noise levels in areas where people live and work, the effects of pesticides on health conditions and food supplies, and the amount of radioactivity and other fallout that occur or are present in various areas near nuclear fueled power plants and nuclear test sites. It is the responsibility of these protective groups to select methods and instrumentation measurements needed to establish controls to either limit pollution to tolerable limits or to eliminate it.

An evaluation of the conditions in different types of areas, namely, the industrial, the urban, the suburban, and the rural farmland, can be made with existing types of instrumentation, although improvements in techniques and selectivity would be helpful in many cases. In order to establish whether contamination levels generated by pollutants are safe for sustained human existence, the EPA has to make measurements and establish levels that appear to be tolerable now. These measurements and their associated instrumentation are discussed first, and then the instrumentation and control devices needed by industry to meet the arbitrarily established tolerable levels are considered.

12.1 METEOROLOGICAL INSTRUMENTATION AND MEASUREMENTS

Wind speed and direction are needed to determine the rate at which pollutants are dispersed into the air and the direction from which they are released. At times it is also important to know the turbulence present in the airstream and the amount of pollution per unit volume of air sampled. Wind speed and directon can be measured simultaneously with the instrument shown in Figure 12.1. Wind direction and velocity transmitters are located outside the measuring center on towers, and send their signals to recorders which maintain a continuous record of their outputs. Wind direction is measured in 0 to 360°, and wind velocity is given in miles per hour. Wind speeds and gusts of up to 90 mi/h can be measured and recorded. If there is *turbulence in the airstream* it has to be measured with a separate unit known as a UVW anemometer which is

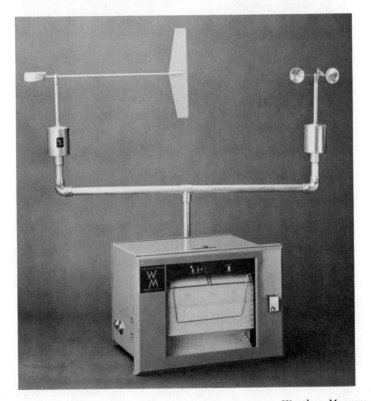

Figure 112.1 *Wind direction and velocity meter. Courtesy Weather Measure Corp.*

shown in Figure 12.2. The turbulence is integrated over a small time span of a minute or more to give a vector output. An air volume and speed anemometer is shown in Figure 12.3. With measurements of direction, speed, and volume, it is quite simple to determine the concentrations of the pollutants described in the following sections.

The *relative humidity* of the atmospheric air is important as a comfort factor, and it is a measure of how many airborne particulates are held in suspension where we can take them into our lungs as we breathe. Relative humidity can be determined by a relative humidity measuring device such as that shown in Figure 12.4, by means of *dew point* measurements, or by use of the wet bulb temperature method. Dew point temperatures

Figure 12.2 *UVW anemometer. Measures directly the three orthogonal vectors of the wind. Courtesy R. M. Young.*

Figure 12.3 *Air volume and speed anemometer. Courtesy Davis Instrument Manufacturing Co.*

range from −40 to 90°F. Dew point measurements are helpful in determining the levels of sulfur dioxide (SO_2) and ethylene (CH_4) in the atmospheric air.

To determine the contribution of *solar energy* to the *ambient temperatures* measured on the earth's surface, we make use of special pyrometers. The amount of solar energy absorbed or reflected by a specific area of the earth's surface can be measured as total energy, with the contributions in respect to wavelength, ranging from the ultraviolet to the infrared. These measurements are usually made on a mV/(mcal)(cm²)(min), mV/(cal) (cm²)(min), mW/(cm²)(min), or mW/(in²)(min) basis. The ultraviolet energy is measured by means of a photocell. This energy passes through an opaque quartz window and a bandpass filter which limits the spectral response to the wavelength interval from 295 to 385 nm and provides close adherence to the Lambert cosine law. (If the normal to a surface makes an angle with the direction of the rays, the illumination is pro-

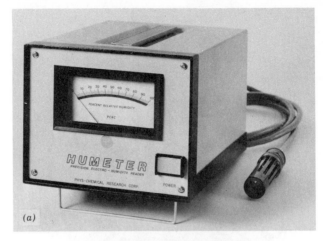

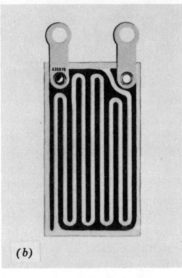

Figure 12.4 *Relative humidity meter (a) and sensor (b). Courtesy Phys-Chemical Research Corp.*

portional to the cosine of that angle.) A typical unit is shown in Figure 12.5. This unit has a sensitivity of approximately 0.2 mV/(cal)(cm²)(min), with a ±2% linearity from 0 to 0.1 cal/(cm²)(min).

The total sun and sky radiation energy is normally measured by means of a differential thermopile with blackened hot junction receivers and whitened cold junction receivers. The energy passes through an optical

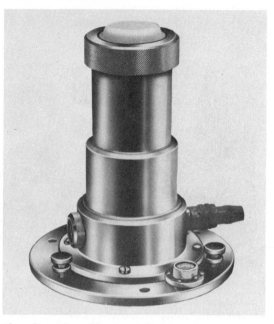

Figure 12.5 *Eppley ultraviolet radiometer (photometer). Courtesy Eppley Laboratory, Inc.*

glass hemisphere (Schott WG7) which is transparent from 280 to 2800 nm. The sensitivity of these units ranges from approximately 2.5 to 7.5 mV/(cal)(cm²)(min), with a linearity of ±1% from 0 to 2.0 cal/(cm²) (min). These units are referred to as black-and-white pyranometers or pyroheliometers. A typical unit is shown in Figure 12.6.

Another type of total sun and sky radiation energy measuring unit is exposed through a variety of hemispherical filters on a time sharing basis. In this unit the spectral range covered is from 285 to 2800 nm in four steps, using Schott optical filters GG-14 (approximately 500 m), OG1 (350 m), RG2 (630 m), and RG8 (700 m). Other filters are under development to cover the entire 300 to 3000 m solar spectral range. This unit has a sensitivity of approximately 5 mV/(cal)(cm²)(min) and a response that is linear up to 4 cal/(cm²)(min).

The infrared radiometer (pyrgometer) shown in Figure 12.7 is a typical instrument for measuring terrestrial long wave radiation in the infrared region from 4 to 50 nm. This is accomplished by the use of a KRS-5 hemisphere whose inner surface is a vacuum deposited interference filter. This unit has a sensitivity of approximately 5 mV/(cal)(cm²)(min) and a linearity of ±1% from 0 to 1 cal/(cm²)(min). Such units are intended for

Figure 12.6 *Eppley black and white pyranometer. Courtesy Eppley Laboratory, Inc.*

Figure 12.7 *Eppley precision infrared radiometer (pyrgometer). Courtesy Eppley Laboratory, Inc.*

unidirectional operation in the measurement, separately, of incoming or outgoing terrestrial radiation as distinct from net long wave flux.

Pyranometer units provide a means of measuring the ambient temperature contribution from the solar source and separating it from the other sources of energy making up the total ambient temperature measured. This is an excellent method of establishing the distribution and the variation of incoming, outgoing, and net radiation, either directly from the sun or radiated by the sky. Reflected and diffused sky radiation is due to the scattering of solar radiation by water vapor, dust particles, and clouds in the path of direct solar radiation. The air layer near the earth's surface also prevents the escape of reflected radiation.

Rainfall measurements are made to determine the amount of moisture falling on a given area, and the residue collected from the water provides a good measurement of the particulates and other pollution the rain washed out of the air as it descended. Of course the water soaks into the ground, and as it does so it leaves the collected residue behind. Not all the water is soaked up by the soil; the runoff carries part of the residue along and in time distributes some of it in the stream bed it traverses. The amount of residue carried or deposited enroute depends on the flow volume and velocity and the weight of the accumulated residue. Solubles are carried along as far as the water travels and are deposited only in the stream, lake, or ocean bed when the liquid evaporates.

A typical rain gage is shown in Figure 12.8. An accumulation record is kept so that the annual rainfall for a given area can be measured and recorded to be used in statistical evaluations with respect to agriculture, forestry, and marine products. Rainfall data are also useful in predicting the level of the water table, the height of feeder streams, reservoir storage rates, and rates of soil erosion in the rainfall area.

The *ambient temperature* of an area can be measured with a mercury-in-glass thermometer, a bimetallic thermometer, a resistance thermal detector (especially for air inversion measurements), an electric thermometer, or a thermistor. Usually, a thermistor and a good bridge circuit are fast and accurate enough for this application. When precision air temperature measurement is required to determine air inversion, use can be made of two resistance thermal detectors. One is placed on the top of a tower 30 to 100 m, high, and the other at the base of the tower or at the measuring station. When the air at the top of the tower is at a higher temperature than the ground level temperature, we have an inversion in which the warm air cannot rise and disperse into the atmosphere. This mean that any stack pollutants are held close to the ground. Under air inversion conditions we experience smog in cities and heavy industrial

Figure 12.8 *A typical rain gage. Courtesy Weather Measure Corporation.*

areas. There is usually a high fallout of airborne particulates during inversions, and the concentrations of the particulates often exceed established tolerable levels. Air inversion measurements are required when stack outputs are dependent on air currents to disperse the pollutants emitted so that the pollutant level is maintained at a value below the designated tolerable level.

Atmospheric pressure is measured by means of a *barometer,* the output of which is usually indicated, recorded, or transmitted to a central station.

12.2 AIR POLLUTANT MEASUREMENTS

Air pollutants consist of gases, vapors, aerosols, and particulates. They can occur singly or in combinations. Combinations are most common, for we seldom if ever find a perfectly dust-free atmosphere. All these types of pollutants are found as exhaust emissions from various industries, combustion type vehicles, or even home heating plants.

Gaseous pollutants include sulfur compounds (such as sulfur dioxide, hydrogen sulfide, and sulfur oxides), nitrogen compounds such as nitrous oxide and nitrous dioxide, carbon monoxide, carbon dioxide, ozone, hydrocarbons, ethylene, and other mixtures. Vapors include water and other solvents with measurable vapor pressures. Aerosols include man made dispenser types and very small bacterial types which have to be trapped in special filters.

Sources of sulfur and sulfur compounds are fuel oils, combustible fossil fuels (such as coal, coal dust, charcoal, coke, and peat), and by-products of the paper manufacturing process and petroleum refining. The exhaust fumes of certain combustion engines generate sulfur and nitrogen products through imperfect combustion of the basic fuel.

When only one type of pollutant has to be measured, it is highly desirable to have an instrument that measures that pollutant accurately and excludes the effects of other pollutants that may be present. Where redundant measurements are required, one should be very selective, so that there can be no doubt that the element is present and that it can be measured accurately. This is equally desirable in measuring the source of pollution or in monitoring the air into which the pollution may be discharged.

Sulfur and sulfur products (SO, SO_2, H_2S) concentrations can be measured by wet chemical sampling, chromatography, electrical conductivity, nondispersive infrared, and spectrometric techniques. Probably the best method for continuous sampling simulation is the chromatographic technique. The words continuous sampling simulation are used because chromatography measurements involve sampling and separation techniques. The system has to be flushed after each sample, and so-called zero air must be introduced to ensure zero concentration without a sample present. This calibration check is accomplished by the introduction of pure gas from cylinders or by means of permeation tubes. The use of permeation tubes requires the employment of very closely controlled oven temperatures, because the change in concentration of the gas emitted from the permeation tube is highly temperature dependent. The ratio is about 10:1. A 1°C temperature change causes a 10% change in permeation tube gas concentration. The block diagram of Figure 12.9 shows a chromatograph with sulfur and nitrogen product permeation tube sections, a clean air system, and a hydrogen generator section with appropriate switching mechanisms to sample, flush, calibrate, measure, and purge. Helium is used as the carrier gas in the sulfur system, and pure filtered air is used with the nitrogen permeation system. As shown, the instrumentation can measure both sulfur and nitrogen compounds, but the measurement must be made separately on a time sharing basis. Both the helium and air are thoroughly dried before they enter the chromatograph system. The sample injection and stripper valve, select

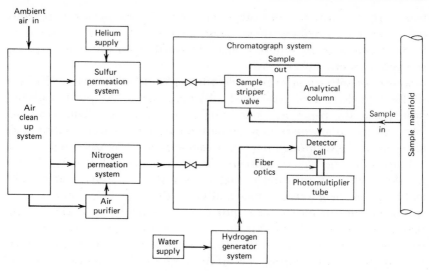

Figure 12.9 *Chromatograph and auxiliary operating components. Courtesy Bendix Corp. Process Instruments Division.*

the sample to pass through an analytical column whose output passes through a detector cell where it is mixed with hydrogen gas and burned as a hydrogen-rich mixture. The flame is monitored by a photomultiplier tube, and the gas constituents are discharged through vents to the atmosphere. The sampe is drawn from a type of manifold system. This manifold may be especially constructed for a remote monitoring station or system, or it may be part of an exhaust stack on a neighborhood industry. Chromatographic units are available with minimum sensitivities of 0.005 ppm sulfur or nitrogen and cover the range up to 1 ppm for sulfur and nitrogen products.

Where concentrations of sulfur remain consistently higher, it is also feasible to make these same measurements using nondispersive infrared techniques. Such a system is shown in Figure 12.10, and the infrared analyzer section is shown schematically in Figure 12.11. This system indicates that an ultraviolet source can be substituted for the infrared source, or that the system can employ both types of sources in a two-gas system, depending on the application. (See Figure 12.12). Nondispersive infrared systems generally have a higher minimum sensitivity than chromatograph systems.

Ozone has been established as a gaseous air pollutant. One can smell it after lightning strikes close or when there is a corona discharge. Too large an exposure may lead to severe headaches, and plant life may suffer considerably. Ozone measurements can be made by the chemiluminescence technique. The principle involves the photometric detection of a flameless phase reaction of ethylene gas with ozone. This is a selective and valid measurement process, since this reaction results exclusively from a combination of ozone and ethylene gas. The ethylene gas and sample flow path diagram are shown in Figure 12.13. The air passes through a chemical filter to remove moisture and particulates, and the sample passes through a Teflon filter to remove particulates above 5 μm in size. The stream selector solenoid puts either the output of the ozone generator or the sample through the reaction chamber. The ozone input is used to calibrate the system. This calibration gas can be generated in a system similar to that shown in Figure 12.14. This is a sampling system in which the ozone and the sample are time shared in the reaction chamber. Adequate ethylene pressure is required to provide the best flow for maintaining the sensitivity required in the specific application. Care must be exercised, because the ethylene–air mixture is highly combustible. Measurements are usually made on a parts per million or parts per billion basis, and instruments are available in ranges from 0 to 1 ppm with a minimum sensitivity of approximately 0.001 ppm.

Nitric oxide (NO), nitrogen dioxide (NO₂), and other *nitrogen compounds (NO$_x$)* can be measured adequately by the chemiluminescence principle, discussed under ozone, and by the infrared/ultraviolet two-gas analyzer discussed under sulfur products and shown in Figure 12.12a. Using the chemiluminescence technique, the lowest detection limit is 0.005 ppm with a precision of ±2%, which is adequate to meet present EPA standards. A *colorimetric* method of continuously measuring nitric oxide or nitrogen dioxide is available. In this method the air sample is scrubbed with a continuously flowing stream of Griess-Lyshkow reagent. The color formation that takes place depends on the level of nitrogen dioxide contamination. The colorimeter uses two matched flow cells, a reference cell through which the clean reagent flows, and a detector cell through which the colored reacted reagent flows. A common light source is used to illuminate the two flow cells, and the transmitted light is detected by two cadmium photoresistors.

The light transmission from each cell is compared by a sensitive Wheatstone bridge circuit, and the bridge produces an electrical signal which is proportional to the nitrogen dioxide concentration in the atmosphere. The signal is displayed as a digital indication of the concentration. The output is in binary coded decimals for computer input, and is reproducible to ±1% of full scale. The instrument sensitivity is in the order of 0.02 ppm or better.

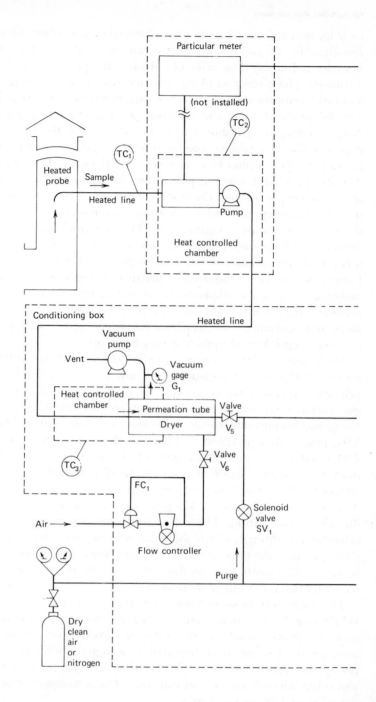

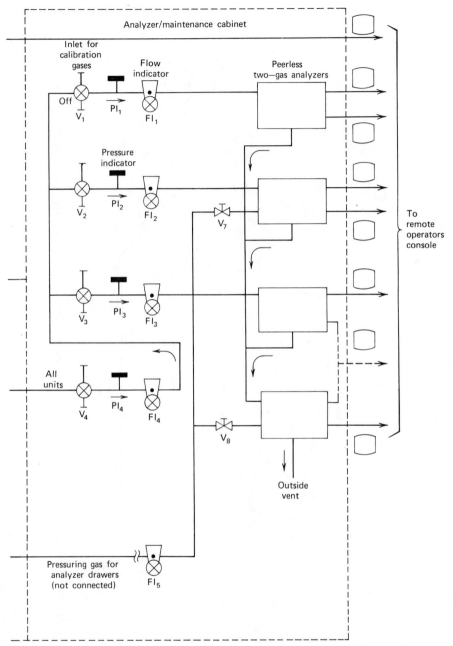

Figure 12.10 *Nondispersive infrared/ultraviolet sulfur and nitrogen products analyzer. Courtesy Peerless Instrument Co., Inc.*

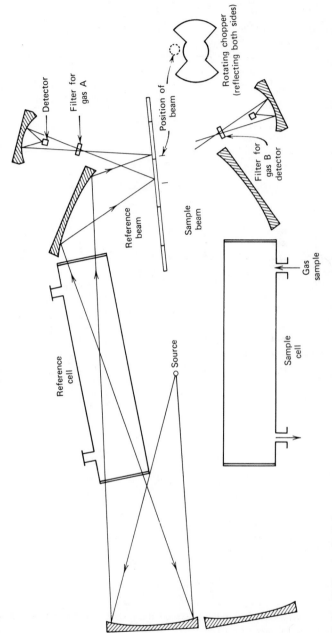

Figure 12.11 *Analytical system for a nondispersive infrared gas monitoring system. Courtesy Peerless Instrument Co., Inc.*

Detector

Filter for gas A

Position of beam

Rotating chopper (reflecting both sides)

Reference beam

Sample beam

Filter for gas B detector

Reference cell

Source

Gas sample

Sample cell

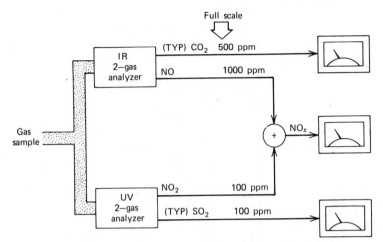

Figure 12.12 *Infrared/ultraviolet two gas analyzer. Courtesy Peerless Instrument Co., Inc.*

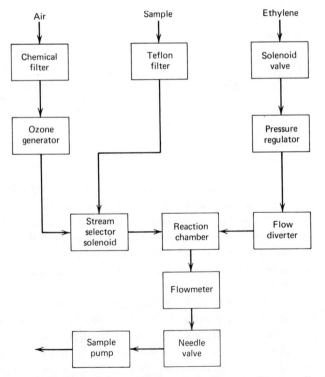

Figure 12.13 *Ethylene gas and sample flow path diagram. Courtesy Bendix Corp., Process Instrument Division.*

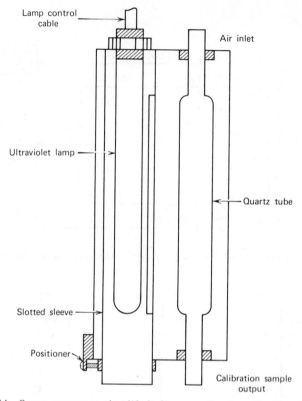

Figure 12.14 *Ozone generator simplified diagram. Courtesy Bendix Corp., Process Instruments Division.*

Carbon monoxide and *hydrocarbons* of the methane and nonmethane groups can be detected and measured by several methods including non-dispersive infrared, gas chromatography, and optical techniques. It is customary to measure the gases separately, either with different instruments or separately with the same instrument as total hydrocarbons, methane, and carbon monoxide, when the methane and carbon dioxide are below 1 ppm.

The portable carbon monoxide measuring instrument shown in Figure 12.15 operates on widely recognized electrochemical principles. Ambient air is drawn through the detector cell behind the diffusion type catalytically active electrode, where any carbon monoxide present is oxidized to carbon dioxide. The rate of oxidation is related to the concentration of the carbon monoxide, and is read directly as parts per million. This

Figure 12.15 *Portable carbon monoxide analyzer. Courtesy Energetics Science, Inc.*

catalytic action is not affected by water vapor or relative humidity, and a ±1% full scale accuracy can be maintained over a temperature range of 0 to 40°C. This instrument has a minimum detectable sensitivity of 1 ppm and is available in ranges from 0 to 50 ppm to 0 to 2000 ppm.

A total hydrocarbon, carbon monoxide, and methane air monitor system block diagram is shown in Figure 12.16. In this system an integral pump extracts a sample from a sampling manifold and introduces it into the chromatograph which performs an analysis every 5 min. The analysis is stored in a memory system so that a continuous readout is available to a central control computer through a telemetry line. This chromatograph system employs a *hydrogen flame ionization detector* (HFID) to measure the low concentration of the air pollutants in the atmosphere. A very sensitive detector is needed. This chromatograph uses a coiled capillary packed with granular material, some of which is coated with a non-volatile liquid in one column and some of which is adsorptive material in another column. The carrier gas is dry nitrogen, and pure air is furnished by the catalytic oxidizer shown in Figure 12.17. The separation columns have been selected to give rapid and complete separation of methane and carbon monoxide. Normally, the ambient air contains more than these two pollutants, so a backflushing technique is used on the columns to prevent other types of pollutants from reaching the detector. At a particular point during the analysis cycle, the flow is reversed in one of

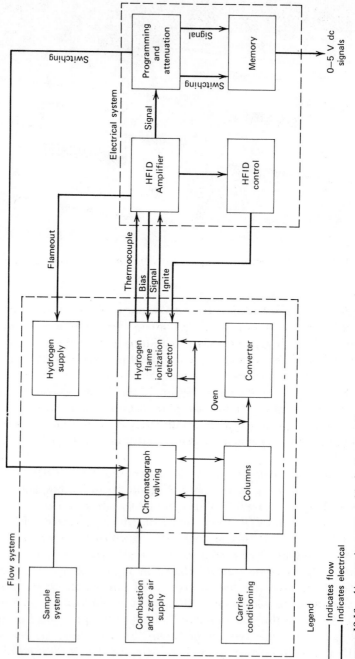

Figure 12.16 *Air monitor system block diagram. Courtesy Mine Safety Appliances Co.*

520

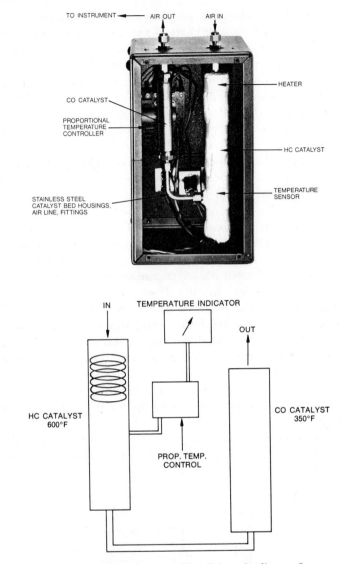

Figure 12.17 *Catalytic oxidizer. Courtesy Mine Safety Appliances Co.*

the columns to purge the unwanted pollutant components from that column, while permitting the continued forward flow through the other column to the detector.

The HFID operates on the principle that a flame of pure hydrogen burning in air produces a very small number of ions, but when carbon molecules are added a large number of ions is produced. It turns out that the number of ions produced is proportional to the number of carbon atoms combusted. Since ions are charged particles, regardless of how they are produced (ions are produced by radioactive particle collisions in the Geiger-Mueller tubes mentioned in Chapter 10), they may be collected by an electric field. This can be accomplished by using the setup shown in Figure 12.18 in which the flame is placed in an electric field and the ions are collected by the collector ring and measured as a current. The jet is connected to a bias voltage to set up a voltage gradient to direct the current flow.

The detector just described detects only hydrocarbons, yet the chromatograph system was selected because it has the capability to measure carbon monoxide, methane, and total hydrocarbons. This means that the carbon monoxide has to be converted to methane to be detected and measured. The carbon monoxide is converted to methane in the catalytic converter shown in Figure 12.19. This is accomplished by injecting the carbon monoxide and pure hydrogen gas on a continuous basis into one side of an electrically heated converter containing a granulated catalytic bed through which the gas mixture flows. The mixture is catalytically converted to methane and water vapor. When methane passes through the converter, it is not affected by the catalyst and emerges as methane. The detector thus sees both methane and carbon monoxide as methane and measures them as methane, but at separate times as they are eluted by the separate columns.

For the total hydrocarbon measurement, the columns are bypassed, and the sample is injected directly into the detector without going through the converter.

Pure air is supplied by the catalytic converter to make the zero base reading. The air emitted by this converter does not contain detectable

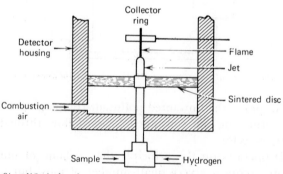

Figure 12.18 *Simplified sketch of a HFID. Courtesy Mine Safety Appliances Co.*

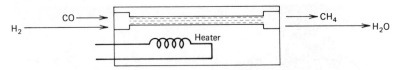

Figure 12.19 *Simplified sketch of a catalytic converter. Courtesy Mine Safety Appliances Co.*

amounts of any of the pollutant gases measured by the monitoring equipment. A mixture of this pure air and hydrogen is used to establish the zero reading for each sampling cycle.

Vapor measurements include fog, haze, smog, clouds, and other liquid vapor phase conditions that may hinder visibility, cause breathing problems, generate toxic environments, or affect plant life. Water vapor from steam, heavy fogs, and various types of clouds, fumes from chemical reactions, acid vapors from plating and stripping operations, vapors from paints, varnishes, paint removers, thinners, and lacquers, and mercury vapors are all undesirable and are definitely considered pollutants for one reason or another. Depending on the nature of the vapor pollutant, it may be measured optically, chemically, or electrochemically in much the same manner as gases.

Fog, haze, smog, precipitation, high concentrations of dust, and other light adsorption vapor conditions that hinder or block normal visibility can be measured by means of the photometric technique shown in Figure 12.20. In the fog Visiometer the air sample is illuminated by a pulsed xenon lamp and light scattering from the sample volume is measured by a photomultiplier and associated solid state electronics circuitry.

Another technique is to study plumes from acid plants by means of time lapse photography. An instrument setup is shown in Figure 12.21, and the results are shown in Figure 12.22. This unit is useful for producing evidence of any type of stack emission or dust condition where there is enough light to take pictures.

Airborne particulates include aerosols as well as dust, dirt, bacteria, and other airborne solids.

Aerosols are very fine particles, and are becoming more of a problem as more products are packaged with aerosols such as Freon used as the dispensing agent. Aerosol dispensers are used for hair sprays, shaving creams, deodorants, disinfectants, paints, lacquers, varnishes, insecticides, fertilizers, and food items such as quick whipped cream.

In general, airborne particulates are collected on filters, and their types and sizes measured. Various types of so-called air samplers, air monitors,

Figure 12.20 *Fog visiometer. Courtesy Meteorology Research, Inc.*

filter systems, and collectors are used to procure the samples from the air. In some cases evaluations are made on the basis of density for an exposure time of 15 min to a whole day or longer. The filter system may be multiple-screen, and filter devices set up in stages so that the first stage collects specimens of 20 μm or larger and the last stage collects particles 0.2 μm in size. For units of this type the airflow should range from 50 to 20 ft^3/min. A typical continuous flow unit with a tape filter for sample collection and measurement by the density technique is shown in Figure 12.23. A multiplate filter high volume air sampler is shown in Figure 12.24. These samplers are normally used outdoors, but they can be used indoors, for example, in grinding shops, cement processing areas, and fossile fuel grinding and pulverizing areas. Such units can also be used to an advantage in fossil fuel burning areas where pulverized coal and rock dust are injected into high pressure boiler fireboxes, or in the open hearth or quenching areas of steel mills where heavy smoke or vapor laden air is filled with lung penetrating or irritating particulates. Other areas of concern are rock crushing operations, clay grinding and screening processes, pulverizing of fertilizers, and mine screening operations.

Where dust particles or aerosols are extremely small and well dispersed, use can be made of an integrating nephelometer such as the one shown in Figure 12.25. This is an optoelectrical unit which measures the particulate concentration as a function of scattered light reflections from the particles present in the measurement volume. This volume is changed by pulling air through it from a dispensing manifold at a

Figure 12.21   *Time lapse photography equipment. Courtesy Meterology Research, Inc.*

prescribed rate. The integrating nephelometer operates on a comparison principle. This unit draws a continuous air sample through a chamber where it is illuminated by a pulsed flash lamp. The scattered light is detected by a photomultiplier tube focused on the illuminated air sample. The photomultiplier output is averaged and compared with a reference voltage from a second photomultiplier tube focused on a clean air sample illuminated by the same flashing lamp. A calibration reference consists of light scattering from clean, filtered air and from Freon-12. The nephelometer measures the same wavelength as the human eye, and the local visual distance is indicated in miles. The range is from 0.3 mi to infinity.

The nephelometer can be operated both as a stationary detection device or in a mobile capacity in an automobile or aircraft. The nephelometer is employed to monitor air quality for day-to-day variations, to evaluate the effectiveness of regulatory measures, to make aerial surveys of pollution, in aerosol research, for measurement of automotive and jet plane engine emissions, and for visibility measurements for flight operations, photography, outdoor testing, and transportation.

12.3   ENVIRONMENTAL AND AIR MONITORING SYSTEM EQUIPMENT FOR REMOTE AND CENTRAL STATION LOCATIONS

The parameters to be measured in an environmental and air monitoring system may consist of any or all of the items covered so far in this chap-

Figure 12.22 *Time lapse photographs showing the intervals of pollution. Courtesy Meteorology Research, Inc.*

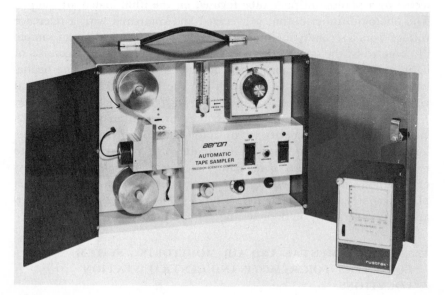

Figure 12.23 *Aeron automatic tape air filter. Courtesy Precision Scientific, Subsidiary of GCA Corp.*

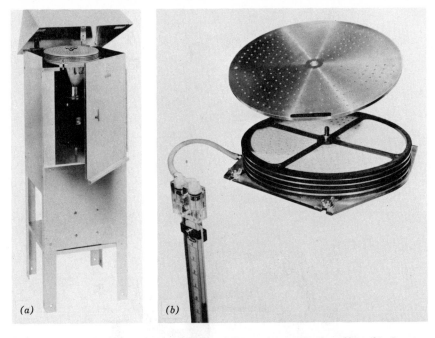

Figure 12.24 *A high volume air sampler (a) using a multiple plate filter (b). Courtesy General Metal Works, Inc.*

ter. A total typical system is shown in Figure 12.26. There is no need for a central station or home base unless there are two or more remote stations. These stations are usually of the mobile type, so that they can be moved from one area to another. The number of remote stations should not exceed 40 to 50, primarily because there is too much data for reasonably rapid comparisons and evaluation by telemetry on a continuous basis using a minicomputer. Larger computers can be employed to increase the number of reporting stations or the number of measurement and monitoring functions. The central station should have all the facilities shown, and the computer should be programmed to obtain the data in sequential order and initiate commands to have special tests and/or calibrations performed. The central station should also be equipped with instruments such as atomic mass spectrometers and spectrum analyzers, or other highly flexible analytical equipment, to verify or supplement the measurements made by instruments at a remote station on any type of critical contaminating pollution that may be detrimental to human life.

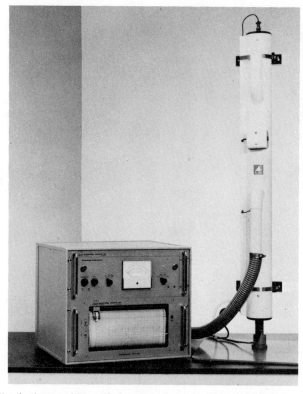

Figure 12.25   *An integrating nephelometer. Courtesy Meteorology Research, Inc.*

12.4   INDUSTRIAL AIR MEASUREMENT AND CONTROL

Industries that have a potential for contaminating the air with gaseous, vapor, or airborne particulate pollutants that can be detected by the EPA or other government agencies with air monitoring equipment need equipment as good as or better than that used by the EPA.

In some instances stack control equipment is adequate to solve the industrial problem. One way of solving the sulfur pollutant problem is to remove the sulfur from the fuel or process producing the stack exhaust. Where this is not possible, one method is the fume and solvent air pollution control system shown in Figure 12.27. This system is effective in meeting Office of Safety and Health Administration (OSHA) standards for combustible organic gases, fumes, and particulates such as organic solvents, phenols, aldehydes, oil mists, sulfides, thinners, mercaptans, rendering and sewage odors, and aromatic hydrocarbons. This system

Control module
Magnetic tape
Printer
Computer, mini
Wet chemistry
Auxiliary equipment
Display

Central
control

2 to 50

Remote
stations

Each functional measuring device to have an interface
with the computer to transmit and receive signals. Measuring instruments output to be

Environmental measurements: bcd compatible with a 0 to 5 V signal or to have signal conditioning
units to match the interface. Signals to be transmitted for out of
tolerance measurements, high or low, on
critical flows, temperature controls,
concentrations, and unit malfunctions on
a digital telemetry system

Wind direction: 0 to 360° from a north or south bearing
Mean wind velocity: 0 to 90 m/h in three ranges
Mean wind turbulence: azimuth vector
Relative humidity: 0 to 100%
30 meter tower temperature: ambient
Ground temperature: ambient
Barometric pressure: inches of mercury
Dew—point: −40°F to 150°F
Rainfall: inches of water
Solar temperature: visible band, total or by selected bandwidth
Solar temperature: ultraviolet
Solar temperature: infrared

Air sampling:

High volume air sampler — single filter, 50 ft³/min
High volume air sampler — fractional filters, 20 ft³/min
Radioactivity air sampler — alpha, beta, and gamma
Tape filter air sampler — 2 to 10 ft³/min

Air contaminates or pollutants:

Sulfur products	ppm	Chromatography, nondispersive infrared, wet chemistry
Nitrogen products	ppm	Chemiluminescence
Carbon monoxide	ppm	Chromatography, nondispersive infrared, optical
Methane	ppm	Chromatography, nondispersive infrared, optical
Total hydrocarbons	ppm	Chromatography, nondispersive infrared, optical
Nonmethane hydrocarbons	ppm	Chromatography, nondispersive infrared, optical
Ozone	ppm	Chemiluminescence

Figure 12.26 *Environmental and pollutant monitoring system equipment.*

529

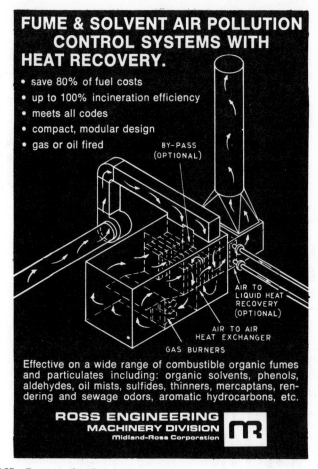

Figure 12.27 *Fume and solvent air pollution control system with heat recovery. Courtesy Ross Engineering, Machinery Division.*

draws the fumes or other combustibles through a gas or oil fired incinerator chamber maintained at a high enough temperature to burn all the pollutants to nearly 100% efficiency. Provision can be made for recovering most of the heat for a preheat section of the system in front of the combustion chamber. Temperatures range from 1100 to 1400°F.

When it is inadvisable to use the combustion system, or when the pollutants are not combustible, a scrubber/separator system can be used. Fume scrubber/separator units such as the one shown in Figure 12.28 remove up to 99% of the particulates from incinerator fumes or electro-

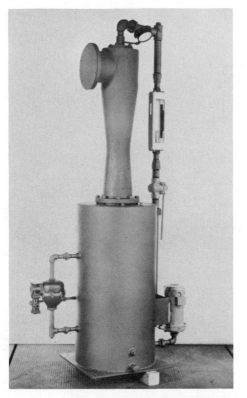

Figure 12.28 *Fume scrubber system. Courtesy Croll-Reynolds Co., Inc.*

static precipitator units containing fly ash dust and similar abrasive and corrosive constituents. The unit shown is cast of ductile iron, a tough and durable alloy which provides high abrasion, corrosion, erosion, and oxidation resistance and is stable under fluctuating temperatures up to 1100°F. Although the airstream is clean enough to exhaust into the atmosphere, the pollutant-contamination problem does not end there. The precipitated solids and the scrubber fluids, which have to be treated, are bulk requiring disposition. In some cases the heated airstream has to be cooled before it can be released. These systems are expensive to buy, operate, and maintain. Therefore it is almost mandatory that offenders be placed in a position where its more expensive to pay the fines for disregarding the law than it is to buy, operate, and maintain the proper control equipment. However, the management of a particular offending industry should know its problem and how to control it the least expensive way.

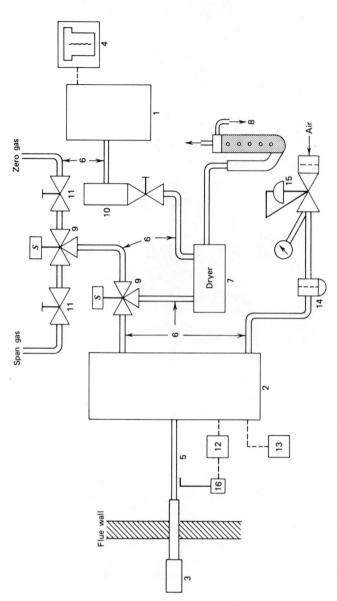

Figure 12.29 *A typical stack monitoring system. 1. Infrared analyzer or infrared/thermal conductivity analyzer. 2. Automatic blow-back gas sample conditioner. 3. Heated probe with primary conditioner. 4. Recorder. 5. Heated line and termination kit. 6. ¼ in. teflon tubing. 7. Refrigeration dryer. 8. Condensation trap, water leg. 9. Two three-way solenoid valves. 10. One flowmeter with needle valve. 11. Two needle valves. 12. One auto transformer, 120/240 V. 13. One timer for automatic blow-back. 14. One air line filter and trap. 15. One air pressure regulator. 16. One thermostat for use with #5. Courtesy Leeds and Northrup.*

The ideal way is to set up a control system which removes the pollutant or ecology problem and provides a salable product from the waste. For example, the smoke and fumes generated by a coke producing furnace provide by-products that are worth far more than the coke itself.

A nondispersive infrared stack monitoring system is shown in Figure 12.29. The sampling equipment withdraws and conditions a portion of the stack effluent and presents a clean, dry sample to the analyzer. Infrared radiation is directed onto the clean sample. A simple beam splitter produces a reference beam and a measuring beam. A differential thermopile measures the radiation traversing each beam and produces a voltage signal which represents the difference in the energy levels of the two beams. This difference signal, which can then be related to the concentration of the gas of interest in the sample, is amplified in the analyzer and transmitted to a recorder where the information is graphically displayed. This system can be made to interface with any type of data handling equipment using conventional current outputs to drive the recorder.

The optical components are housed in a temperature controlled cabinet to ensure stable performance. Units of this type are capable of handling as many as three infrared sampling cells, so that simultaneous readings can be made for three combustion gases such as sulfur dioxide, carbon dioxide, and carbon monoxide.

The ideal sampling point in a power plant for such a monitor is in the ductwork immediately following the gas cleaning equipment. In some cases this point is not readily accessible, so a compromise has to be made so that maintenance can be readily accomplished. In some cases the compromise point may be in the ductwork immediately following the induced draft fan.

Airborne dust and dirt are usually some type of abrasive material, and if in the 0.2 to 2 μm size range, particles can clog breathing passages and cause lung damage. In many cases the dust is actually part of the industrial product escaping, so it is essential that it be collected. Vacuum systems are employed to collect the airborne abrasives created by most grinding and pulverizing operations. Two of the highest dust producing industries are cement and fertilizer manufacturers. Other industries producing a high level of particulate matter are smelting, sugar refining, foundries, die casting, and textiles. Usually, dust or particulate collecting systems remove up to 99% of the harmful particles. These particulates can be transferred directly to a material storage bin as in the case of cement, ceramic clay, or fertilizer dust, or they can be collected in hoppers, such as those shown in Figure 12.30, for later disposition. In some

Figure 12.30 *Dust collecting systems. (a) Collector for a rock products plant. Courtesy Sly Manufacturing Co. (b) Torit 140-H dust collector with bagging outlets. (c) Torit PIC dust collector. Courtesy Torit Corporation.*

cases the particles are placed in bags and disposed of as solid waste; in other cases they may be pelletized for storage or disposal, or recycled as in the glass industry where soda ash and lime dusts are collected during furnace charging and mixing operations and placed in furnace charging hoppers.

The effectiveness of dust collecting systems can be measured by means of a high volume air sampler, with particulate sizing filters, to evaluate both the size and volume of the particulates. Instruments such as a nephelometer are used only to test the effectiveness of filter systems for clean room operations or for the evaluation of aerosol filters and control systems.

If any radioactivity is involved, a tape air sampler should be used and the collection spots surveyed or measured for radioactive particles with the radiation measuring instruments discussed in Chapter 10.

The size and type of control unit depend on the type and volume of pollutant or contaminate to be removed or destroyed. In most cases common parameters such as volume, flow, pressure, temperature, and weight are measured with the instruments used in the process or analytical applications discussed in the preceding chapters. The only difference that might be found is that a signal conditioner is used to convert a physical motion into an electrical or electronic signal employed in a computer monitoring or command circuit.

In the majortiy of cases in which flame photometry, chromatography, chemiluminescence, colorimetry, densitometry, ultraviolet, or infrared analytical techniques are used, there is a prescribed wet chemistry method that can be used to verify any measurement of which there is serious doubt or where there may be severe consequences from faulty unverified measurements. These measurements are slower and may take from several hours to a day to complete; they usually require the services of highly trained chemists or chemical technicians.

12.5 NOISE MEASUREMENT

Noise is generally considered sound without harmony, which is annoying and in some cases detrimental to human hearing. There is no generally accepted agreement as to exactly how much noise, what type of noise, or what length of exposure to how much of what type constitutes a health hazard. The Walsh-Healey Public Contracts Act has established a legal definition in the form of safety regulations, issued by the U.S. Department of Labor as Section 50-204.10 in 1969, for noise limits beyond which employers must take steps to protect employees. The provisions of this act now cover all industrties engaged in interstate commerce, but factually should apply to all groups generating detrimental noise. This is now OSHA Public Law 91-596 which also encourages industrial plant management to purchase noise measuring equipment for in-plant noise measurements and monitoring. It also encourages audiometer tests for workers employed in noisy work areas. The noise exposure limits recommended by OSHA are shown in Table 12.1.

When noise has various levels and frequencies during the work day, or a worker performs his duties in areas having different noise levels, the total is then considered a combined effect which can be calculated by the following equation:

$$C_T = 100 \left(\frac{C_A}{6} + \frac{C_B}{4} + \frac{C_C}{3} + \frac{C_D}{2} + \frac{C_E}{1.5} + C_F + 2C_G + 4C_H \right) \qquad (12.1)$$

where C_A = total time of the noise level of band A

C_B = total time at the noise level of band B, and so on

C_T = total combined exposure for the time period under investigation

Table 12.1 Noise Exposure Limits per 8 h Day

Band	Noise Level [dB(A)]	Limit (h)
—	Less than 90	Unlimited
A	90	8
A	90–92	6
B	92–95	4
C	95–97	3
D	97–100	2
E	100–102	1.5
F	102–105	1.0
G	105–110	0.5
H	110–115	0.25
—	Over 115	0

As shown in Table 12.1, each noise level in the range that can cause induced hearing loss is shown as an A weighted value in dB(A) because A weighted levels have been found to correlate well with hearing loss. It now appears that sound in excess of 65 dB in the 500 to 2000 Hz frequency range is the most critical to noise induced hearing loss. The limits shown in Table 12.1 are based on tests indicating that each 5 db(A) increase in sound intensity required a 50% reduction in exposure time to assure no increase in noise induced hearing loss.

The decibel term for acoustical power is $10 \log_{10} P_1/P_2$, where P_1/P_2 is the ratio of any two different acoustical powers producing the decibel value.

$$N_{dB} = 10 \log_{10} \frac{P_1}{P_2} \tag{12.2}$$

where N = number of decibels

Noise measurements can be spot checks of noisy areas, continuous measurements integrated on a per hour or per day basis, or continuously monitored on a real time basis.

Noise levels of concern may be listed as definitely dangerous to humans when they cause vibration of cranial bones, blurred vision, or weakening of body muscular structure. Some of the sources of noise considered dangerous in causing hearing loss, which generate 155 dB or above, are rifle blasts, jet engine exhausts and sirens; shotgun blasts as heard by the hunters, and drag strip motor exhausts as heard by drivers and pit crews, generate 140 db or above; jet airports, some electronic music, and pneumatic drills drilling hard rock generates 120 db and above.

Noise levels that probably cause hearing losses if endured for a suffi-cient time period are 115 to 125 db generated by such equipment as drop hammers in steel mills and chipping hammers used in casting and weld cleaning areas; 100 to 115 dB generated by such units as wood working planers and routers and sheet metal speed hammers; 95 to 100 db created by subway cars, paper making machines, and high speed weaving looms; and 90 to 95 dB emanating from screw machines, punch presses, riveters, cutoff saws, motor generator welding units, and air grinders.

Noise levels that could possibly cause damage to hearing are in the 80 to 90 dB range and are generated by spinners, looms, lathes, heavy traffic, diesel motors, or plate mills in steel making facilities. Generally, any noise of less than 90 dB is not considered for industrial facility measure-ments. Just to orient ourselves with respect to the noise levels we hear daily, a stenographic room with a group of typewriters in use generates 65 to 75 dB. Areas that are considered quiet and comfortable for low conversation are at the 45 to 50 dB level. A quiet city apartment or a level good for sleeping is in the 20 to 30 dB range, and a very quiet area where one can hear a pin drop or a leaf rustle is in the 15 dB range. Normally, the human ear does not distinguish a sound level change of less than 3 dB.

Noise Measurements.   Spot check or short duration noise level measure-ments are usually made by means of portable equipment, and the size of the equipment may range from the small hand held unit shown in Figure 12.31 to the deluxe noise level and analysis unit shown in Figure 12.32.

Where the sound level is over 40 dB and less than 140 dB, and a micro-phone is desirable for a completely hand held unit, the one shown in Figure 12.31 meets OSHA noise and U.S. Bureau of Mines gaseous mines safety requirements. It has A, B, and C weighting, all solid state cir-cuitry using FET and integrated circuit design. It has an input imped-ance of 13 MΩ at 15 pF. The circuit has a 1.2 V rms output into a 620 Ω meter movement at full scale, or equivalent impedance analyzers or re-corders for monitoring applications. This portable instrument can be calibrated at 125, 250, 500, 1000, and 2000 Hz.

For personnel monitoring, the system shown in Figure 12.33 is a com-bination of dosimeter and readout unit. The pocket noise exposure moni-tor shown in Figure 12.33a is worn by a worker, and it moves with him throughout his normal workday. The monitor uses a microphone detec-tor whose output is weighed and accumulated to obtain the total noise exposure for the workday, based on OSHA criteria. If the 115 dB level is exceeded, the exposure level and time are stored for separate readout. The monitor readout is made on the readout unit shown in Figure

Figure 12.31 *40 to 140 dB hand held portable sound level meter. Courtesy General Radio Co.*

12.33*b*. This readout unit is designed to display the accumulated percentage of noise exposure levels from the individual noise exposure monitors worn by personnel working in noisy areas. This unit registers from 1 to 999% of legal limits established by OSHA. A light indicates if the 115 dB(A) level is exceeded during the monitored workday. This monitor covers only the 90 to 115 dBA range and has an accuracy of ±1 db over a 60°C temperature span.

In addition to reading the accumulated percentage of noise a worker has experienced, the indicator also has provisions for checking the monitor battery condition, resetting the monitor to zero for the next day's use, and verifying acoustic calibration of the monitor.

The unit shown in Figure 12.32 covers the 24 to 150 dB measurement range, and responds to frequencies from 20 Hz to 20 kHz. The indicating meter is calibrated for a 16 dB span from −6 to 10 dB. This meter is used with an attenuator calibrated in 10 dB steps from 30 to 140 dB above 20 $\mu N/m^2$. The instrument output is 1.4 V behind a 7000 Ω meter movement at full scale. The input impedance is 25 MΩ with 50 pF of capacitance.

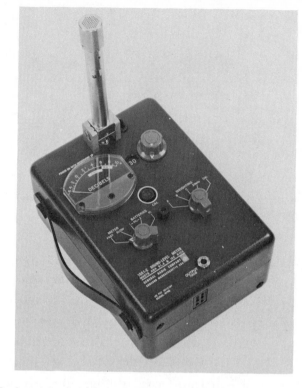

Figure 12.32 *Sound level meter. Courtesy General Radio Co.*

This sound level meter can be used alone or as a detector and pre-amplifier for spectrum analyzers. Its output can also be used to drive headphones, oscilloscopes, graphic level recorders, and magnetic tape recorders. It can operate in ambient temperatures from 0 to 60°C and at relative humidities from 0 to 90%.

When noise is generated by impact, the noise signal can be best measured by means of an impact noise analyzer which can measure electrical and acoustical noise peaks. The two most significant characteristics of impact sound appear to be the peak amplitude and the duration, or decay time. The instrument shown in Figure 12.34 measures the maximum peak value reached by the noise, a continuous indicating measure of the high levels reached just before the time of indication and a time average measure of the average noise level over a predetermined period of time. This unit measures and analyzes the output of the meter shown in Figure 12.32.

Figure 12.33 *Noise dosimeter system. (a) Personnel pocket meter. (b) Noise exposure indicator. Courtesy General Radio Co.*

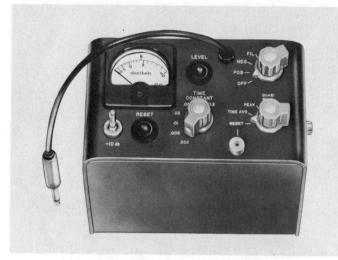

Figure 12.34   *Impact noise analyzer. Courtesy General Radio Co.*

In the measurement of noise generated by a product or facility, three properties can be detected, measured, and analyzed. They are amplitude, frequency, and time distribution. Of coures it is always ideal to stop the noise at its source but, if it cannot be stopped, it must be controlled or confined if possible. A sound level meter is always a good first choice, but its primary function is to measure amplitude. However, the output of the sound level meter can be fed into an analyzer unit if a compatible sound level meter is chosen. The sound level meter must have the right output to feed into a standard octave band analyzer unit with center frequencies from 31.5 Hz to 16 kHz.

Mil Standard 740B requires continuously timed one-third and one-tenth octave analysis features on product noise analyzers. Most product noise problems require analysis in one-third octave bands. Units are now available that provide fully automatic stepped one-third octave analysis while minimizing test time. Some units offer real time analysis by providing a complete, new, one-third octave spectrum analysis as often as every $\frac{1}{8}$ s. These new units use a digital detection scheme which provides true rms measurements of band levels, wide dynamic range, and eliminates the spectrum smearing problems associated with the usual analog approaches.

A new generation of real time analysis equipment has been made possible by the development of fast Fourier transfer algorithms. This development allows the spectrum to be divided into as many as 8192 segments which can be updated every 12.2 ms (for 1024 lines). The design of such analyzers embodies arithmetic operations to perform amplitude, time, and frequency analysis. Trade-offs can be made among several versions for processing speeds versus cost. In general, the faster the processing speed the greater the cost.

For the ultimate in acoustical analysis in the narrowest frequency bands to establish the exact frequency content of an acoustical signal, use is made of wave analyzers with a 1% constant percentage bandwidth. This type of equipment is needed only where closely spaced low frequencies are combined with high frequencies and both low frequency and high frequency resolutions are required. Some of these units can identify up to the fiftieth harmonic. Such a system consists of a wave analyzer, a graphic recorder, attenuating potentiometers, and drive units to provide a synchronized automatic spectrum change over the range of the attenuator and frequency band.

In the analysis of a noise problem, one should not use more equipment than is absolutely necessary, and the choice should be based on the need to solve an existing problem or to prevent one from developing. Where noise creates a safety hazard, or is loud enough to be detrimental to human hearing, the noise generating device should be isolated by sound absorbing covers, mounting pads, or enclosures. If the noise maker cannot be made to be self-contained, personnel working on the equipment and in the immediate vicinity must be fitted with earplugs or special headphones or earmuffs to minimize noise induced hearing losses.

Prime areas for analyzing sounds that are known or suspected causes of hearing loss should be continuously monitered so that tolerance levels can be established. Then the number of hours exposure without danger of causing hearing loss, and the precautions to be practiced in the event this duration is exceeded, can be posted. Monitored values are normally recorded to provide legal evidence or proof that levels may or may not exceed established tolerances, and are used to set up exposure limits for areas that cannot be otherwise controlled.

12.6 LIQUID POLLUTION MEASUREMENTS AND CONTROL

Water is one of the prime liquid requirements for sustaining all types of life, so we have to be concerned about its availability, both in quality and quantity. This concern then leads us to examine closely the bodies of water from which we draw our supplies and those into which we dump our excess effluents. These excess fluids may be chemically clean and pure legally allowable effluents, or contaminated with pollutants that exceed lawful tolerance limits.

If the body of water from which drinking water is drawn is not pure and clean, it is necessary to purify the water of microorganisms, chemical and particulate pollutants and contaminants before it can be used. This purification is accomplished by filtration and chemical treatment. Filtration may be hastened by the use of coagulants such as alum to remove the large majority of suspended particles. In some cases the water may have to have most of the dissolved minerals removed and the hydrogen ion concentration (pH) adjusted so the solution is nearly neutral (pH value of 7.0, as discussed in Chapter 9). Another method of measuring the amount of dissolved matter is by the electrical conduction technique. A pure, double-distilled sample of water has a practically infinite resistance to current flow, meaning that the liquid allows very little current to pass through it. Such measurements are usually in megohm or higher resistance values. This means that there are few ions free to carry current through the liquid, and the liquid is considered clean. However, if the electrical conduction is high, it is certain that a large number of ions is present to carry the current, and the liquid may very well be contaminated with pollutants.

In the majority of large water purification systems there is a chemical treatment station such as the one shown in Figure 12.35 where chlorine is automatically fed into the water stream to kill bacteria, viruses, and microorganisms that could cause human disease. Many times the water is also treated with fluorides to aid in the reduction of tooth decay.

Water treatment and storage is a big business, and may be either a private enterprise or a municipal government function. In either case the water has to be collected and stored in large reservoirs which may cover acres of land. In some cases the surface of the water is used for recreational purposes, so long as no wastes are permitted in the watershed area. This water requires final treatment before release for use. Large storage areas are needed to provide an adequate supply in case of drought or large fire emergencies. This means that pipeline sizes are important to provide adequate flows. These flows and reservoir levels have to be measured, so that the head depletion is known and the volume can be replaced by pumping or diverting stream flows from auxiliary water sources.

The contents of any body of water that supports plant and animal life, such as a lake, river, or feeder stream have to be sampled and analyzed for pollutants that are detrimental. Measuring pH values indicates if the

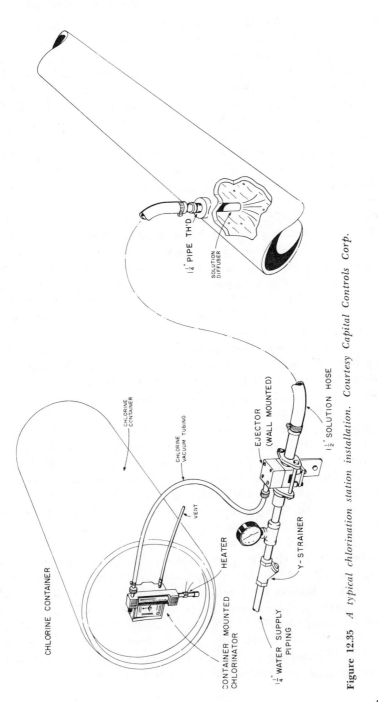

1¼" PIPE TH'D

SOLUTION DIFFUSER

CHLORINE CONTAINER

CHLORINE VACUUM TUBING

VENT

HEATER

CONTAINER MOUNTED CHLORINATOR

CHLORINE CONTAINER

EJECTOR (WALL MOUNTED)

1½" SOLUTION HOSE

Y-STRAINER

1¼" WATER SUPPLY PIPING

Figure 12.35 *A typical chlorination station installation. Courtesy Capital Controls Corp.*

sample is acidic, neutral, or alkaline. Such a measurement in itself does not indicate what type of acid or alkali is present. The sample then has to be analyzed to determine if the acidity is due to highly ionized strong acids such as nitric, hydrochloric, and sulfuric acids; poorly ionized weak acids such as acetic, carbonic, and tannic acids; or hydrolyzing metallic salts such as sulfates or chlorides of iron and aluminum. Alkaline solutions must be analyzed to determine if the alkalinity is due to strong alkalies such as sodium and potatssium hydroxides or to weak alkalies such as carbonates and bicarbonates. Alkalinity may also be caused by hydrolyzed metallic salts such as sodium acetate and sodium cyanide. The different compounds mentioned as acids or alkalies can be identified by chemical reactions and spectrometric techniques. Some can be identified by colorimetric techniques, and others by ion exchange or deposition.

Water samples may also yield biodegradable pollutants, and in such cases the oxygen demand of the sample is usually studied for a 5 day period to determine the biochemical oxygen demand (BOD). The demand is measured in milligrams of oxygen per liter of sample required to support the bacteria in breaking down the material into simpler substances such as carbon dioxide and water. These simpler substances are easier to change into nonpollutants.

Water or other fluids being discharged from a manufacturing facility or process can, and by law must, be analyzed and purified before being allowed to enter any feeder stream in a watershed area.

Most industries that create fluid pollutants know what type they are creating and in general do only as much in the way of control as the law requires and, where there were no laws, there were few if any controls. In the past 5 or 6 years this situation has changed drastically, and today industries show what they are doing for improved ecology to improve their corporate image. Offenders are also being fined heavily, so that it is becoming mandatory to stop polluting in order to stay in business.

If liquid waste has to be stored for long periods of time, it can be retained in sealed pits, lagoons, or other lined structures where the contents can settle, be leached, aerated, agitated, or just stored for future treatment. These structures are lined with leakproof plastics so that the liquid cannot soak into the ground and contaminate it. After treatment has been completed, the liquid can be pumped or drained off, and the solid, slurry, or other debris can be incinerated, compacted, buried, or used as sanitary landfill. For example, if the solution is acidic and an alkali is added to neutralize the acid and form a neutral salt, the neutral salt may be dried for use in some other process, slurried or mixed with other material, or incinerated to form carbon dioxide, water, and mineral ash. The liquid has to be analyzed for clarity, pH, and biological organisms before it can be released.

In the case of oil and grease wastes, many cities have problems created by automotive service shops and service stations. When some service stations change crankcase oil, the oil is supposed to be collected, but in some cases it is drained into a service drain and eventually finds its way into a storm sewer or sewage sewer line. From then on it is part of the city sewage problem. The second and most common problem is generated by the gunk material that is used to clean the greasy, oily floor in the grease pit and the repair area. This material makes greases and oils water soluble and high potency detergents are employed to complete the cleaning job. The oil and grease are now emulsified with a phosphate that encourages heavy organic plant growth in any stream into which it might be released. Service stations are supposed to dump the used oil into recovery tanks so that at least a portion of it can be recovered and used for other lubricating applications such as glazing oil for dies used in the extrusion of clay products to prevent tears on square corners. It may be necessary to incinerate the residue portion, or it may become part of an asphalt or tar product.

Liquid waste products from chemical processes and scrubbers used to remove heavy airborne fumes and odors must be analyzed and treated before they are released. It is not improbable that the liquids are highly corrosive and require extensive and expensive treatment before they can be safely released. In such cases the liquids are collected in corrosion resistant lined vessels and transported to a remote treatment area for leaching, filtering, and dilution so that the prescribed amount can be legally released into a feeder stream in a watershed area. It is quite possible that other pollutants are absorbed by rainfall in the area and that the total stream pollutant level can be exceeded, so it is necessary to maintain constant monitoring of the stream. Such measurements are the basis for controlling the fluid release from the treatment station.

Perhaps one of the most prevalent liquid pollutant problems arises from improper sewage treatment and disposal. Raw sewage has been dumped into streams and delta areas for years. Sewage problems involve both liquid and solid wastes. Untreated sewage presents problems involving decaying organic matter, biodegradable materials, microorganisms, and disease generating viruses.

Where possible, municipalities should maintain separate sanitary and storm sewer systems. However, there are too few separate systems in use today. Where separate systems are in use, the storm sewer effluent receives little or no treatment before release into watershed streams, rivers, or other bodies of water. Quite often storm sewer wastewater is used to dilute the liquid effluent released from an Imhoff tank or other sewage treatment facility. An Imhoff tank is a facility for treatment of sanitary sewage and biodegradable waste solids. Liquid is separated from suspended solids in the top portion of the tank. Solid wastes enter the bottom portion of the tank by gravity, where anaerobic digestion takes place.

In conventional sewage treatment plants, the primary treatment is basically a separation operation involving a sedimentation step in which the solids settle out of the waste fluid and the liquid effluent is treated for removal of microorganisms and biodegradable pollutants. A secondary treatment removes the organic wastes by filtration or by aeration. Aeration cultivates bacteria which consume the pollutants. At the end of such treatment, the effluent is about 90% pure and requires a tertiary treatment for obtaining what is considered safe drinking water. The tertiary system or step removes microscopic contaminates, usually with activated carbon. The tertiary treated effluent is generally fit for drinking after chlorine treatment.

One of the latest innovations in the treatment of liquid effluents to obtain the highest purity is to treat the liquid with ozone in the presence of ultrasonic vibrations. There are still some unknowns in this latest technique of treating sewage, but the EPA is backing studies to determine if there are any major problems associated with ozonized materials. At least one company is now constructing a unit which is envisioned to be effective as a combined primary, secondary, and tertiary system. Such a system is considered no more expensive by the developing firm than present secondary systems. EPA contractors who have built and operated pilot plants feel that the effluent is as clean as can be obtained short of distillation.

In effect, ultrasonics are used to break up pollutants into very small sizes that are readily acted upon by ozone to form harmless oxides. Ozone is a highly effective oxidant which produces carbon dioxide and water as end products in the tertiary treatment system.

12.7 SOLID WASTES AND POLLUTANTS

Solid wastes come in all varieties and sizes. They can be small enough and light enough to be airborne for miles, or they may be heavy and drop out very soon after they are generated, as discussed for cement, glass, clay, and other types of grinding operations producing pollutants that may or may not be considered waste. However, some articles considered waste but not necessarily contaminants are bottles, cans, scraps of wood, abandoned cars, tires, paper, and leaves.

Solid or semisolid wastes from sewage treatment plants can be treated so that they are not pollutants, but if not properly treated and controlled they may present health hazards, especially if they are dumped in open landfill areas and not covered, or at sea where marine life may become contaminated by consuming any biodegradables that were not completely consumed.

Garbage is another waste product that can become a pollutant if not properly handled and treated. In some cases it is finely ground and treated much the same as sanitary sewage solids. In other cases it is compacted to remove the majority of the liquid content, and the solid is incinerated. In other cases it is finely ground and taken out to sea and dumped. If stored for too long or dumped too close to shore, it will become a pollutant. Under proper conditions clean garbage wastes can be dumped at sea for consumption by marine animals, or can be used as forage by swine and other farm animals.

Solid waste materials from slaughterhouses, stockyards, and poultry farms create odor problems as well as disposal problems. In general all three sources of waste end up in different types of fertilizers which are deodorized, pulverized, and pelletized for spreader operations and nitrogen release control.

With conglomerate wastes it is customary to separate reusable materials such as metals before incineration or burial of the major mass. Incinerator ash also must be disposed of as a type of sanitary landfill. Many communities use old mining pits, abandoned quarries, ravines, or other excavations as landfill areas. Stringent control has to be exercised in such areas so that they do not become infested with rodents and have the general appearance of filth and disorder.

Compaction of trash and household wastes is a growing trend. Individual home compactors are available, and most refuse collectors have compactor trucks which reduce the volume by at least 20 to 1. The general idea is to make the volume as small and as dense as possible, so that storage space is not wasted when the mass has to be stored. This factor is becoming more important, because space is becoming a scarce commodity in the vicinity of the biggest producers of waste materials. This type problem is not limited to the United States, it is worldwide. Some large cities are already leasing landfill acreage and transporting both sanitary and garbage wastes to the fill site. These operations are expensive and are likely to become more expensive as the energy crisis drives prices upward.

Whether the waste collection is for a small or large area, many handling and disposal problems are similar and call for stringent control. There is no one answer to satisfy all the requirements, mainly because there is no acceptable answer as to what constitutes a totally realistic requirement.

Some waste materials are very hard to reduce to reclaimable or compactable solids. Automobile and truck tires are an example, and they are banned from landfills. Several types of reducing and disposal units are now available, and one such unit is shown in Figure 12.36. In this system tires are reduced to ashes and gases in a furnace. The ashes drop into the ash pit and the gases are scrubbed; a portion of the heat from the discharge gases is recovered in a heat recovery boiler and used in the furnace heating cycle. In the scrubber system airborne particulates are washed out, and gases pass through a desulfurizing unit for removal of sulfur dioxide. The scrubbed and desulfurized gases are then discharged

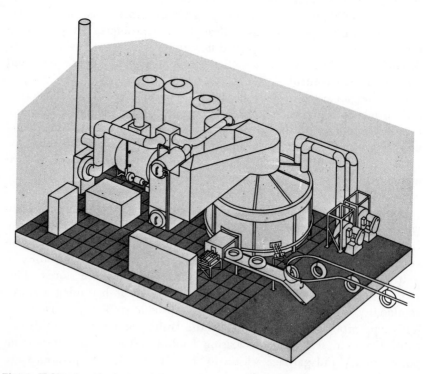

Figure 12.36 *A cyclonic waste furnace for the disposal of tire carcasses. Courtesy Flour Utah, Inc.*

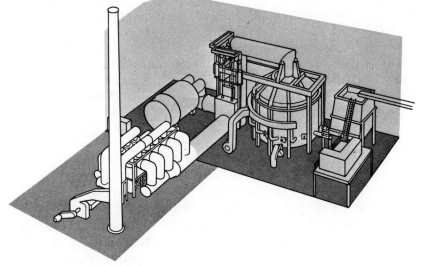

Figure 12.37 *A cyclonic furnace for the incineration of sludge waste materials. Courtesy Fluor Utah, Inc.*

through the stack. Automotive bodies, engines, and frames, minus the tires, can be compacted and incinerated to remove plastic, leather, and rubber parts so that the scrap metal can be recycled by the iron and steel industries. A sewage sludge incinerator for general sludge disposal is shown in Figure 12.37. In this system the sludge is fed into a premix and storage hopper from which a feeder system regulates the sludge flow into the furnace. The sludge is incinerated, and the ash portion falls into a waterseal trap and is dredged away. Any airborne particulates and combustion gases are passed out through heat exchangers and wet scrubbers. The particulates are collected in the base of the scrubbers, and the gases are passed out the stack.

Pollution and waste controls are only as successful as people make them. There is no surer way for an excellent control program to fail than lack of interest in its success. At present, public opinion has made this an area of immediate attention. Energy shortages will provide some curb on the effort, and continued lobbying by opponents of the effort may slow implementation, but long range controls are needed and must be supplied for good ecology practices.

Under present conditions our wastes are either burned into nonpollutant substances, buried, or just dumped into landfill areas, rivers, lakes, or oceans. The rules of disposal are becoming more strict, so it is mandatory to measure the wastes we dump in our environment and keep the concentrations below what is now considered a tolerable level, or public opinion will force the establishment of such stringent controls that industries will either break the rules or go out of business. We need a means of establishing controls that will safeguard human and plant life without strangling the industries we need to make living worthwhile.

12.8 SUMMARY

Each pollution problem requires an individual solution, at least, until some realistic controls are established by a nonpartisan group to cover the specific types of pollution. The established levels should be stated in concentration levels on a basis that can be understood and measured. Standards must be established, and they must be reliable and usable, at least by competent technical personnel. While we have progressed in cleaning up our environment, we have only scratched the surface in reliable and continuous measurements. The technology is still quite new, and we rely heavily on laboratory instrumentation to collect data for evaluation. Since no other reliable equipment is available, engineers should develop more definitive and reliable instrumentation that can be used to detect the generation of pollutants or to remove them from the air and water we need for daily living.

Bibliography

Capital Controls Corporation, Division of DART Industries, Inc., *Chlorination,* Bulletin 922, Pub. No. 572-3, Colmar, Pa.

Fluor Utah, Inc. A Subsidiary of Fluor Corporation, *A Revolution in Sludge Disposal,* San Mateo, Calif., 1973.

Fluor Utah, Inc., A Subsidiary of Fluor Corporation, *A Revolution in Tire Disposal,* San Mateo, Calif., 1973

General Radio Company, *General Radio Catalog,* 73, West Concord, Mass.

IGY Instruction Manual, Part VI, Radiation Instruments and Measurements, Pergamon Press, no date.

Leeds and Northrup Company, *Stationary Source Monitoring for SO_2 on Fossil-Fuel-Fired Combustion Process,* L & N Application Bulletin E1.1301-AB 80-672, North Wales, Pa.

Meteorology Research, Inc., A Subsidiary of Cohu, Inc., *Fog Visiometer,* MRM Ca 77/82, Altadena, Calif.

Meteorology Research, Inc., A Subsidiary of Cohu, Inc., *Integrating Nephelometer,* MRM Ca 81/22, Altadena, Calif.

Meteorology Research, Inc., A Subsidiary of Cohu, *Time Lapse Photography*, MRM Ca 61/112, Altadena, Calif.

Peerless Instrument Company, *Two-Gas Analyzers 209/210*, Elmhurst, N.J., 1972.

Precision Scientific, A Subsidiary of GCA Corporation, *Aeron Air Quality Instruments*, Bulletin 680A, Chicago, May 1970.

Telecommunications Industries, Inc., *Is Ozone the Way to Treat Sewage?*, TII Ecology Division, Sunnyvale, Calif., 1972.

Weather Measure Corporation, *Weather Measure Instruments*, Catalog 1272, Sacramento, Calif.

R. M. Young Company, Product Bulletin, Traverse City, Mich., November 1972.

Index

Accelerometer characteristics, capacitance, 359
 charge sensitivity, 358
 construction, 348
 cross axis sensitivity, 358
 directional sensitivity, 353
 frequency response, 358
 natural frequency, 359
 phase shift, 359
 temperature response, 359
 voltage sensitivity, 358
Accelerometer forms, 348
Accelerometer mounting techniques, 352
Accelerometer shapes, 353
Accelerometer types, bender, dual output, 349
 compression, 348
 linear, 347
 NRL, 349
 piezoelectric, 346
 seismic, 347
Accuracy in measurement, 291
Acoustical signature, 494
Aeron type air filter, 526
Air control, 527
Airborne particulates, 523
Air filters, 526
Air measurements, 528
Air pollutants, aerosols, 510
 gases, 510
 mixtures, 510
 nitrogen, 513
 ozone, 512
 particulates, 510
 sulfur, 511
 sulfur products, 511
 vapors, 511
Air sampler, high volume, 275
Alpha particulates, 404
Ambient temperature, 502, 509
Amplifiers, direct current (DC), 427

linear, 427
 logarithmic, 426
 negative pulse, 425
 positive pulses, 424
 transistor type, 428
 tube type, 425
Analytical instruments, 323
Anemometers, UVW, 503
Annubar flow element theory, 253
Appliance industry, 20
Atomic mass spectrometers, 527
Automatic control, basic characteristics, 307
 floating, 307
 proportional, 307
 proportional plus reset, 307
 rate, 307
 two position, 307
Automatic measurement, anticipatory, 290
 basic characteristics, 291
 feed forward, 290
 feedback, 290
 theory, 290
Automotive industry, 15

Backscattered radioactivity, 444
Bailey meter system, 71
Barometer, 65, 510
Barrier photovoltaic cells, 178
Beer's law, 374, 376
Bell gages, balanced lever gage, 66
 beam gage, 66
 spring balance, 68, 69
Bellows, 76
Bennoulli's theorem, 240
Beta particles, 404
Bin-O-Matic Level Gages, 223
Black body, 172, 173, 177, 190
Bloom mill, 9
Bolometer, 174
Boron trifluoride gas, 414

Bourdon Tube, 81, 123, 124
  C-type, 82
  Helix, 82
  spiral, 82
Bridge, Wheatstone, 48
Brightness temperature, 190
Bulb thermometer responses, 125
Butt weld, thermocouples, 157, 160

Calibration, parallel, 336
Calibration standards, 2, 28
  Deadweight, 100
  LVDT, 370
  manometers, 99
  pressure gages, 99
  radiation pyrometers, 186
  radioactive sources, 49
  standard test gages, 101
  temperature, 31
  ultrasonics, 477
  vacuum, 101
  x ray, 463
Cam, 84
Carbon monoxide, 518
Case compensation and capillary, 124, 125
Catalytic oxidizer, 521
Cells, barrier, 178
Cells, photovoltaic, 178
Ceramics industry, 2, 3
Chamber, ionization, 406, 408
Charge amplifier systems, 355
Charge sensitivity, 358
Charles's law, 123
Chromatography definitions, 396, 397
Chromatography system, 396, 397
Chromatography theory, 395
Chromatography types of packing, 395
Chemical industry, 10
Cipolletti notch, 282
Circuit analysis, ac, 332
  bridge, 331
  dc, 331
  dynamic, 336
  full bridge, 332
  half bridge, 332
  multiple bridge, 332
  R_{cal} values, 340
  strain gage, 331
  Wheatstone, 331
  with load, 341

  without load, 341
Circular magnetism, 458
Colormetric detection, acid, 513
Compensation, reference junction, 168
Computer control system, business area, 316
  design, development, research, 316
  On-line process, 316
  time sharing, 317
Conductance probes, 221
Conductivity, eddy current, 477
Contact meter, 164
Control, floating, 309
  proportional, 310
  proportional-plus-rate, 314
  proportional-plus-reset, 312
  proportional-plus-reset-plus-rate, 315
  rate action, 313
Control function, floating, 309
  proportional, 310
  two position, 308
Count rate meters, 433
Counters, binary, 428
  boron trifluoride, 415
  decimal, 431
  electronic, 428
  Geiger-Mueller, 417
  neutron, 415
  scintillation, 417
Cross axis sensitivity, 358
Curie point, 351
Cyclonic furnace, 550, 551

Dead time, 23, 26, 293
Dead zone, 293
Decibel (db), 537
Densitometer, 407
Density, depth probes, 447
Density measurement, radioactivity,
    backscatter technique, 444
  moisture interference, 444
  system concept, 444
Density surface probe, 446
Detecting junction, 151, 152
Detector, ionization, 411
  neutron, 415
Dew point sensor, 394
Dew point techniques, 504
Differential magnetic reactor analyzer, 483
Dust collecting, 535

Dye penetrants, 459
Dynamic error, 291
Dynamic response, First order type
    instruments, 296
  Second order type instruments, 300

Eddy current material properties, ratios,
    477
Eddy current testing, differentially, 483
Eddy current testing theory, 477
Eddy current test setups, 479
Electrical industry, 12
Electrical signature, 494
Electrodes, pH, antimony, 383
  glass, 381
  hydrogen, 383
  quinhydrone, 383
Electromagnetic flowmeter, 269
Electronic counters, 428
Electronic thermometers, 143
Elution, 396
Emission spectra, 377
Emissivity, 173, 190, 191
End device, 1
Environmental concentrations, 502
Environmental protection agency, EPA,
    502
Eppley ultraviolet radiometer, 507
Error, dynamic, 291
  random, 23, 24, 25
  static, 291
  systematic, 25
Extension wire, thermocouple, 151

Factor, gage, 324
  transfer, 325
Film, photographic, 405
  x-ray, 461
Flow coefficient, 248
Flow meters, head, 240
Flow meters, diaphragm, 279
  differential pressure, 254
  electrical, 256
  electromagnetic, 267
  electronic integration, 259
  fixed volume, 269
  Mass, 269
  mechanical, 255
  nutating piston pumps, 275
  Open channel, 281

  peristaltic, 279
  piston pumps, 275
  positive displacement, 274
  rotary pumps, 278
  Turbine, 272
  variable area, 261
Flow nozzle, 243
Flow pressure transducers, 258
Flow rates, 248
Flumes, Capacities, 286
  Parshall flume dimensions, 285
Flumes, types, 283
Fluorescent spectrometers, 377
Folkers bond tester, 497
Food processing, 11
Four lead method, 133
  differential system, 136
Fractionating tower, 13
Furnace, blast, 7
  Open hearth, 9
  refractory, 4

Gage factor, metal strain, 324
  semiconductor strain, 325
Gages, balanced lever, 66
  beam, 67
  bell, double, 69
  bell, single, 66
  bellows, 76
  bourdon tube, 81
  deadweight, 38, 39
  diaphragm, 74, 78, 79
  draft, 63
  Dubrovin, 70
  elastic membrane, 76
  electromechanical, 85
  inclined manometer, 63
  U-type manometer, 61
  well manometer, 61
Gamma backscattering, 444
Gamma rays, 405
Gamma sources, 466
Gas amplification, 410
Gas expansion factor, 248
Gas thermometer calibration, 125
Gear sector pinion, 83
Geiger counter, 404
  end window, 417
  side window, 417
Geiger Mueller tube, 417

Generators, frequency, 55
  pressure, 53
  radiation, 55
  temperature, 31, 53
  time, 54
  vacuum, 54
Griess-Lyshkow reagent, 513

Hall detectors, 482, 485
Head flowmeters, 240
Helicoid movement, 84
High volume air samplers, 275
Holography, 343
Homogeneity, eddy current, 477
Honeywell Radiamatic low range unit, 184
Honeywell Radiamatic standard, 178
Hot junction, 151
Hydrocarbons, 518
Hydrogen flame ionization detector, 519

Ice point, 151
Imhoff tank, 548
Impact noise, 542
Indicators, diaphragm, 220
  electrical, 221
  liquid level manometer, 216
Infrared detectors, 375
Infrared NDT tool, 487
Infrared pyrometry, 174
Infrared spectrometry, 374
Insert venturi tube, 243
Instruments, analytical, 323
Ionization chambers, 406
Iron manufacturing, 6

Kiln, continuous, 3

Lamb waves, 470
Lambert's cosine law, 505
Laplace transforms, 304
Lead sulfide detectors, 174
Ledoux bell meter, 256
Leeds and Northrup rayotube, 178
Level measuring systems, ball float
    mechanism, 211
  bubbler system, 216
  cage floats, 215
  capacitance type, 228
  conductance type, 225
  displacement float, 213

  float, 209
  magnetic ball float, 214
  nuclear type, 231
  paddle type, 223
  pressure drop, 215
  rod type, 207
  sight glass, 208
  sticks, 207
  ultrasonic type, 229
  weighing, 232
Level system comparisons, 235
Linear variable differential transformer,
    LVDT, 359
Liquid pollution, 543
Liquid waste, 547

Magnetic comparison testing, 487
Magnetic particle testing, 457
Magnetic Reaction Analyzer, 481
Manometers, inclined, 63
  mercury, 64
  plastic, 66
  ring balance, 65
  U-tube, 36, 53, 62, 63
  well type, 36, 62, 63
Mass flowmeter, axial flow, 271
  operation, 269
McLeod gage, 92
Measurement, radiation, thermal, 172
  thermal radiation, 172
Mechanical level instruments, 207
Mechanical signature, 494
Millivoltmeter pyrometer, 157, 158, 160
Moisture measurement, nonradiation, 393
  gamma ray, 449
  neutrons, 449

Nephelometer, integrating, 525
Neutron, 405
Neutron radiography, 468
Neutron sources, 468
Newtonian fluid, 390
Nitric acid, colormetric detection, 513
Nitrogen compounds, 513
Noise amplitude, 536, 537
Noise distortion with time, 536
Noise exposure limits, 537
Noise frequency, 537
Noise levels, 537
Noise measurement, 536

Nondestructive testing, 457
Nondestructuve tool, microwaves, 489
Non-dispersive infrared detector, 515
Non-dispersive ultraviolet detector, 515
Notch, Cipolletti, 282
  rectangular, 282
  trapezoid, 282
  V, 282
Nuclear reactors, 13
Null balance, 168

Offset error, 330
Ohmmeter, 48
Open channel flowmeters, flumes, 283
  Weirs, 281
Open hearth, 8
Optical pyrometry, 192
Orifice plates, Conventional, 244
  eccentric, 243
  segmented, 245
Orifice plates, definition, 241, 243
OSHA, 536

Paper manufacturing, 11
Particles, alpha, 404
  beta, 404
Peerless Instrument Co., 515, 516
Peltier emf, 152, 153
Penetrameters, 463, 465, 466
Permeability, eddy current, 477
Petroleum industry, 10
pH definition, 377, 378
pH, antimony electrodes, 378
  chemical indicators, 379
  glass electrodes, 381
  hydrogen electrodes, 383
  Measuring cells, 381
  measuring system, 384
  meters, calibration, 385
  meters, direct reading, 384
  meters, null type, 384
  quinhydrone, electrodes, 383
  reference cells, 379
  salt bridges, 380
Piezoelectric materials, 346
Pipe taps, 245
Pitot tube, 250
Photocells, 178
Photographic film, 405
Photomultipliers, tube, 419

Photoplastics, strain, 344
  stress, 344
Phototubes, 178
Photovoltaic cells, 178
Poiseuille's law, 387
Positive displacement flowmeters, 274
Potentiometer, 47
Potentiometric pyrometers, 165
Power supplies, constant current, 439
  constant voltage, 436
  fullwave rectification, 438
  halfwave rectification, 408
Preamplifiers, linear, 424
  trigger, 424
Pressure gages, barometer, 64
  bellows, 76
  draft, 63
  inclined manometer, 63
  mercury manometer, 64
  ring balance, 65
  strain gage, 88
  U-type liquid column, 61
  well type liquid column, 62
Pressure gage calibration, 99
Pressure taps, 245
Process characteristics, basic, 301
  capacitance, 302
  capacity, 302
  load changes, 301
  process lag, 301, 302
Proportional counters, 412
Psychrometer, 393
Pulp preparation, 11
Pump type flowmeters, Diaphragm, 279
  nutating piston pumps, 275
  peristaltic, 279
  piston pumps, 275
  rotary, 278
Pyranometer, 508
Pyrgometer, 507, 508
Pyroheliometer, 507, 508
Pyrometer range, 153, 167, 168
Pyrometer sensors, 203
Pyrometer span, 167, 168
Pyrometers, industrial thermal radiation,
    173
  infrared, 174
  Leeds and Northrup automatic, 200
  Leeds and Northrup optical, 196
  millivoltmeter, 157

optical, 192
Pyro microoptical, 195
Pyro optical, 193
Radiation, 171
Pyrometry, 151
    millivoltmeter, 157
    optical, 192
    thermal radiation, 177
    two-color, 189

Radiation liquid level measurement, 451
Radiation pyrometers, 171
    applications, 189
    calibration, 186
Radiographs, 465
Radiometer, infrared, 507
Rain gage, 509
Rate meter, linear, 435
    logarithmic, 435
Ratio of size, diameter, mass or thickness
    to eddy current, 477
Rayotube, response curve, 183
    temperature range, 183
    time constant, 182
Rays, gamma, 404
Reactor analyzer, absolute magnetic, 483
    differential magnetic, 483
    effective permeability plane, 485
    impedance plane, 485
    magnetic reaction, 483
    magnetic vector plane, 485
    voltage plane, 485
Reactor radiation industries, 444
Reactors, nuclear, 13
Readout instruments, 424
Recorders, liquid level, 220
    diaphragm, 220
    electrical, 221
Rectangular notch, 282
Reference junction, 151, 152, 168
Relative humidity, 374, 504, 506
Reproducibility, 292
Response, First order system, 297
    frequency, 26
    Second order system, 306
Resistance rods, 222
Reversible motor, 223
Reversible motor level system, 223
Reynolds number, 248, 249, 390
Rheometer, 388

Roentgen, 414
Roller, 84
Ruska air piston, 71

Schott filters, 507
Schott WG7 optical glass, 507
Scrubber, fume, 530
Seebeck, 152
Sensitivity, cross axis, 354
    directional, 353
Sector gear, 83
Semiconductor detectors, 422
Sewage treatment, 547
Sewage treatment plants, 548
Sight glasses, 208
Signature analysis, 493
Snell's law, 471
Snubbers, 85, 87
Solar energy, 505
Solid level weighing, 232
Solid waste, 548-549
Solid waste disposal, 550
Sound level, 537, 538
Source follower, 355
Spectrometer, atomic mass, 527
    emission, 377
    flourescent, 377
    infrared, 374
    ultraviolet, 376
    visible light, 373
Spectrometer, Operating principles, 373
Spectrometer schematic, 272
Spectrometry, absorption, 370
    transmittance, 373
Spectrum analyzer, 527
Speed of response, 294
    Delay in response, 294
    dynamic response, 298
    lag, 294
    mass of system, 295
    temperature, 296
Stack monitoring system, 532
Standards, ac current, 46
    ac voltage, 48
    ac and dc voltage, 46
    dc current, 44
    dc voltage, 44
    deadweight gages, 38
    decade resistors, 50
    flow, 40

frequency, 51, 55, 56
manometers, 35
power, 46
precision resistance, 49
pressure, 32
primary, 29
radiation, 49
secondary, 29
speed, 49
temperature, 31
time, 42
velocity, 49
weight, 41
Standardizing, 108
Standing wave, 490
Standing wave ratio, 490
Static error, 292
Steel manufacturing, 6
Stem correction, 30
Step analysis, single-time constant, 305
   two-time constant, 306
Step change, 292
Stephen-Boltzmann law, 173, 177, 488
Strain gage characteristics, calibration, 334
   circuits, 331
   construction, 327, 329
   dummy, 330, 333
   forms, 326
   gage factors, 325
   metal, 324
   offset error, 330
   resistivity, 325
   semiconductor, 325
   sensitivity, 335
   sizes, 334
   transfer factor, 324
Strain gage pressure measurements, 88
Strain gage types, biaxial, 334
   rosettes, 339
   triaxial, 334
   uniaxial, 334
Stray magnetic field effects, 368
Strobotac, 49, 52

Tachometer, 51
Taps, pipe, 245
Taut band meter, 84
Temperature scales, 109
Thermister, 137
   applications, 145

infrared detectors, 175
typical shapes, 141
Thermocouples, chromel-alumel, Type K,
   154
copper-constantan, Type T, 154
high speed, 157, 159
iron constantan, Type J, 154
platinum-platinum-rhodium, 155
type R, 155
type S, 155
Thermocouple, arrangements, 161, 162
   Materials, 153
   protective tubes, 156, 157
   response, 182, 183
   sensitivity, 157, 161
   systems, 157
Thermocouple insulators, 155
Thermocouple wells, ceramic bonded
   silicon carbide, 156
Fyrestan, 156
high speed response, 159
inconel, 156
iron, cast, 156
iron, wrought, 156
nickel, 156
seamless steel, 156
stainless steel, 156
Thermometer materials, noble metal, 34,
   153, 155
precision glass, 53
resistance, 34
Thermometers, types, Angle, 114, 115
bimetallic, 126
gas, 122
liquid-in-glass, 109, 111
liquid-in-metal,
   resistance, 131, 135
vapor, 120
Thermopile, differential, 506
Thevenin's theorem, 342
Thickness by radiation, 450
Thomson emf, 152, 153
Three lead method, 131
Three-wire industrial system, 134
Throttlers, 85, 87
Time lag, 23, 25
Time lapse photography, 526
Tolerance limits, 2
Tolerance uses, 17
Tower, fractionating, 13

Townsend avalanche, 412
Transducer characteristics, 344
Transducer operating principles, 345
Transducers, capacitive, 90
  inductive, 89
  oscillatory, 91
  piezoelectric, 90
  piezoresistive, 345
  pressure, 85
Transfer device, 1
Transfer factor, 325
Transfer function, 304
Transformers, LVDT, 359
  application, 361
  calibration, 370
  core position, 361
  form, 361
  frequency response, 363
  linear range, 367
  linearity of response, 365, 367
  mounting, 364
  phase, 361
  operation, 360
  output, 361
  sensitivity, 365
  stray magnetic fields, 368
Transverse strain effect, 329
Trapezoid notch, 282
Traveling wave system, 490
Turbine flowmeter, 272
Turbulence, 503
Twisted pair, 157, 160
Two-lead method, 131

Ultrasonic attenuation, 472
Ultrasonic bond testing, 497
Ultrasonic energy transfer, 470
Ultrasonic holographic equipment, 496
Ultrasonic Measurement system, Angle
  beam technique, 474
  A-scan, 476
  automatic operation, 476, 477
  B-scan, 476
  bubbler technique, 475
  C-scan, 476
  delta technique, 474
  pulse echo technique, 473
  reflection technique, 473
  through transmission, 471
Ultrasonic sound waves, 468

Ultrasonic transducers, 469
Ultraviolet spectrometers, 376

Vacuum gages, Alphatron, 99
  Dubrovin, 70
  ionization, 98
  Knudsen, 97
  McLeod, 92
  Phillips, 98
  Pirani, 97
  Thermocouple, 97
Vacuum thermocouple, 174
Variable area flow meter, 260
  inductronic, 265
  potentiometric, 265
  variable head, 262
Vector relationships, 482, 483, 485
Vena contracta, 244
Venturi tube, 241
Venturi tube insert, 242
Viscometer, 391, 524
Viscosity, Coefficient of, 386
  measurement of, 388
  theory of, 386
V-notch, 282
Voltage amplifier systems, 354
  zener reference, 170
Voltage Standing Wave Ratio (VSWR), 490

Walsh-Healey Public Contracts Act, 536
Water treatment, 544
Wavelengths, 177, 178
Wein's equation, 190
Wein-Planck law, 190
Weir, 281
Weir capacities, 284
Wind direction, 503
Wind speed, 503
Wheatstone bridge, 170
Windings, resistance thermometers, 130
Wire covering, alumni, 155
  asbestos braid, 155
  ceramic beads, 155, 156
  ceramic tubing, 155
  extruded nylon, 155
  glass fiber braid, 155
  Heat resistant enamel, 155
  Heat resistant rubber, 155
  high temperature silica braid, 155
  molybdenum oxide, 155

silicon impregnated asbestos, 155
silicon impregnated glass braid, 155
teflon-glass braid, 155
waxed cotton braid, 155
Wire, extension, 151, 152, 153

X ray beams, 461
X ray calibration standard, 463
X ray equipment, 462
X ray film speed, 465

Zener diode, 170

DATE DUE

MAY 3 1			
July 6/83			
AUG 2 3 1983			